Greening Affordable Housing

An Interactive Approach

Editors

AbdulLateef Olanrewaju
Universiti Tunku Abdul Rahman
Jalan Universiti, Bandar Barat
Kampar, Perak
Malaysia

Zalina Shari
Department of Architecture
Faculty of Design and Architecture
Universiti Putra Malaysia
Serdang, Selangor
Malaysia

Zhonghua Gou
Griffith University
Gold Coast, QLD
Australia

CRC Press
Taylor & Francis Group
Boca Raton London New York

CRC Press is an imprint of the
Taylor & Francis Group, an **informa** business
A SCIENCE PUBLISHERS BOOK

Cover illustration reproduced by kind courtesy of Jian Zhang, a PhD student at Griffith University

CRC Press
Taylor & Francis Group
6000 Broken Sound Parkway NW, Suite 300
Boca Raton, FL 33487-2742

Printed on acid-free paper
Version Date: 20181130

International Standard Book Number-13: 978-1-138-10260-6 (Hardback)

Visit the Taylor & Francis Web site at
http://www.taylorandfrancis.com

and the CRC Press Web site at
http://www.crcpress.com

Foreword

We can't afford to not build green. That was the mantra in 2007 when I was editing the *Blueprint for Greening Affordable Housing*. With the green affordable housing movement now past the two-decade mark, it is deeply inspiring to see this new book make a thoughtful, diverse, and inspired contribution to the small but growing canon that is exploring the intersection of green strategies and community development.

The fundamental building blocks of the green building argument—energy and water cost savings, reduced exposure to environmental toxins, increased durability, stronger connections to the surrounding community—are as relevant today as then. This book, *Greening Affordable Housing*, expands on that foundation by offering a diverse perspective that goes well beyond the conventional green building topics, showing both curiosity and a willingness to tackle emerging questions of governance, the role of affordable housing in shaping sustainable neighborhoods, the potential flexible design and construction techniques, and need for policies and market mechanisms that can sustain green practices.

The *Blueprint* arose out of a desire to share the potential to merge a green perspective with the passion and optimism of community development. Case studies were popping up around the country of family housing, co-housing, tribal housing, public housing, housing for the formerly homeless, senior housing, and sweat equity projects - all using the green building process to integrate smart design strategies with innovative systems and healthy, environmentally responsible materials. Once the concept of green affordable housing was proven in practice, doing anything else seemed irresponsible; to tenants, society, and the planet. I, my colleagues at Global Green, and our collaborators across the country wanted to share what we had learned, so that green affordable housing would become the expectation, rather than the exception.

In the decade since the *Blueprint for Greening Affordable Housing* was published, I've helped over 30 affordable housing developments to earn LEED for Homes or Enterprise Green Communities certification. Each a microcosm of urban sustainability, these projects reemphasize how a spirit of collaboration and commitment to integrated design are essential to achieving a transformation of practice. Policy is another powerful lever for change. What was a nascent effort in 2007 by groups including Global Green, New Ecology, and Enterprise Community Partners to incorporate green building standards into the low-income housing tax credits (LIHTC), has resulted in green criteria now featured in the LIHTC criteria of nearly all 50 states.

Today the discourse on green affordable housing is turning toward resilience.

Affordable housing is often the foundation of economic stability, physical and mental health, and social cohesion. The buildings themselves can be resilience hubs; familiar community gathering places that feature enough stored renewable energy to support several days of refrigeration for medicine, cell phone charging, lighting, and heating and cooling. Furthermore, each green affordable housing development has the potential to be a catalyst for equitable neighborhood revitalization by seeking ways to improve access to parks, stormwater management, mobility, and local sources of fresh food.

As we strive to confront the many serious challenges facing urban areas—adapting to climate change, providing shelter for the homeless and refugees, and ensuring equitable access to opportunity—the fundamentals of green affordable housing holds true: collaborate, seek design strategies that provide multiple benefits, stay mindful of the needs of the residents, seek opportunities along the entire arc of a project from land acquisition to building operation, and leverage real-world success through policy.

By addressing emerging issues and closing the gaps between knowledge, assumptions and real world experience, *Greening Affordable Housing* is enabling better, more thoughtful, and more impactful decision making by architects, developers, public policy makers, and all the actors that are involved in designing, building, and operating affordable housing.

March 2018
Venice, CA

Walker Wells
AICP, LEED AP, EcoDistricts AP
Executive Director Global Green
Lecturer UCLA Urban Planning

Green affordable housing is a very difficult domain to address, namely because there are so many 'pull factors' that need to be resolved simultaneously. A green affordable housing is not just about putting a roof over those that can least afford it. It is also about creating an environment where the cycle of poverty can be broken, and the stigma of a low-cost housing removed. It also has to be able to instill pride and responsibilities toward shared common facilities, promote happiness and social upward mobility for those that lived in these housing schemes.

Not only it has to be structurally safe, it has to be socially safe and secure for the young, old and weak, to be able to socialize happily in an environment without a fragment of fear of being robbed or kidnapped any time of day or night. It also has to provide a healthy environment that promotes walking and exercise, with good access to daylight and fresh air, using low polluting materials; while at the same time, it has to be well-connected for cheap and easy access to workplaces, schools and health care. Only then, the typical green requirement of comfort, efficiency of energy and water; and reduction of waste be addressed. Finally, the most important requirement of all - such a housing scheme must be cheap and abundance.

Fortunately, these requirements are not mutually exclusive of one another, even though, many in the industry does think so. This is the challenge that we need to overcome if we are to build a better world tomorrow.

We need a total rethink for an integrated Green Affordable Housing development, where waste are generated into wealth, sustainability is creating jobs, energy in its various forms are reused again and again until it is completely exhausted, water scarcity is recognized and a system of using it efficiently and reusing it again and again before it is finally discarded into the drain, river and sea has to be discovered; and at the same time, it must be promoting health, well-being and happiness.

A book like this is an important step forward, for all of us in the building industry to start talking, discussing and debating about the direction that we all must take today. The policy makers, developers, architects, engineers, quantity surveyor, real-estate agents, building owners and even the home owner/buyers, need to ask ourselves, what do we do next? To put us and our next generation towards a path of better quality of life, not just for ourselves but for everyone on this planet and beyond.

With this I would like to congratulate all the writers that have contributed to this book. A job well done. It raises the right questions, but the solution(s) is far from certain. It is a long difficult road ahead for all of us, to develop this planet right, for a future that we can be proud of.

March 2018
Kuala Lumpur, Malaysia

Tang Chee Khoay
Honorary Secretary
Malaysia Green Building Confederation

Preface

This book "Greening Affordable Housing" was written in response to the emerging need to address green building and housing affordability in an integrated approach. Although, affordable housing is the largest types of buildings, a book that specifically addresses itself to the issues and context of green or sustainability is conspicuously lacking. The industry and academia have seen an increase in texts in this field. A comparison of contents of the representative books indicates that the emphasis is on materials, design, site activities and green tools In particular, Blueprint for Greening Affordable Housing, edited by Walker Wells (2007), is an earlier book that looks into making affordable housing green. This book intends to take a comprehensive approach to the supply of sustainable affordable housing by looking at the various issues in greening affordable housing from different cultural perspectives. This book looks at the emerging issues in greening affordable housing as new requirements, constraints, techniques and methods. The emerging issues are becoming more critical as housing the homebuyers are becoming more demanding and complex.

Research on the need for the sustainable built environment is growing drastically and at the same time, research shows that housing plays a significant role in achieving sustainable built environment. According to data published by Demographia (2017), housing is far from being adequate in most developed countries including the USA, the UK, Australia, Singapore, Canada and New Zealand. No doubt, housing is even more inadequate in the developing and underdeveloped countries across the world. In most of developing countries, the number of houses per 1000 population is less than 200. In many cities around the world, families have to keep more than one job to afford a decent accommodation. A great amount of city residents have to forgo medical treatment, education, social life and clothes to afford a home (Tesco Bank, 2017; and Bank of America, 2016). Housing is synonymous with development and growth. It is interesting to know that the 2008 global meltdown or financial crisis was triggered and denigrated by the housing sector. The impacts are not over yet, and there is no evidence that the same crisis will not reoccur in the nearest future. Many households especially the US continues to be affected by the 2008 foreclosure crisis and many homeowners are increasing becoming renters (Nagy and Goldberg, 2017).

Besides severe housing inadequacies around the globe, the housing sector presents a challenge and opportunity to the environmental sustainability considering the amount of energy and resources being used in construction and maintenance (UN, 2016). During the Paris Climate Conference (COP21), 195 countries agreed to keep the global average temperature to well below 2°C and aimed to limit the increase to

1.5°C. The Kyoto Protocol, Rio and the South Africa Conventions discussed the same requirements. At the micro level, the concern for sustainability, climate change, global warming, flood, drought, pollution and resource depletion are now part of homebuyers' and households' decisions making (Olanrewaju et al., 2016; Xie et al., 2017). As such, housing developers have started to opt for sustainable design strategies in housing developments in order to increase revenues and profit margins. Many stakeholders are involved in sustainability for altruism purposes. However, the recent research shows that the level of awareness among housing developers, contractors and designers (i.e., architects and engineers) on the concept of sustainability is low (Shari and Soebarto, 2012; 2013). This has implications on their level of preparedness and ability to incorporate green features in the housing planning, designing, construction and maintenance.

The processes and techniques involved in the design, construction, maintenance, operation and demolition of affordable housing have a profound influence on the built environment, house owners, occupants and other stakeholders. As such, a global green building movement is happening with the proliferation of various green building rating tools proposed and developed by government bodies, professional bodies, corporate organisations or university researchers (Gou et al., 2013). Examples of such rating systems include US Leadership in Energy and Environmental Design (LEED), UK's Building Research Establishment's Environmental Assessment Method (BREEAM), Malaysia's Green Building Index (GBI), Japan's Comprehensive Assessment System for Building Environmental Efficiency (CASBEE), Singapore's Green Mark, Australia's Green Star and Hongkong's Building Environmental Assessment Method (HK-BEAM). Most of rating tools emerged as a response to the local climate and policies in their respective countries of origin (Gou and Xie, 2017). However, not all of these rating tools take into account the availability of affordable, sustainable housing, and there remains until today, very limited rating tools (if any) that are specifically developed to assess the sustainability performance of affordable housing projects. The requirements of sustainability in the housing delivery will lead to a significant energy optimization, enhancement of the housing durability, minimization of waste, improvements in social impacts and overall users' satisfaction on indoor environmental quality, reduction in pollutions, consideration of project life-cycle costs, and creation of more inclusive living environment. Integrating sustainability criteria into affordable housing delivery enables stakeholders to address the environmental components that have not traditionally been seen as an integral part of affordable housing development process.

The motivation behind this book is to address three main issues. Firstly, greening the affordable housing still presents a unique challenge to the industry today, simply because of the negative attitudes and perceptions that green housings are expensive and not profitable. Affordability and sustainability are generally considered incompatible; hence, it is of paramount importance to demonstrate that much can still be achieved despite the constraints and negative perceptions. Secondly, we are becoming more aware of the need to increase our measures to achieve environmental and social sustainability. Homebuyers are now demanding for sustainable homes and questioning the sustainability credentials of developers and contractors. Thirdly, the market for green homes, though currently still low at about 5%, will grow steadily in the near future as technologies improve and homeowners become more informed. Developers and homebuyers require simple yet practical evidence to facilitate their decision makings.

References

Bank of America. 2016. Homebuyer Insights Report. Retrieved January 20, 2018, from https://info.bankofamerica.com/homebuyers-report/?cm_mmc=CRE-HomeLoans-_-vanity-_-CA01VN0065_homebuyersreport-_-NA.

Demographia. 2017. 14th Annual Demographia International Housing Affordability Survey: 2018 Rating Middle-Income Housing Affordability. Retrieved January 20, 2018 from http://www.demographia.com/dhi.pdf.

Gou, Z. and X. Xie. 2017. Evolving green building: Triple bottom line or regenerative design? Journal of Cleaner Production 153: 600–607.

Gou, Z., S. Lau and D. Prasad. 2013. Market readiness and policy implications for green buildings: case study from Hong Kong. Journal of Green Building 8(2): 162–173.

Hilber, C. A. and W. Vermeulen. 2016. The impact of supply constraints on house prices in England. The Economic Journal 126(591): 358–405.

Olanrewaju, A., S. Y. Tan, L. T. Lee, M. Ayob and S. Ang. 2016. Investigating the compatibility of affordable housing with sustainability criteria: A conceptual framework. pp. 228–240. *In*: Proceedings of the Putrajaya International Built Environment, Technology and Engineering Conference (PIBEC2016), 24–25 September 2016, Bangi, Malaysia. ISBN 978-967-13952-8-8.

Olanrewaju, A., S. Y. Tan, L. L. Tat and N. Mine. 2017. Analysis of homeowners' behaviours in housing maintenance. Procedia Engineering 180: 1622–1632.

Nagy, C. and L. Goldberg. 2017. Affordability at a Cost: What We Can Learn from Mobility Patterns. https://shelterforce.org/2017/12/19/affordability-cost-can-learn-mobility-patterns/.

Shari, Z. and V. I. Soebarto. 2012. Delivering sustainable building strategies in Malaysia: Stakeholders' barriers and aspirations, ALAM CIPTA. Int'l Journal of Sustainable Tropical Design Research and Practice 5(2): 3–12.

Shari, Z. and V. I. Soebarto. 2013. Investigating sustainable practices in the Malaysian office building developments. Construction Innovation: Information, Process, Management 14(1): 17–35.

Tesco Bank. 2017. The UK Home Buyers Survey 2017. Retrieved October 30, 2017 from http://www.tescobank.com/guides/buying-a-house/home-buyers-survey-2017/.

Global Green USA. 2007. Blueprint for Greening Affordable Housing. Washington, DC: Island Press.

Xie, X., Y. Lu and Z. Gou. 2017. Green building pro-environment behaviors: Are green users also green buyers? Sustainability 9(10): 1703.

Acknowledgements

We will like to extend our since thanks to our contributors who accepted our criticisms and suggestions and remained with us through the lengthy production period. We are equally thankful to our esteem referees for accepting our requests and providing useful comments and suggestions to authors and editors.

We are humbled and honoured by the esteemed contributors for accepting our critics and comments on the submissions and for their patience throughout the editing and publication stages.

Contents

List of Referees

Associate Prof. Ghada Ragheb
Pharos University Egypt

Associate Professor Dr. Zan Yang
Tsinghua University, Beijing

Dr. Radzi Ismail
Universiti Sains Malaysia

Dr. Tan Hai Chen
Heriot-Watt University Malaysia

Dr. Noor Suzilawati Binti Rabe
International Islamic University Malaysia

Ar. Dr. Loo Kok Hoo
Universiti Tunku Abdul Rahman Malaysia

Dr. Lim Poh Im
Universiti Tunku Abdul Rahman Malaysia

Dr. Mohd Hafizal B. Ishak
Universiti Tun Hussein Malaysia

Associate Prof. Dr. Zulhabri Ismail
Universiti Teknologi MARA (UiTM)

Ir. Dr. Lee Wah Peng
Universiti Tunku Abdul Rahman Malaysia

Prof. Dr. John Kurian
Universiti PETRONAS Malaysia

Dr. Trivess Moore
RMIT University

Dr. Khairusy Syakirin Has-Yun Bin Hashim
International Islamic Universitiy Malaysia

Introduction

This book proposes strategies for affordable housing delivery that complied with the sustainability requirements at individual, organisational, and policy levels. This book showcases the validated strategies that will deliver sustainable affordable housing through significant, persistent, and measurable reductions in resource consumption and waste generation. From the peer-reviewed chapters, the book identifies relevant strategies for various types of housing developments that could be implemented by individuals, groups of individuals and institutions. Furthermore, the book identifies the gaps in the current literature and approaches and provides some topics in greening affordable housing for future research.

Greening Affordable Housing is designed for policymakers, housing developers, architects, urban planners, engineers, contractors, place managers, quantity surveyor, third-party agencies and other professionals and scholars concerned with sustainability, housing, whole life appraisal and building maintenance. In particular, the book addresses the concern of all stakeholders that have interests in affordable housing. It focuses on approaches, methods, indicators, processes that are central to the delivery, operation and maintenance of sustainable affordable housing. This is achieved through collections of the latest thinking in affordable housing planning, design, construction and operation. The book is also presented in a form that is useful to undergraduate and graduate students that have interest in the sustainable built environment. It presents various techniques that are required for outcome through operation phases of housing in order to deliver residential buildings that complied with sustainability requirements. It also provides a platform for debate and intellectual discourse on measures and policies to deliver sustainable buildings and built environment. A short summary of each chapter is presented in the following text.

Chapter 1, by Olanrewaju and Tan, provides a summary of activities that each player should execute based on the RIBA 2013 plan of work. While misconceptions and myth are prevalent among members of the housing supply about the compatibility of affordability with sustainability, the major problem noticed was issue appertaining to the roles they perform in the design and constructions of sustainable affordable housing. They argue that the major reason for the lack of sustainability compliance in the design and construction of affordable housing is that there is a knowledge gap amongst team players. For that reason, the chapter presents a framework that aims to serve as a reference for design professionals, government officials and local communities to help making the incorporation of sustainability in affordable housing more appealing for the growing market.

In Chapter 2, Huston and Baines argue that the UK has not been able to meet its housing need and the government intervention has not been able to resolve the affordability crisis. The chapter calls for holistic integration among the various government agencies in the UK in order to increase affordable housing delivery. In many developed countries and the UK in particular, the major causes of high housing prices are due to the restrictive planning system. The chapter finds that cost is the primary reason why affordable housing cannot be sustainable. The chapter discusses affordable housing in the different contexts as the basis for the development of a framework for each group of developing and developed countries.

Khor, in Chapter 3, emphasizes that the collaboration among housing designers, city planners and other key players are essential to deliver sustainable affordable housing. The argument within this chapter is that sustainable housing requires a multidisciplinary approach. As such, the author identified ten sustainable management multi-steps to be incorporated into sustainability management system and sustainable housing development delivery. The first step deals with the identification of the scope and purpose of the housing project that sets the stage for the other activities. It is important that stakeholders are familiar with the project's brief and corporate objective, a failure of which is detrimental to the entire sustainable development process.

Kozlowski's review in Chapter 4 focuses on the creation of green neighbourhoods based on case studies from Denmark, Sweden and Singapore. He shows that these cities have met the requirements for green neighbourhoods such as vibrant physical housing, high connectivity, less use of cars, and passively designed houses for energy efficiency and good thermal comfort. He also examines and explains the roles of urban design as a tool in the creation of a sustainable built environment as well as the roles of ecological urbanism in the integration of natural and built environments. Constraints posed by the governments towards achieving sustainable neighbourhoods are also discussed. This chapter should be read in conjunction with Lin and Qin's analysis of energy efficient designs for affordable housing and Chakraborty's discussion on passive design strategies for affordable housing in a tropical climate, in Chapter 11 and 12 respectively.

In Chapter 5, Yusoff and Abd Rashid focus on the life-cycle assessment (LCA) approach to justify the need for standards and policies to advance the implementation of sustainable housing. They argue that if LCA is used appropriately, it is able to identify building element that poses threat to the environment or sustainability. The difficulty of implementing the LCA in buildings is also identified and discussed. The chapter presents a framework for LCA specifically for the Malaysian building industry. The chapter concludes that it is imperative for green building rating systems, particularly Malaysia's Green Building Index (GBI), to integrate an LCA approach towards providing sustainable built environments, in line with other major rating systems in many parts of the world.

Green building policies is the topic covered by Liu and Gou in Chapter 6. They argue that the lack or weak government policies are the major constraints for the incorporation of green building features in many countries. Without effective governments' supports together with lack of widespread evidence that green design will increase developers' profits, developers would not take the initiatives to supply green affordable housing. Furthermore, most developers are not convinced that homebuyers will be able to afford the additional prices of green buildings in the face of increasing housing prices and homelessness. In line with this realization, the authors emphasise that governments should play the role of an enabler, a provider and a regulator as well as provide a framework that includes five main requirements to increase the uptake of green affordable housing in the market.

In Chapter 7, Mohd Noor examines the integration of flexible housing with the circular economy principle based on experiences from the Japanese housing industry to advance the supply of affordable housing in Malaysia. Mohd Noor has developed a business model called flexZhouse (or flexible house) that emphasises the innovative partnership between housing developers, manufacturers and homebuyers to support the principle of rent-then-own. A guideline on required new technologies and how they can support the flexZhouse model is provided in the chapter. Mohd Noor uses the fundamental economics principle to show how housing developers could reduce the prices of housing. He envisions that the flexZhouse model could potentially encourage mass production and customisation because it makes use of the principle of economies of scale.

In Chapter 8, Kraatz discusses the shortage of social affordable housing in many developed countries including Australia, Canada, the UK, the US and the Netherlands. Kraatz argues that it is grossly inadequate to define affordable housing in terms of price alone. He first defines and conceptualises affordability by considering non-economic factors such as security, accessibility, location, the health and well-being of occupants, and social and educational factors. She then outlines the main benefits of delivering an affordable housing that considers the social and environmental aspects in the supply chain. The 'composite to return on investment approach' (CROI) is then proposed to enable housing providers to justify their investments in affordable housing developments.

In Chapter 9, Shi looks at landscape (e.g. public parks and gardens) as the most commonly neglected issue in affordable housing design and planning. Expenditure on the landscape and other external amenities are seen as an additional burden to developers as it might reduce their profit margins. The author distinguishes the importance of green network among various concepts and tools in balancing urban development and the environmental need. Shi argues that since affordable housing developments often involve mass development of large pieces of lands, opportunities abound to integrate green network are convincible to the policymakers and housing developers. Shi, however, cautions on the effects of such integration in places like Hong Kong, Monaco, Bahrain, Malta and Singapore where cities are densely populated and lands are exceptionally expensive.

Mohd Noor once again shares his research findings in Chapter 10. This time he advances the flexZhouse model presented in Chapter 7. The model benefits from container technology that is adaptable and flexible to accommodate the changing lifestyles of homebuyers and contribute to the reduction of waste generation. Mohd Noor anticipates that the flexZhouse model will extend the building lifespan due to its design that encourages refurbishment and conversion with minimal waste generation. Mohd Noor's study finds that potential homebuyers are in favour of the flexZhouse concept as it will allow them to make future changes without incurring too much additional costs. Each module could be developed and manufactured according to the client's requirements in terms of the number of rooms, space configuration, size, internal fittings and external amenities.

In Chapter 11, Lin and Qin discuss how energy efficient features can be implemented in affordable housing design and construction. The authors argue that energy management should be a primary concern of affordable housing design professionals in Hong Kong. The chapter focuses on operational energy and passive design and explains various design techniques that are relevant for affordable housing. Although the authors acknowledge that passive design may increase the initial construction cost, they posit that the resulting lower operating costs would make it more desirable and profitable to the homeowners in the long run. Reduced operational energy would also lead to a reduction of negative environmental impacts.

Chakraborty, in Chapter 12, extends the discussion on passive design for an affordable housing made in Chapter 11 by focusing on tropical climate like in India. Chakraborty defines affordable housing as a home for the poor in India, though the housing needs to be supported by basic amenities and facilities for the occupants' comfort and well-being. Chakraborty stresses on the roles of passive design in reducing energy cost and environmental degradation, hence supporting Lin and Qin's proposition. However, since the level of awareness on the need to integrate passive design features into affordable housing is low, the author calls for all design team members to start thinking about passive approach and be actively involved from the beginning of the design process.

Shearer, in Chapter 13, presents the tiny house concept as a measure to increase housing affordability and at same time, address some social and environmental concerns. Shearer argues that the increasing land price and regulations have led to the expansion of tiny house markets around the world, especially in Australia. With the increasing number of homelessness among singles and new couples, tiny houses are said to offer a viable alternative for shelters. Shearer provides an elaborate context for tiny houses to justify its appropriateness in terms of sustainability and affordability. Despite certain benefits associated with tiny houses, the author also presents some barriers to its implementation in many capital cities. These include planning restrictions, transportation policies as well as safety and security risks.

Olanrewaju et al., in Chapter 14, investigate the possibilities of implementing green features in affordable housing in Malaysia by using the principle of critical success factor. Through a survey questionnaire involving 121 housing professionals, the results reveal nine groups of green features, namely operating cost, energy, pollution, water, green, location, ecology, social and place. Interestingly, none of the green features is required to be traded-off for one another. This is contrary to existing knowledge that shows the indicators for incorporating sustainability in building developments required some trade-offs. The chapter contributes a sustainable affordable housing framework to serve as a reference tool for government, industry, academics and housing developers in producing sustainable housing for the Malaysian market.

In Chapter 15, Ogunsemi and Ade-Ojo highlight the importance and method of estimating green affordable housing by using the Nigerian context. The authors justify a reliable cost estimation of green housing by emphasing the fact that cost or price of housing constitutes a major decision making. In fact, cost is often the most often cited reason for the low uptake of green housing. In Nigeria, housing is mainly unaffordable and the high affordable deficit is associated with high materials costs. With the lack of up-to-date and reliable information, modelling the capital cost of housing has been very difficult. The chapter reviews several predictive modelling tools and subsequently recommends Artificial Neural Network as the most compatible model with the Nigerian housing market.

Nizamani et al. in Chapter 16, look at the foundation design of affordable housing in Malaysia. They argue that the cost implications of safety and structural integrity are often neglected in the design of affordable housing. Cases of public housing collapse due to foundation failure are on the rise around the world. Foundation is perhaps the most critical element in the multi-storey housing and selecting the most appropriate foundation type for a multi-story building is a major concern among design professionals. Therefore, the authors examine the design code for housing with a specific emphasis on pile foundation. The authors highlight the complex issues involved in different lengths for pile foundation designs and discuss the design process in pile foundation using two types of simulation software. The cost comparisons between different pile foundation design alternatives are also provided.

In the final chapter, Chu looks into waste management in housing construction. The author reminds us of the huge volume of waste generated during construction stage and more than 90% of these wastes are inert. As such, Chu argues that an efficient waste management is crucial to help minimizing wastes disposals to landfills; reducing carbon emissions; decreasing the cost of operations; increasing developers and contractor profits; as well as improving the overall environmental quality. Chu suggests that concentrating efforts to reduce waste is not enough. Instead, market for the wastes should be created and adequate information on the existence of these wastes should be made available to project players, if needed. The waste market will involve circular flow of resource, cradle to cradle approach, birth to rebirth and the active participation of all stakeholders.

CHAPTER 1

Roles of Design and Construction Teams in Sustainable Housing Delivery

*AbdulLateef Olanrewaju** and *Seong Yeow Tan*

Introduction

Housing delivery involves many stakeholders from both the demand and supply side of housing delivery. Each party performs specific, but complementary functions at a specific stage. Often, the parties include architects, engineers (especially for services and structure), developers, contractors, suppliers and manufacturers. The scope and nature of the work of the parties depends on many factors, including the procurement systems used, complexity, size and methods of construction, the extent of customization, location, regulations and clients' funding arrangements. Housing has huge implications on sustainability or sustainable development. The market for sustainable housing is increasing. The trend will continue to grow in order to increase housing performance, occupants' and the community's satisfaction levels and to protect the environment. Sustainability matters, particularly with affordable housing, which should address social and environment dimensions. Sustainability and affordability are not reciprocally exclusive. Affordability is far more complex than mere (capital) costs of housing. Affordable housing should address occupants comfort and productivity, the operational costs of housing, householders' relationships with the neighborhood and be socially acceptable for the rest of the neighborhood. Sustainable affordable housing is more demanding. The main goal of the theory of sustainable affordable housing delivery that the present study focuses on is to produce typologies to explain how to implement sustainability in affordable housing delivery. The strategic objectives of this study are to stimulate the supply of affordable housing but also consider environmental,

Unversiti Tunku Abdul Rahman Malaysia.
* Corresponding author

social and economic values of certain stakeholders. In the past, it can be seen that homebuyers have demanded carbon offsetting projects which have resulted in higher overall satisfaction levels with housing performance. Homebuyers have begun to realize that investing a little more to offset the carbon footprints of their activities will contribute to better living. Recent research has found that specifying sustainability requirements during the design and procurement stage results in a higher opportunity to implement sustainability requirements during construction and operational stages of a housing scheme. With this in mind, housing developers and contractors are developing various measures to increase revenues and profit margins. However, a major limiting factor against sustainable housing delivery is a lack of awareness on the roles and responsibilities of design professionals involved in the supply of affordable housing. The sustainability credentials of housing developers, contractors and design teams are being questioned. The governments' agenda and interventions are also being questioned. The government has responded to this by introducing regulations and policies to increase sustainable affordable housing supply. The government, professional bodies, NGOs, academics and various associations are responding with advice, guidelines and certifications on affordable sustainable housing. Nevertheless, research that examines the roles and responsibilities of design and construction teams in sustainable affordable housing delivery is nascent. In order to fill this gap in knowledge, the present chapter examines the roles and responsibilities of design professionals in the supply of affordable housing and whether they are capable of helping households save energy, water, materials, reduce waste, pollution and emissions in houses as well as promote social integration of communities.

Theoretical Framework

Sustainable development or sustainability is a global challenge because it's causes, impact and solutions has an effect on everyone. The origin of sustainable development can be traced back to environmental economics, which dates back to 1960. Environmental economics primarily seeks to measure the implications (i.e., costs) of economic decisions on the external environment and to proffer solutions to mitigate or eliminate these implications. The implications are usually significant but are not often known when economic decisions are taken. Thus, sustainable development goals are future based activities. The topical issues faced in the efforts towards sustainable development are the need to integrate economic, environment and social factors in decision making. For simplicity, but without losing its meaning and generalization, sustainable development places a limit on the use of natural resources. Essentially, the bottom line is, globally, we are currently consuming more than the earth can produce and support. Moreover, we are producing and discharging waste that is far beyond what the earth can accommodate. In simple terms, sustainable development is a practice that integrates various criteria, including energy efficiency, durability, waste minimization, social impact, a good indoor environment, durability, pollution control, life-costs, user-friendliness, user comforts and others.

Achieving sustainability goals involves a huge capital outlay. It is estimated that USD 2 trillion is required to counter the impact of climate changes (UN, 2016).

During the Paris Climate Conference (COP21), 195 countries agreed to maintain the global average temperature to well below 2°C and aimed to limit the increase of the world temperature to 1.5°C. The Kyoto Protocol, Rio and the South Africa Conventions have discussed the same requirements. At a micro level, the concern for sustainability [climate change, global warming, pollution, resource depletion, etc.] is now part of consumers' decisions and plans. Furthermore, firms and corporate organizations have also inculcated the same into their business models. As a result, households now examine the finer details on products and services labels to better understand how harmful or complimentary certain products or services are to the environment and the community. Homeowners now check carbon emissions, water and electricity consumption rates of products and equipment before completing their buying decisions. Moreover, homeowners and home buyers also consider the social and economic implications of the choices they make. The sustainability credentials of firms are being questioned. In response to this, most governments around the world have set targets to address climate change, pollution, resource depletions and energy and water shortages. The prosperity of countries is no longer solely based on Gross Domestic Product (GDP). The Social Progressive Index that measures social and cultural dimensions is now being considered as an alternative measure. Lately, sustainable GDP is also being seriously discussed. A sustainable GDP attempts to deduct the negative externalities from an actual GDP value. The construction and operation of buildings has a huge impact on sustainability goals. Although there are many indicators for sustainable development (Enquist et al., 2007), to simplify, three indicators, economic, social and environmental dimensions, are often commonly cited.

Sustainable Housing

Ever since sustainable housing development has been recognized, the call for sustainably built environments has heightened. Built environments pose the greatest threat to sustainable development goals. The impact of construction on sustainability is huge. Specifically, buildings consume more than 40% of the world energy, release 1/3 of CO_2, use about 25% of harvested woods, release about 50% of fluorocarbons, produce 40% of landfill materials, use 45% of energy in operations, emit 40% of Green House Emissions and use 15% of the world's usable water (see; Killip, 2006; UNEP-SBCI, 2014). Building interiors contain five times more pollutants than outdoor air . When other CO_2 emissions attributable to buildings are considered, such as the emissions from the manufacture and transport of construction and demolition materials, the impact is even higher. The energy that buildings consume is even higher if we take into account other energy used for buildings. For example, the embodied energy in a single building's envelope is around 8–10 times the annual energy used to heat and cool a building. Energy is also used due to transport related activities for buildings. People use cars because of the distance from their houses to places of work, childrens' schools, and markets. Cars also play a significant role in emitting pollution. Most houses spend up to 20% of their household income on transportation. From these statistics, it can clearly be seen that buildings are part of the threat to sustainable development. Moreover, affordable housing plays a significant part in this threat.

Housing development involves resources such as energy, water and raw materials. The production, process and use of these resources depend on the environment and community structures and organization. Construction of housing construction generates wastes and emits potentially harmful substances. Without a doubt, all these contribute to climate changes. The concerns for climate change, population explosion, environmental degradation, pollution, and resource depletion has led to the quest for sustainable buildings. This chapter proposes that housing is a major contributor to the climate change phenomenon. The requirements of sustainability on housing delivery will lead to significant energy optimization, the enhancement of housing durability, waste minimization, improvements on social impacts, good indoor environments, a reduction in pollution, the consideration of building project life cycle-costs, active user-friendliness, user comfort and total satisfaction of users. Therefore, there is a need for sustainable buildings.

Following on the definition of sustainable development, sustainable housing would imply "housing that meets the needs of the present without compromising the ability of future generations to meet their needs". But the flaw with the above definition is that it is rather philosophical and is not scientific. It is an elusive definition because it is difficult to measure. It is ontologically subjective. For this reason, sustainable housing is defined as housing that strives to improve the capital outlay, operational performance and costs, not be harmful to the environment and communities and is accepted by its users. It could also be defined as housing that strives to preserve the economic, ecological, spiritual, social and cultural requirements for societal well-being while at the same time preserves the natural environment for future use (Olanrewaju and Kafaya, 2008). These definitions allow for the easy identification of indicators and criteria to measure sustainable buildings. The relevant stakeholders will thus be aware of what they need to do. Therefore, housing development strategies that contribute to attain these objectives should have some measurable indicators. In the present chapter, terms including eco-housing, livability housing, green housing, natural housing, and sustainable housing are synonymous with sustainable affordable housing and they are sometimes used interchangeably. The common features found in most guidelines for sustainable homes from many countries include:

- Energy reduction
- Energy management
- CO_2 reduction
- Efficient use of water resources
- Use of local construction materials sourced without destroying habitats
- Use of durable construction materials
- The collection of surface water
- Efficient waste management
- Minimizing all forms of pollution, including noise
- Maximizing the health and well-being of those using the building
- The fostering and care of local ecology

- Location
- Occupants' satisfaction levels and productivity
- Neighborhoods
- Site management

Sustainable Affordable Housing

There is not a clear definition on what affordable housing is, in any part of the world. Nevertheless, one common basic trend with all the different definitions is that it is a measure of the affordability of homes for lower and middle income earners. While some classifications and definitions focus on market affordability, others dwell on individual affordability. The factors that characterize the various interpretations and meaning of affordable housing are whether households can rent, mortgage or buy a house using income as a major denominator. The term affordable housing has been used quite loosely to represent all types of housing developed by governments, its agencies or partnerships with developers or social landlords. It seems that the meaning of affordable housing is more easily understood than defined. However, the problem with this common definition is that it is more concerned with class than with performance requirements. As Jewkes and Delgadillo (2010) explained, most definitions for affordable housing are based on a housing price in relation to income affordability. The primary factor that is used to determine the affordability of a home is the disposable income of a household. More specifically, the amount of money that households have available for spending and saving after income tax and other mandatory charges have been deducted. Thus, housing is 'affordable' if the rental cost or mortgage repayment cost does not exceed 30% of a household income for households in the lowest 40% of the income distribution range (U.S. Department of Housing and Urban Development, 2006). Based on this definition, housing is affordable if the housing does not cost more than three times a median income over a period of three years. For rental purposes, the 30% would include utility bills, which includes electricity, water, gas, sewage and garbage collection. In case of a mortgage, the amount includes the actual payment, tax, insurance, utilities and maintenance costs. This definition is referred to as the 30/40 rule. The Demographia compares the median house price to the median household income to measure affordable housing based on an annual international Housing Affordability Survey. However, both of these definitions are acceptable internationally. Based on either of these definitions, housing is not affordable to more than 90% of the world population. In the OECD, the average price of housing, water, electricity, gas and other fuels, furnishings, household equipment and routine maintenance is 28% of a household's consumption (OECD, 2018). However, if transportation costs are included, households spend more than 40% of their expenditures on housing in OECD. However, defining affordable housing in terms of financial affordability is misleading. Affordable housing should account for homebuyers/homeowners' value systems, location, price, distance and size of the house. The criteria within homebuyers' value systems include comfort, satisfaction levels, quality, preferences and tastes. If homebuyers sacrifice their value systems to occupy a house for financial reasons, the house is not affordable in the long run.

The implication of defining affordable housing financially is that the prices of houses become the only denominator or standard of comparison.

Despite the fact that numerous research has shown that sustainable buildings perform better compared to standard buildings (buildings designed, constructed and operated using standard regulations and specifications), issues regarding compatibility between sustainability and affordability of houses is high. Many members of design and construction teams are skeptical about the compatibility between affordability and sustainability of housing. Research on sustainable affordable housing is grossly inadequate. In the US, it was found that the assimilation of the green building concept on the "realm of affordable housing remains a work in progress" (The Urban Land Institute, 2008). But the institute noted that despite the challenges, "the future of green affordable housing is promising". Sustainable affordable housing supply has remained the biggest challenge in the housing sector in Australia (Susilawati and Miller, 2013). Despite the fact that there is not a classic or acceptable definition of sustainable affordable housing, the present chapter has defined it as affordable housing that strives to preserve the economic, ecological, spiritual, social and cultural requirements for societal well-being, while at the same time preserves the natural environment for future use (see Olanrewaju and Kafayah, 2008). Recent literature has shown that the major obstructers for the delivery of sustainable affordable housing include:

- A lack of awareness on the supply side of the concept of sustainable development
- A lack of understanding on the benefits of the supply of sustainable affordable housing
- A lack of awareness on the indicators of delivering sustainable affordable housing
- A lack of awareness on the duties and responsibilities of the supply side
- A lack of willingness to deliver sustainable affordable housing
- A lack of expertise (i.e., competency) and technology (i.e., plant, and equipment)
- Restrictive government regulations and policies
- A lack of market demand
- A lack of availability of sustainable materials and components.

The present chapter addresses itself to the duties and responsibilities of design and construction teams. A lack of awareness of the housing industry on the concept of sustainable built environments and, in particular, sustainable housing is a major limitation. Therefore, cognizance of the roles and responsibilities of the teams in the design and construction of sustainable affordable housing is fundamental and critical.

The Complexity of Sustainable Housing

The supply of sustainable buildings is fraught with various complexities. An example of this is a timber processing company that cuts down too many trees in order to meet demand. Most certainly, for the company and its employees, the large number of trees that are cut makes business sense, because they would benefit from economies of scale, which would make their timber production cheaper, etc. In addition to this, their customers would be happy because they would be able to buy quality timber at a cheaper rate. However, if the tree cutting continues without proper checks, the trees would soon disappear and the company would close and thereby terminate the jobs of its employees (i.e., the locals), thus increasing unemployment rates. Apart from the direct implications on the company, other businesses that depend on the materials would close shop and the price of timber would increase unnecessarily. For example, the price of timber for the construction industry would increase and would also lead a shortage of labor for other businesses, because the company that produces timber is willing to pay higher wages to meet its production requirements. The excessive cutting of trees would also lead to environmental degradation and social unrest in the long term. Other things to consider are the increased demand for energy such as crude oil, and electricity, due to an expansion of the industrial and transportation sector. Despite the fact that the expansion is required in order to increase the economic prosperity of a country, it will lead to massive pollutant emissions into the atmosphere. Consequently, this deteriorates the air and water quality and has a negative impact on people's health and lives. Expenditure on medical treatment would increase, population growth may be affected, and expenditure on treating air and water pollution and erosion would also increase. Cement production is another common construction material. Its production and use has significant implications on the environment and on social organizations. For instance, cement is an essential input for the production of concrete and a primary building material for the construction industry. However, cement production has an impact on the environment and on human health due to CO_2 emissions produced in its production process.

Professionals in Affordable Housing Delivery

The concept of sustainable affordable housing is only beginning to be explored by practitioners and researchers. There is a lack of understanding among design teams on sustainable strategies for affordable housing. The present study explores the following questions. What specific functions or duties can stakeholders perform at a particular phase? The aim of this study is to produce a matrix for affordable housing delivery that is capable of producing sustainable affordable housing. Stakeholders in affordable housing delivery comprise of public and private stakeholders (Olanrewaju et al., 2016). Private stakeholders include material suppliers, contractors, developers and designers (architects, engineers and quantity surveyors). Professionals in housing delivery include architects, engineers, and quantity surveyors. Although developers also employ quantity surveyors, architects and engineers (i.e., structure, services), the stakeholders addressed in this research are the consultant architects, quantity surveyors and engineers. However, issues in housing operation and housing demolition were also identified. The architects specialize in the development of housing briefs, concepts and designs for building schemes. During a construction phase, the architects continuously revise plans, drawings, and specifications to meet the requirements of clients and statutory regulations. For housing projects, civil engineers are mostly concerned with road designs and the construction and structure of high rise housings. The building engineers are involved in calculating the strength and force of proposed housing schemes. They also prepare structural drawings and specifications from architectural drawings and other relevant contract documents. They work to ensure that the buildings can carry and withstand the loads it will encounter during a housing operation. The mechanical and electrical engineers are often segmented together as services engineers and are an important aspect of modern housing constructions, especially for work relating to lifts, escalators, electrical work, plumbing and air conditioning systems. The duties and responsibilities of service engineers are crucial, especially in multistoreyed design and the construction of buildings. Quantity surveyors deal with cost management, procurement and contractual issues involved in the delivery of housing schemes. They monitor and update initial estimates and contractual obligations as the construction progresses based on additional work and variations.

Framework for Duties and Responsibilities of Sustainable Affordable Housing

In building terminology, a framework is a supporting structure of a building. It is also defined as the skeleton of a building. A framework is an abstract presentation of real life issues. However, it is important to note that it carries different meanings in other disciplines. For instance, in computer language, it is defined as a structure that shows the types of programs that can be designed, as well the interactions among the programs. To clarify, a framework provides structure and consistence. It allows a process to achieve a particular set of purposes or objectives. In general, a framework is an outline, rules, guidelines or a skeleton of an idea or concepts formulated, designed or proposed toward achieving a particular set of objectives. A framework can be graphical, tabularized, conceptual, statistical or mathematical. However, some will disagree with these classifications, because they believe that, a mathematical expression involves a model. The present chapter focuses on a conceptual framework or guideline. The framework developed in this chapter aims to facilitate designers to incorporate sustainability features into the planning, designing and construction of affordable housing. It is important for roles and duties to be identified as the roles and duties of design professionals are currently based on anecdotal evidence, instead empirical facts, as there is no established approach available in academic and practical literature. Chapter 14 presents the results of a study that explores the identification and establishment of critical success factors in the design, construction and operation of affordable housing. The three main components of sustainability are economic, environment, and social factors. Economic sustainability in housing entails balancing capital and operating costs of housing over its lifespan effectively and efficiently. The price of owning a house should not unnecessarily deny a household the opportunity to better health, education, a social life and other relevant factors. Traditionally, cost was the only priority of housing, which neglected the other two dimensions of sustainability. The decision to design and construct a house has always been on one dimension. However, there has been a continuous realization on the impacts of a decision that takes into account only costs. Indeed, decisions have been broadened to embrace a multidimensional approach. Over the last five decades, there has been a paradigm shift from making housing decisions solely based on capital to include both capital and operating. However, a building may be affordable if it takes into account life cycle costs, yet be socially and environmentally unsustainable. Hence, there is a need to consider the other two dimensions of sustainability. Environmental sustainability of housing implies its design, construction, operation, maintenance (and if necessary its refurbishment, extension and conversion) and demolition should not contribute to waste generation, pollution, climate changes, floods, heat and other environmental degradations. To this effect, affordable housing should be designed, constructed, maintained and operated to use renewable energy, reduce fossil fuel consumption and emissions, reduce deforestation, use recycled materials and components for construction and maintenance (i.e., refurbishment, extension), and have superior waste management. The social sustainability factor in affordable housing ensures that housing does not threaten social orders and community values, does not lead to neighborhood instability, nor increase crime rates or discrimination. The housing should therefore, be designed, constructed, operated and maintained to ensure peace, social justice and have initiatives that promote social equity. The roles and duties of members of the design professionals are contextualized into five different stages. The details are contained in Table 1.1. A framework was established through the research which attempted to address the knowledge gaps identified in the literature review (i.e., there were clear no roles or responsibilities of design teams and construction stakeholders in affordable housing). At a macro level, the five phases were plotted against three dimensions of sustainability.

Much of the responsibilities and duties in the conceptual framework might be common to most efficient managed construction delivery processes. However, for affordable housing delivery, most of the responsibilities and duties are only implicitly considered. Because most affordable housing projects are located in towns and at the outskirt of cities, waste management is not considered. Whole life appraisals of housings are not considered, foreign labor is mostly used and abuse occurs on sites. National and state level guidelines, regulations and policies on housing construction are not followed. Due to policies or regulations, or both in most cases, affordable housing schemes are provided at the outskirt of cities or far away from city centers. This is largely because there is a scarcity of land or it is part of an urban design or development project. Whatever be the case, affordable housing is constructed on untouched land. While this might open up the towns and villages to development, it will involve destroying untouched lands, increase deforestation, increase pollution, and disturb the ecosystem. Erecting housing on untouched land means there will be a reduction in agricultural land, which ultimately implies that the next generation will faces a scarcity in food production and vegetation. The above are some factors that lead to climate changes such as heat waves, deadly floods and storms. In most cases, affordable housing occupiers or users work in the cities due to poor transportation systems, commute daily to their places of work, which ultimately increases carbon emissions, worsens already depleted fossil fuels and increases pollution levels. Transportation costs take up a large part of the incomes of affordable housing occupiers. The quality of material and components used for affordable housing construction are usually of poor quality, which leads to an increase in operation and maintenance costs. Poorly maintained and operated houses will increase operational costs and even generate more waste. Such houses will be unsuitable for occupation due to safety and security issues. In most developing countries, affordable housing is designed and constructed as semi-detached, detached, flats, apartments or bungalows, hence using more land and other natural resources. Energy and water efficiency and

Table 1.1. Sustainable affordable housing matrix.

Stage	Environmental	Economic	Social	Main parties	Comments
BRIEF	• Evaluate source and nature of environmental degradation • Evaluate different kinds of pollution • Conduct an impact assessment on housing • Conduct site investigation and have environmental sustainability as objectives • Identify environmental objectives • Set environmental sustainability as housing objectives • Provide answers on how selection of a site can reduce environmental degradation • Provide answers on how housing orientation can prevent environmental degradation • Provide answers on how design and construction teams reduce environmental degradation. • Prepare a site to reduce traffic	• Produce a feasibility study report • Identity green financing methods • Prepare a life cycle cost model • Conduct a feasibility study to reduce housing life cycle costs • Analyze and recommend method of finance to reduce down payments and purchasing prices • Develop housing objectives that will reduce capital costs • Propose housing close to material sources to avoid long haulages • Prepare proposal to reduce maintenance costs	• Identify social values • Conduct neighborhood analysis • Avoid 'ghettoization' of affordable housing • Select a site that will increase community, and households' wellbeing • Select a housing site that will reduce crime rates • Select a housing orientation that improves community and house wellbeing • Aim to reduce conflict and litigation • Locate housing in a mixed community • Provide recreational facilities to promote integration • Parking should be located to prevent congestion and to promote contact among households	Architects, engineers, contractors, quantity surveyors/cost specialists' subcontractors, suppliers, clients/homebuyers	The core objective of this phase is to set sustainability criteria as a housing development objective, i.e., the housing should aim to reduce pollution, waste, costs, CO_2 and promote cultural and neighborhood integration. The key aspects to look into are: how selection of a site, orientation, design and construction teams and methods of finance can reduce pollution, waste generation, overcrowding, maintenance costs, reduce crime and promote cultural harmony, friendships, household wellbeing, peace and happiness. While this stage is critical, sustainability is not well addressed. The traditional procurement strategy is cited as a major problem responsible for this neglect. This problem is further compounded in affordable housing delivery as homebuyers are not even known during the design and even during the major part of a construction phase.

DESIGN	• Develop designs that reduce carbon emissions • Prepare designs that reduce waste generation • Develop proposals that consume less water • Develop proposals to reduce energy consumption. • Develop proposal based on renewable energy • Specify natural materials • Source for materials locally • Specify biodegradable material • Specify recyclable materials • Specify reusable materials • Update designs to reduce waste generation • Perform eco-audits • Prepare design to accommodate the impact of floods, earthquakes • Specify materials and components that are reusable or can be recycled • Specify materials that have recyclable or reusable packaging. • Specify materials that do not require finishes. • Prepare designs that reduce pollution	• Review and develop conceptual budgets • Develop reliable preliminary cost information • Prepare concepts to reduce variations • Prepare concepts to reduce pre-occupancy obsolesce • Prepare flexible designs to reduce costs of reconstruction, refurbishment and extensions • Prepare affordable design • Evaluate life cycle of the design phase • Seek information on homebuyers' finance methods • Review and evaluate methods of finance • Review conceptual budgets • Recommend recyclable materials • Prepare architectural details to reduce costs • Specify materials based on whole life appraisals • Select contractors knowledgeable in value engineering/management • Select contractors that prepare tenders based on localized pricing.	• Prepare designs that reduce conflict among design team, construction team and community where the housing is constructed • Prepare cost information to increase social integration among design and construction teams and owners • Recommend localized workers • Recommend localized contractors • Produce concept designs to include safety measures • Produce designs that consider security of a property and the properties of households and neighborhoods. • Prepare open designs • Prepare designs that encourage cultural integration • Prepare designs that are socially acceptable to the community • Consider safety of households and neighborhoods in the design of the housing	Architects, engineers, contractors, quantity surveyors/cost specialists' subcontractors, suppliers, clients/homebuyers	During the design phase, how to reduce embodied energy, operational energy, waste, pollution, energy consumption, and water consumption in affordable housing? During the design phase, how to economize affordable housing in terms of construction costs, maintenance costs, future modifications (i.e., refurbishments, alterations During an outline proposal for a structural design, building services systems and outline specifications, how to enhance household and community wellbeing and happiness. Energy efficiency components are considered. While sometimes it affects economics, social aspects are not considered Only technologies based solutions are considered

Table 1.1 contd. ...

...Table 1.1 contd.

Stage	Environmental	Economic	Social	Main parties	Comments
	• Prepare designs to reduce waste generation • Update designs to reduce waste • Update designs so they are energy efficient • Select contractors that are sympathetic to the environment • Recommend subcontractors and suppliers that are certified by third party green agencies • Select contractors that have effective site waste management methods		• Consider the security of occupants' and neighborhoods in the design of the housing • Reduce barriers among neighborhoods • Finalize designs to include safety measures • Update designs to protect the security of properties of occupants and neighborhoods. • Provide parking for visitors as well as homebuyers • Parking lots should provide for bicycle users • Specified materials available locally • Select contractors, subcontractors and material suppliers that are familiar with the culture of where s house is located. • Select contractors from the housing community	Architects, engineers, contractors, subcontractors, suppliers, clients	During off-site manufacturing and on-site production, how to economize, socialize and project the environment through material selection, claim avoidance, risk avoidance and reduce variations? Currently, most attention is paid to this stage but even then, the approaches are fragmented.

PRODUCTION	• Prepare a site waste management plan • Construction teams to set requirements for the efficient use of materials • Design teams to monitor waste plans • Include waste reduction instructions or standards in contracts • Recommend suppliers that have waste minimization/ environmental plans or credentials • Prepare energy manuals • Prepare maintenance manuals • Use prefabricated elements and components • Access environmentally certified materials • Ensure that gums, adhesive and paints are environmental friendly	• Conduct waste audits • Contractors to prepare estimates and targets • Contractors to have strategies to reduce waste • Contractors to have strategies to maximize recyclable residue waste. • Engage a local workforce • Source material locally and especially within a host community • Contractors to produce accurate waste estimates • Source and use recycled or reused materials • Separate waste streams on site • Conserve on-site biodiversity	• Engage local work forces • Reduce conflict among construction teams and those in the community where the housing is situated. • Patronize local suppliers • Ensure that gums, adhesives and paints used are odorless, and non-irritable		
COMMISSION	During handover and commissioning, prepare a commissioning plan that clearly defines the roles and responsibilities of households, prepares a critical equipment list, integrates testing, safety and security and emergency	• Prepare handover notes during the design phase and modify it accordingly during the construction phase • Prepare handover notes based on building information	• Involve all parties in preparing the handover notes	Architects, engineers, contractors, subcontractors, suppliers, clients	This is seldom identified as an opportunity to reaffirm sustainability in housing delivery

...Table 1.1 contd.

Stage	Environmental	Economic	Social	Main Parties	Comments
	preparedness, operation manuals, training, reduces pollution, waste, energy reduction and consumption. • Educate homebuyers on methods of energy management • Train plant operators on plant operations • Engage experts on the commission process				
OPERATION	While a house building is in operation, how to reduce energy consumption, CO_2, operational energy, waste, pollution, maintenance, water consumption through usage and through maintenance, retrofitting, extension and refurbishments? • Use less energy • Switch off lights when not in use • Make use of openings (i.e., windows) for ventilation • Use water efficiently	While a house building is in operation, how to economize operations through maintenance, retrofitting, extension and refurbishments, energy consumption and water consumptions? • Avoid excessive usage of elements • Use experienced maintainers • Use recycled components • Use local materials and components	• Engage locals for maintenance • Employ plant operators from the local host community • Employ locals as security personnel	Maintenance organizations, specialist subcontractors, suppliers, homebuyers, home occupants	This stage is very critical, while new affordable housing may appear to meet sustainability regulations on paper, the performance of houses are not met once they are in operation. Hence, the need arises to bridge the gap between *as designed* and as *operated*. However, mostly the gap is created by various factors such as design, briefs, expert advice, and consideration for capital, nature and interest of clients. To improve, systemic rethinking is required by both clients and developers and more awareness and knowledge needs to be

DEMOLITION	• Dismantle rather than demolish. • Use deconstruction techniques to remove individual buildings • Engage specialist deconstruction contractors • Recycle waste from the demolition process that cannot be salvaged for reuse. • Avoid noise pollution	• Protect materials and salvage them • Sell salvaged materials	• Identify market for salvaged material locally • Use local labor • Avoid harmful discharge to the community • Avoid foul odors	Maintenance organizations, specialist subcontractors, homebuyers, home occupants, local authorities	created. In fact, recent research emphasized that sustainable housing needs sustainable occupants to close the loop. During this stage, sustainable behaviors of home occupants are very crucial. Home occupants' maintenance approaches must be sustainable.

Modified after Olanrewaju et al. (2016).

the conservation of materials and components are not used during their construction. Hence, those houses could be cheap but the operational and maintenance costs are often higher compared to high end housing. The design and socio–economic profile of the occupiers does not encourage social integration among the occupiers of affordable housing. Moreover, even when there are building management frameworks, but which in most cases is not even available, they are always weak and ineffective. Affordable houses are often over populated or there are a higher number of occupiers per housing unit, this normally provokes or increases the deterioration of buildings and common facilities, and instead promotes social interactions which sometimes lead to security issues. A noise and poor air quality problem is rampart in affordable housing units. Unlike the construction of high-end properties in cities, waste management during the construction of affordable housing in the outskirts is poorly managed. Indeed, there is indiscriminate dumping of waste on open fields. Poor monitoring of the construction of affordable housing by relevant government agencies results in high cases of health and safety problems. Furthermore, site operators are poorly treated or abused when compared to those working in cities on high end properties.

Conclusion and Summary

The present chapter has attempted examine preliminary sustainable affordable housing to guide housing stakeholders and especially design and construction teams to provide affordable housing that is compatible with sustainability criteria. Although the matrix or typology within this study borrows its general framework from the RIBA 2013 Plan of work, it also draws upon a vast body of theories and empirical evidence. This indicates that the achievement of sustainability in housing delivery is multidisciplinary in nature. Indeed, many processes, procedures and stakeholders are involved in different stages of a housing development right through the operational to the disposal stages. Explicit in this study is the compatibility between affordability and sustainability in the delivery of housing.

Based on a literature review, this study has identified the duties and responsibilities of a number of stakeholders in the delivery and use of affordable housing. The duties and responsibilities are based on the construction process of a building. In the present study, we have looked at a number of key concepts in examining the compatibility of housing affordability and housing sustainability. Both the production and operation of affordable housing poses a threat to sustainability. It has since been recognized that the earth is unable to repair itself due to over exploitation of man. The present chapter illustrates some of the duties and functions of stakeholders in affordable housing delivery with regards to sustainability. There is a need for a shift in the affordable housing delivery process towards adding value to the sustainability paradigm. When making decisions on sustainable housing, further emphasis and attention is needed during the briefing and design phases. The initial hypothesis that necessitated the study was that affordability and sustainability are complementary but not mutually exclusive. An important consideration in any sustainable affordable housing delivery is to initially clarify the perception that both sustainability and affordability are mutually exclusive. The government needs to increase awareness of sustainability. However, in the formulation of policies, the engagement of householders and manufacturers is actively required and feedback should be incorporated. Future research may aim at examining the opinions of these stakeholders and whether they can actually perform the outlined roles and if not, to analyze the obstacles and further roles and duties other than those outlined in the present study. The framework is not prescriptive and should be adaptable to different housing projects.

Acknowledgements

The research presented in this paper was supported in full by a grant from the "FRGS"; project: Analytical Investigation of Problems in Housing Supply in Malaysia. Project number: FRGS/1/2015/TK06/UTAR/02/2.

References

Enquist, B., B. Edvardsson and P. S. Sabhatu. 2007. Value-based service quality for sustainable business. Managing Service Quality 17(4): 385–403.

Howes, R. and H. Robinson. 2005. Infrastructure for the Built Environment: Global Procurement Strategies. Oxford: Butterworth-Heinemann

Jewkes, D. M. and M. L. Delgadillo. 2010. Weaknesses of housing affordability indices used by practitioners. Journal of Financial Counseling and Planning 21(1): 43–52.

Killip, G. 2006. The housing maintenance and sustainability debate. *In*: Proceedings of the Annual Research Conference of The Royal Institution of Chartered Surveyors, London.

OECD (Organisation for Economic Co-operation and Development, 2018) Housing expenditure as share of final consumption expenditure of households, 2000 – 2013. OECD Annual National Accounts Database; Eurostat Annual national accounts database. www.oecd.org.

Olanrewaju, A. A. and S. T. Kafayah. 2008. The Need to maintain our buildings: Sustainable Development. In Proceedings of PSIS Enviro (The 1st National Seminar on Environment, Development and Sustainability. Politeknik Sultan Idris Shah, PSIS, Sabak Bernam Selangor Malaysia.

Olanrewaju, A., S.Y. Tan, L.L. Tat, F.m Ayob and S. Ang. 2016. Investigating the compatibility of affordable housing with sustainability criteria: a conceptual framework. pp. 228–240. *In*: Proceeding – Putrajaya International Built Environment, Technology and Engineering Conference (PIBEC2016), 24–25 September 2016, Bangi, Malaysia. ISBN 978-967-13952-8-8.

RIBA (Royal Institure of the British Architects, 2–13) The RIBA Plan of Work 2013.

Susilawati, Connie and Miller, Wendy F. 2013. Sustainable and affordable housing: a myth or reality. pp. 1–14. *In*: Kajewski, Stephen L., Manley, Karen and Hampson, Keith D. (eds.). Proceedings of the 19th CIB World Building Congress, Queensland University of Technology, Brisbane Convention & Exhibition Centre, Brisbane, QLD.

The Urban Land Institute. 2008. ULI Community Catalyst Report Number 7: Environmentally Sustainable Affordable Housing. Washington, D.C.: ULI–the Urban Land Institute.

Thomas, M. J. and J. S. Callan. 2013. Environmental Economics and Management: Theory, Policy, and Applications 6th Edition. 6 edition, South-Western College Pub.

Tony, D. 2006. What is Delivery Management? http://it.toolbox.com/blogs/delivery-doctor/what-is-delivery-management-10341.

U.S. Department of Housing and Urban Development. 2006. Housing impact analysis. Washington, DC: U.S. Government Printing Office.

UNEP-SBCI (United Nations Environment Programme). 2014. Greening the building supply chain. DTI/1753/PA.

CHAPTER 2

Sustainable Management of Affordable Housing

*Simon Huston** and *Richard Baines*

Introduction

In 2008, for the first time in human history, more people lived in cities than the countryside. Inevitably, as human populations burgeon beyond a projected 9.2 billion by 2050 (UNFPA, 2007), affordable housing pressure will increase. Affordable housing is but one of several basic needs essential for empowerment, dignity and eudemonic wellbeing (Sen, 1999; Wadley, 2010) but its sustainable provision involves multiple factors at different spatial scales. The management of sustainable housing cannot be isolated from broader development issues, involving food security and utilities (water, waste, energy). Whilst developing and developed countries face somewhat different challenges, common concerns are land, poverty and power (Niemietz, 2015). Rapidly growing cities need to secure food by formalizing peri-urban settlements and transforming slums. The need to increase and upgrade housing is urgent. In Africa, almost 1 billion people are likely to migrate to urban centres by 2050 pushed by rural poverty and pulled by the prospect of city jobs (UN Habitat, 2016). In developed nations, like the UK and Holland, depleted local governments implemented technical and market-oriented housing solutions. According to the New Economics Foundation (2017), vested land interests and impoverished political economy debate, have spawned spatially-myopic land and fiscal policies. As the artificial financial and rentier economies expanded, private landlords replaced local government social provision (Fernandez and Aalbers, 2017; Hochstenbach, 2017). The 'unearned increments' of London's housing boom were sequestered in overseas tax havens and left behind a flotsam of disenfranchised renters or regional homeowners (Wetzstein, 2016; Arundel, 2017). Reflection on sustainable management of affordable housing is, therefore, timely. After reviewing the affordable housing backdrop, the chapter outlines a

The Royal Agricultural University, Cirencester, UK.
* Corresponding author

Sustainable Management for Affordable Housing ('SMAH') framework and then investigates developing and developed country contexts.

For the purposes of this chapter, affordable housing covers the full gamut of relatively cheap dwellings, including socially rented and community owned accommodation. According to Manchester City Council Planning Guidance (2017: 15), 'housing is unaffordable if it costs more than 3.5x a single or 2.9x a joint gross household income'. In contrast then, affordable housing should consume around a third of a family's income but also meet minimum standards for human dignity and flourishing (Sen, 1999; Johnson, 2007; Maliene and Malys, 2009). However, significant structural factors limit it and undermine the realization of housing dignity. For the poor, cognitive dissonance compounds supply restrictions (Oxoby, 2004). Household financial impediments include transfer costs and mortgage limits. Land banking, planning bureaucracy and lobbying by affluent communities restricts long-run housing supply. Tax shortfalls, distort metropolitan administrations priorities. In the United Kingdom after Thatcher, government subsidized housing construction collapsed whilst planning flux increased (Crook and Whitehead, 2002). Despite privatization, many ex-social or subsidized housing estates remain socially alienated. For new developments, UK planning policy mandates affordable housing quotas but the approach can, at best, only achieve incidental social integration until income, health and education inequality are seriously tackled (Tiesdell, 2004; Wilkinson and Pickett, 2009).

Sustainable Management for Affordable Housing ('SMAH')

To be effective, the SMAH must involve a holistic systems approach with operational and supporting constituents at several spatial scales: national, metropolitan/district/local and at the project level (Fig. 2.1). The system involves top-down, horizontal and bottom-up interactions to coordinate, design, plan, finance, construct or manage appropriate and affordable housing projects. Some aspects are managed centrally whilst others are devolved locally. It has national (legislative, administrative, judicial), local (town planning) and private sector institutional dimensions:

- Executive: political economy direction, land policy and taxation regime
- Legislative: uphold rule of law, and capture unearned increments
- Judicial: land-use and development conflict resolution mechanism
- Administrative: Land Administrative Systems to record transactions, provide reliable spatial data to internalize externalities and attenuate spatial injustice
- Urban executive: shape land-use and planning, development approvals or restraints
- Private sector: design, finance, development, marketing and maintenance.

The SMAH is participatory, responsive to diverse local contexts and stakeholders. Clearly, sustainable management involves both teleological (consequential) and deontological (procedural or moral) considerations (Hausman and McPherson, 2006). Multiple consequentialist assessment criteria include price and quality of housing outcomes. Leasehold or freehold, affordable dwellings are both relatively cheap but also 'fit for purpose' as defined by accessibility, design, construction, and safety as well as energy efficiency criteria. Deontological considerations turn on fair, transparent and

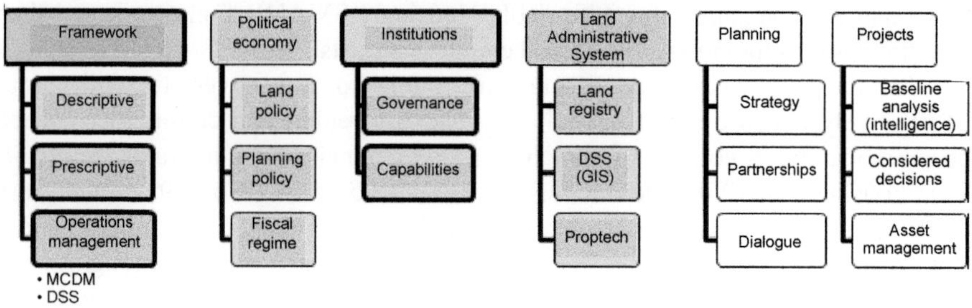

Figure 2.1. Operational and supporting SMAH constituents.

inclusive administrative procedures, based on rule of law. However, since housing is built within or on the periphery of fragmented urban settlements, no system can ever eliminate policy contention or land use conflict (Owens and Cowell, 2011). Sustainable management calls for simplified and agile systems. Likely measures for reform include strategic reviews, bureaucratic rationalization, audits, LAS implementation and controlled delegation of powers to local authorities or communities. In contrast, weak planning systems are corrupt, authoritarian and unlikely to consider indigenous or other local rights. To counter this, SMAH is characterized by local governance framework to strengthen institutional capacity as a prelude to participatory engagement with local communities and businesses.

Key: Shaded—National Scale

Informed policy, dialogue and feedback between the multiple constituents drive system management. The system monitors output quality, learns from mistakes and listens to local communities and property residents. Leasehold or freehold, affordable dwellings are both relatively cheap but also 'fit for purpose' as defined by accessibility, design, construction, and safety as well as energy efficiency criteria (see Fig. 2.2 below). The SMAH confronts a complex mishmash of spatial hierarchies involving the political economy, institutions and innovation (Storper, 2013) and so offers no 'cookie-cutter' solution, only reflection, debate and evolution. Traditional housing management systems guidance comes in descriptive, prescriptive or operations management packages (Johnson and Heinz, 2006). Descriptive affordable housing management investigation is based on case studies of successful projects in various locales. Prescriptive policy research like Maliene and Malys (2009) or a recent New Economic Foundation paper (2017) draw on academic planning literature to formulate ideal sustainable housing 'toolkits' or policy prescriptions. In the operations management tradition, Mulliner et al. (2013) compare the suitability of three different affordable housing estates in Liverpool using a Multi Criteria Decision Making (MCDM) approach with 20 criteria. The criteria included indicators of affordability, housing, crime, accessibility and environment. Nowadays, operations management necessarily involves technologically-driven or 'smart' Decision Support Systems (DSS) and Proptech (property technology) to monitor, contain or plan land use intensification and improve affordable housing decision making (Morano et al., 2015). High tech data systems can enrich information and help administrators balance development

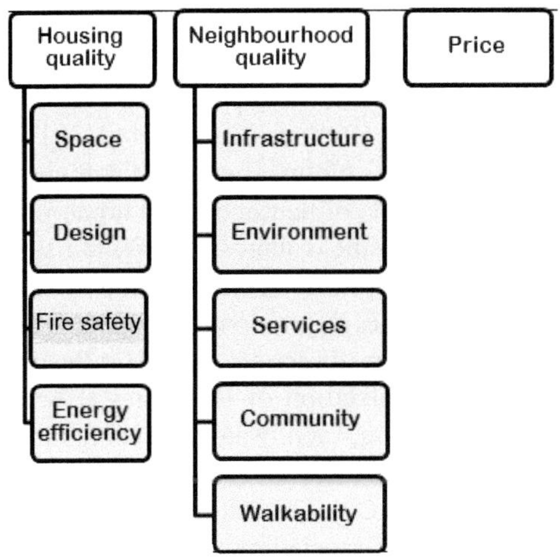

Figure 2.2. SMAH local project design considerations.

with other needs (e.g., for food and environmental protection, tourism and quality of life). Good project management starts with clear objectives: to deliver complex housing projects on time, under budget and at the stipulated quality. Good management ensures the resources are available to successfully complete new social or private sector residential housing projects with appropriate affordable elements. Proper facilities and asset management, involves listening to the needs of diverse stakeholders.

Sustainable housing management increases the supply of affordable dwellings, strengthens community dialogue and multi-criteria reflection. Whether the system is centrally planned, market driven or organized communally, it must be adaptable and involve dialogue between multiple evolving institutions. No system can ever resolve the inherent tensions between informed excellence (top-down), commercial gain (horizontal), local community empowerment and environmental protection (bottom up). Figure 2.2 illustrates SMAH design considerations. At the local project level, decision-makers need to take account of multiple quality dimensions.

Affordable housing which is 'fit for purpose' must pass or meet minimum criteria in terms of space, design, fire safety, and energy efficiency. To untangle the spatial (neighbourhood) aspects of quality, sustainable and affordable housing project managements begin with a baseline analysis. Project design should reflect local geographical and cultural conditions, particularly with respect to flood or other risks and transport connectivity.

Developing Country Focus

Rapid, haphazard urbanization in many developing nations, undermines food security, leading to malnutrition, disease and crime. Sub-Saharan Africa's cities are growing faster than ever before in history and governments struggle to provide housing, infrastructure and employment (Tostensen et al., 2001). Two thirds of the region's urban populations live in slums where poverty, food insecurity and malnutrition, especially among women and children are rife. In developing regions, many practical difficulties hamper sustainable management for affordable housing. First is limited public sector institutional capacity to tackle stressed food-energy-water systems (ibid.). Governments need to integrate science and policy; address cross-scale inequalities and overcome path-dependencies (Romero-Lankao et al. 2017). Rapidly expanding urban areas are challenged in terms of the provision of sufficient affordable food and in dealing with increasing amounts of household and urban waste. Net in-migration exacerbates the situation. The result is almost one billion slum dwellers living in unplanned urban fringe settlement (UN Habitat, 2016). Inadequate energy, water, sanitation, transport and ICT infrastructure lowers productivity and lifespans.

With clear political economy direction, SMAH adjustments for developing country contexts include: formalization of land titles (LAS), urban associations strengthening, cooperative housing, food, water and energy infrastructure provision and tackling waste and pollution (Tostensen et al., 2001). Figure 2.3 below illustrates operational aspects. In this case, a Cities Infrastructure and Growth (CIG) partnership strengthens institutions and urban associations for holistic urban development around food, water, waste and energy. The focus is informal slum settlements and their relationships with planned urban and rural areas. Iterations between the LAS and local communities facilities baseline analysis, reduces land conflict and stimulates engagement. Property rights registration and targeted capital investment help to transform informal settlements into mainstream ones and integrate them with urban centres and surrounding rural communities.

Food security and nutrition for the urban poor are complex and difficult problems dependent not only on food prices and purchasing power but also geography and employment. As food prices climb or incomes falter, families cut back on their purchases of fruits, vegetables and meats and rely instead more on staples, which are cheaper but lack micronutrients. Many are reliant on informal and unlicensed street

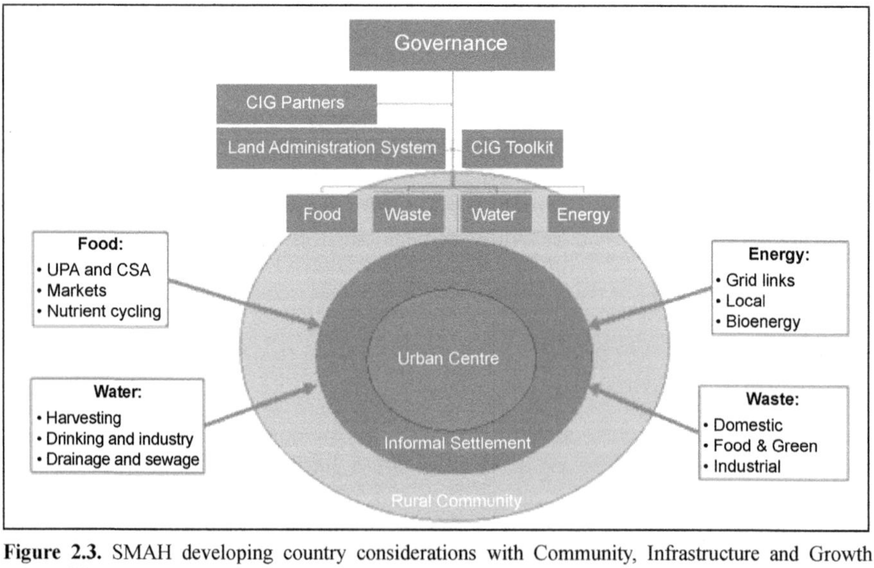

Figure 2.3. SMAH developing country considerations with Community, Infrastructure and Growth partnership and LAS support. Key: UPA - Urban and Peri-urban Agriculture; CSA - Community Supported Agriculture.

vendors whose food is cheaper but generally prepared under unsanitary conditions. It is generally considered that in many African cities, 70% of the calories consumed by the urban poor comes from street food. Formal shopping via regulated outlets is costly for the urban poor as markets tend to be located in the wealthier sections of cities, away from slums and squatter settlements while urban gardens are considered a middle-class luxury. With this shift of food insecurity and malnutrition to the cities, it is incumbent on local administrations, governments and the international community to acknowledge that food poverty is no longer just a rural concern. There are serious urban problems requiring immediate attention, innovative solutions and financial support.

Integrating food systems into urban planning and development has the potential to address food and nutrition security within the urban communities. Furthermore, the surplus food produced from urban and peri-urban agriculture can stimulate economic activity and begin to formalize the food system from a supply and hygiene perspective. Surplus foods and organic wastes can also lead to economic activities to build fertility (composting) and supply green energy (anaerobic digestion) that can maintain food production. Furthermore, such food systems can be integrated with rural food and compost supply chains to stimulate further economic activity and reduce the pressure on in-migration.

Sub-Saharan Africa is one of the world regions with the biggest challenges in water, sanitation and hygiene. Across the continent, 327 million people lack access to safe drinking water, while 565 million must do without a safe and hygienic toilet. Access to water is a human right and access to safe drinking water is the foundation of any health service. In sub-Saharan Africa some 40% of the population does not have access to safe drinking water. There is also a difference between urban and rural access; for example, in 2012 urban access to safe water ranged from 94% in the richest 20% of countries to 64% in the poorest; rural access ranged from 68% to 34% respectively. Urban drinking water supply actually recorded a decline; attributed to the high proportion of slum dwellers in fast expanding urban populations. Rural areas have seen slight progress on sanitation, up from 25% in 1990.

In 2011, African countries reported substantive political commitments to WASH (Water, Sanitation and Hygiene), increasing funding allocations along with leadership and coordination among implementing agencies. The majority of countries have established transparent WASH service provision targets and have put in place supporting policies, and many monitor against these targets. Countries also confirm that the rights to water and sanitation are increasingly adopted in laws or policies. The lack of sanitation is a serious concern because of the associated massive health burden as many people who lack basic sanitation engage in unsanitary activities like open defecation, solid waste disposal and wastewater disposal. The practice of open defecation is the primary cause of faecal oral transmission of disease with children being the most vulnerable. Despite the commitments and technical know-how, there is insufficient domestic financing for WASH overall, especially sanitation. This is exacerbated by a lack of coordination among authorities, stemming from an unclear definition of roles and responsibilities, coupled with lack of harmonization of policies and inadequate staffing in government departments that handle environmental issues.

Participatory planning within informal settlements allied to safe water sourcing (piped and harvested) and the implementation of local WASH programmes can provide the foundation for community managed water utilities. Furthermore, the provision of safe toilets to harvest human wastes can lead to local entrepreneur schemes (as in Kenya) and the recovery of human waste for energy production as well as nutrients for urban and peri-urban food production.

Energy is a critical issue for Africa, where a large number of people do not have access to energy. Energy recovery from waste can play a role in minimizing the impact of Municipal Solid Waste (MSW) on the environment with the additional benefit of providing a local source of energy. Indeed the estimated electricity production from the total waste generated in Africa could reach 122.2 TWh in 2025 which could service the needs of 40 million households. However, waste management is poor. Besides providing an interesting share of gross energy consumption and electricity as a renewable resource, energy recovered from waste could also help minimize the impact of municipal solid waste on the environment. In the developed world there are more than 600 large scale waste-to-energy facilities (mainly incineration) while in Africa, a very limited share of waste is recovered and reused. More recently, there has been increased interest in renewable or green energy sources, especially anaerobic digestion. Often more than 50% of urban waste produced is organic and biodegradable and anaerobic digestion leading to the generation of biogas provides a unique opportunity to fulfil both waste and energy objectives. This approach contributes to improved waste management practices and at the same time fulfils the goals of sustainable energy management. Biogas waste treatment facilities reduce the amount of waste disposed in uncontrolled dumping sites, which if unmanaged, release pollutants into air, water and soil, endangering the environment and contributing to greenhouse gas emissions. Digestate from biogas facilities is a valuable fertilizer for urban and peri-urban agriculture.

As the majority of poor city dwellers lack access to electricity and refrigeration community scale bio digesters could provide the basis for human and municipal organic waste recycling to produce energy for households and for other economic activities. Such facilities could be community owned and managed; furthermore, by harmonizing such systems to that of public energy supply, these local power plants could eventually connect to local, regional or national grids. Attitudes towards solid waste are often contradictory where affluent societies (and districts) often see it plainly as an environmental problem; however, in many cities in Africa it is an important and flexible source of income for a large part of the urban poor. It also provides raw material to many sectors of the economy.

Waste collectors form a vital part of the urban economy. Operating on the streets, kerbsides and dumps, this group of people collects, sorts, cleans, recycles and sells material thrown away by others, therefore contributing to an extent to public health, sanitation and environmental sustainability. Such work is hazardous and it is incumbent on local authorities to work with some of the poorest communities to safely and securely improve their waste management and collection methods. This in turn brings improvement to the health of the slum dwelling families with the creation of safer healthier places to live and work.

Co-ordinating municipal solid waste collection and segregation not only addresses the 'garbage' and health problems but it also creates economic opportunities ranging from local to municipal composting, materials recovery and energy generation. Such services are critical to the quality of life for urban dwellers, especially those in informal settlements who do not have access to even rudimentary waste collection services. Waste collection, segregation and processing also have the potential to stimulate entrepreneurial activities. The Foodscapes model (Fig. 2.4) explores the nexus between primary food production and supply, access to water, urban waste (in particular food and green waste and waste water) and energy supply—all of which can stimulate entrepreneurship if managed sympathetically. It embraces food supply as part of urban planning and economic development along with energy, water and sanitation.

Historically, cities drove economic growth, provided jobs and secured service access. But in many countries today, rapid urbanization today spawns an informal sector, characterized by sporadic and uncertain employment and a lack of access to government services. Unprecedented urbanization in Africa is often typified by unplanned waves of immigration which overwhelm formal and bureaucratic planners. In-migration cannot be stopped, therefore, the only solution is to speed up the planning process. Often, current central government controlled systems are complex, unwieldy and run by different ministries. SMAH calls for a simplified, agile system. Top down rationalization in the wake of the SMAH strategic review (using the CIG toolkit) and LAS enhancements, facilitates controlled delegation (subsidiarity). Often, weak planning systems are corrupted, authoritarian or ignore indigenous informal rights. To counter this, SMAH streamlines the overarching governance framework and strengthens institutional capacity as a prelude to participatory engagement with local communities, informal businesses and urban associations.

Land tenure reform, formalized via LAS, catalyzes local engagement in slum upgrading. Transparency, tenure security and legitimacy de-risks investment, facilitates planning and builds capacity for self-improvement. It transfers power to slum communities and stimulates a virtuous development circle. Dialogue with local slum stakeholders sets regeneration and resilience priorities. When streets and latrines are introduced and power is available to lights in the streets, soon shops will emerge and there will be other economic activities.

Developed Country Affordable Housing: United Kingdom

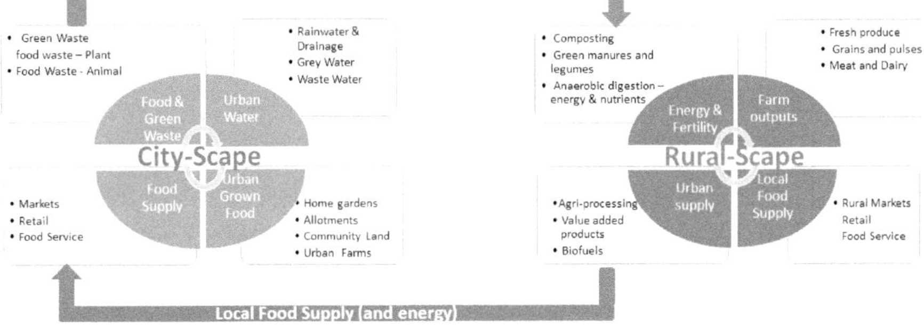

Figure 2.4. Foodscapes model: interaction of city and rural landscapes.

In developed countries like the UK, with advanced agricultural and food distribution systems, the coupling between food security and affordable housing is less obvious. Notwithstanding, the housing affordability crisis is chronic. It underlines inequality and exacerbates deprivation. Its multiple causes include demographic shifts, lax foreign ownership regulations; legitimate and nefarious capital inflows, wage inequality; social housing underinvestment; land supply speculation; planning restrictions and rail policy debacle and mismanagement. One major dwelling demand driver is demographic change, driven by a combination of life expectancy increases, sustained high immigration and the proliferation of single or single-parent households (Goddard, 2016). When house prices escalate, earnings stagnate and mortgage finance tightens, rental tenure remains the only choice for many (Gilmore, 2014). With construction persistently below the estimated 240,000 annual need, government 'Help-to-Buy' schemes have not solved the affordability crisis (New Economy, 2014; Barker, 2004). Opportunistic Private Rented Sector (PRS) housing filled the supply gap (see Fig. 2.5 below). In the last 15 years, the PRS has more than doubled (Scanlon et al., 2015). From being the smallest tenure until 2012, it is now the second largest with over 4 million households. The sector is, however, fragmented and dominated by small unprofessional investors and landlords without training or qualifications in property management (RICS, 2014; RICS, 2015; Crook and Kemp, 2011; DCLG, 2012). Unless developers raise dwelling output by 80% over the next 15 years, Blackburn (2016) predicts 20% PRS growth within five years with prices and rents rising by 25%. The landlord bonanza exacerbated structural inequality, dwelling energy inefficiency and, arguably, abetted illegal immigration, although the 2014 Immigration Act required landlords check tenant status (Goddard, 2016). Increasingly, families are renting

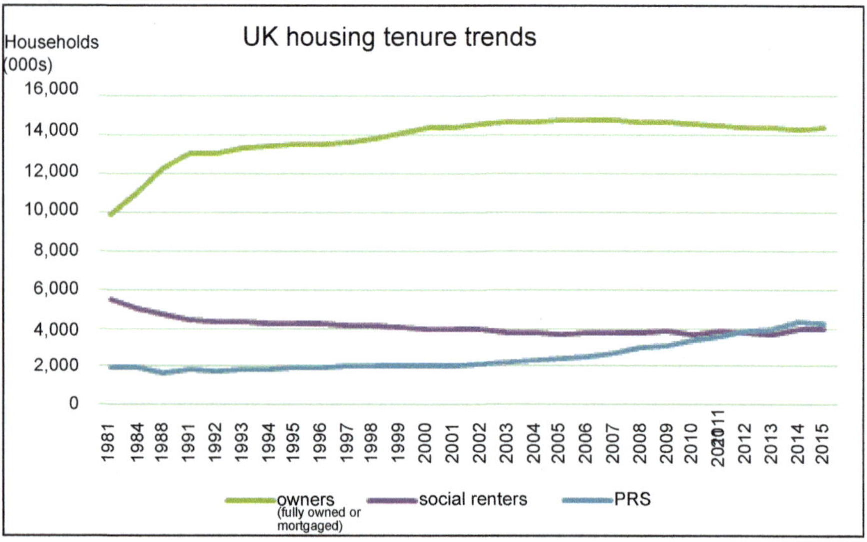

Figure 2.5. UK housing tenure, showing rise of PRS. (Source: DOE Labour Force Survey Housing Trailer; ONS Labour Force Survey; English Housing Survey.)

residential accommodation. The proportion of households in the PRS with dependent children has increased from 30% in 2004 to 37% by 2015 (DCLG, 2016).

If the PRS remains the default replacement for social housing, it requires regulation to tackle quality and energy efficiency (Energy Saving Trust, 2016). Data suggests 13.5% of PRS properties have poor or very poor EPC ratings (F or G), compared to around 9% in the owner-occupied sector (English Housing Survey, 2011–12; BPF, 2013; ACE-EBR, 2015). Despite the obvious benefits of insulation to cut fuel poverty and cold-related illnesses, landlords underinvest. In rental markets, power is asymmetric and often neoliberal competitive assumptions are spurious. Cognitively, many prospective tenants struggle to effectively compare building energy efficiency (Oxoby, 2004; Ambrose, 2015). *Ex post*, mobility is restricted by cost and stress barriers. Tenants fear that informing landlords about energy efficiency or dampness issues could jeopardize lease renewal. Finally, the government passed legislation to force reluctant landlords to act (Middlemiss and Gillard, 2015). By April 2018, it will be unlawful to let dwellings below an EPC of E (Hope and Booth, 2014; Cook and Cross, 2015).

Quality issues plague the PRS which undermines its legitimacy as a solution to the affordability crisis. Ill-considered design and poor workmanship can compromise air quality with a pernicious increase in mould spores. Damp dwellings may trigger asthma and other lung diseases by increasing the incidence of mould and the production mould spores throughout the dwelling (Smith, 2018). Tenants have no powers under s.11 of the Landlord and Tenant Act 1985 to carry out the required works themselves, but the Deregulation Act, 2015 provided some protection for assured shorthold tenants against retaliatory eviction (ibid. 2018). The Tenancy Deposit Protection Scheme (2007) also provided some protection of tenant's deposits.

Unperturbed by either energy or quality failings, neoliberals put their faith in a new breed of institutional PRS investors like pension funds and REITs. However, the Montague Review found two main barriers to institutional residential PRS investment: short leases and low yields (DCLG, 2012). Historically, capital gains, especially in the South of England, have compensated low PRS net yields. However, institutional investors also baulk at complicated tenancy management. Most tenancies in the PRS in the UK are Assured Shorthold Tenancies (ASTs) of 12 months' duration which creates instability and uncertainty for families with school children (Smith, 2018). DCLG (2017) recommends introduction of three-year family friendly tenancies. In the UK, Kemp and Kofner (2010) identified this 'value gap' as the main reason why landlords in England oppose to security of tenure or other rent regulation despite the fact that longer tenancy lengths can reduce voids and tenant search costs (Smith, 2018).

In Germany private landlords take a much longer-term view of residential property investment (ibid. 2018). Open-ended contracts cannot be cancelled arbitrarily. The landlord must have a 'legitimate interest' in the termination such as a breach of obligation, or that the landlord can prove hardship at an eviction trial (ibid. 2018). In relation to rent regulation, landlords are prohibited from increasing rents to a level above the average being a 'reference rent' for the 'community' (local area) for that type of dwelling. German landlords cannot raise the rent by more than 20% in any three-year period (ibid. 2018). In developed countries like the UK, housing affordability remains a chronic problem for a whole host of reasons including income inequality, decades of underinvestment in public infrastructure and unfair land and tax policies. UK interventions, such as *Help to Buy*, housing benefit restrictions, or National Planning Policy Framework have, basically, failed to tackle the affordability crisis (Hills 2001). However, authorities now reluctantly acknowledge the spatial, energy, social, health, community and productivity imposts of dysfunctional housing markets.

Conclusion

Ensuring a sufficient supply of appropriate affordable homes is a central sustainable challenge, given tensions between strategic imperatives, commercial gain, local community empowerment and environmental protection. Whilst sustainable management issues in developing and developed countries differ, their common concerns involve land and fiscal policy, administrative systems and trustworthy planning regimes to balance impartial expertise with local participation. In poorer nations, affordable housing needs to support food security. In developed cities, key affordable issues are value capture, social housing provision and PRS regulatory tightening. In all countries, the sustainable management of affordable housing involves multiple measures at different spatial scales. It calls for political-economy and housing debate, land policy changes, fiscal measures, planning reforms and regulations to actively shape housing markets. A robust Land Administrative System provides data to inform policy, planning and project assessments. Procedural transparency and rich information flows build trust. Locally, professionals manage the system for the Affordable housing system guardians include informed policy debate, long-term planning, law, Independent scrutiny, press freedom and vigilant citizen associations.

References

ACE-EBR. 2015. Cold-man-of-Europe available at: http://www.ukace.org/wp-content/uploads/2015/10/Cold-man-of-Europe update.pdf (accessed 12 March 2013).

Ambrose, J. 2015. Studying the links: The relationship between fuel poverty, energy efficiency and health, RICS Property Journal Nov 2015, pp. 34–36.

Arundel, R. 2017. Equity Inequity: Housing wealth inequality, inter and intra-generational divergences, and the Rise of Private Landlordism. Housing, Theory and Society 34(21): 176–200.

Barker, K. 2004. Review of Housing Supply, Delivering Stability: Securing our Future Housing Needs, HMSO.

Bate, A. 2015. Building the new private rented sector: Issues and prospects (England). Briefing Paper, SN07094, 12 August 2015].

Blackburn, J. 2016. Crisis? What Crisis? RICS Residential Policy, RICS Property Journal, July/Aug 2016, 36–37, (online) available at rics.org/journals.

BPF. 2013. A British Property Federation guide to Energy efficiency and the private rented sector, available at: http://www.bpf.org.uk/sites/default/files/resources/Energy%20Efficiency%20and%20the%20Private%20Rented%20Sector.pdf (accessed 01 December 2017).

Cook, A. and S. Cross. 2015. Minimum energy efficiency standards for UK rented properties, Out-Law.com, Legal news, August 2015, Pinsent and Masons, available at: http://www.out-law.com/en/topics/property/environment/minimum-energy-efficiency-standards-for-uk-rented-properties/(accessed 7 August 2016).

Crook, T. and C. Whitehead. 2002. Social housing and planning gain: is this an appropriate way of providing affordable housing? Environment and Planning A 34(7): 1259–1279.

Crook, A. D. H. and P. A. Kemp. 2011. Transforming Private Landlords. Oxford: Wiley-Blackwell.

DCLG. 2010. English housing survey, Household report, National Statistics, 2008–09.

DCLG. 2012. The Montague Review: Barriers to institutional investment in private rented homes, available at: https://www.gov.uk/government/uploads/system/uploads/attachment_data/file/15547/montague_review.pdf (accessed 6 July 2016).

DCLG. 2017. Housing White Paper-fixing our broken housing market, DCLG, 7 Feb 2017.

DCLG. 2017. Planning and affordable housing for build-to-rent, consultation pap.er, DCLG February 2017 (online) at: https://www.gov.uk/government/consultations/planning-and-affordable-housing-for-build-to-rent.

Department of Energy and Climate Change. 2011. Local Authorities and The Green Deal, Information Note, Nov 2011, DECC, London.

Department for Communities and Local Government. 2016. English Housing Survey Headline Report 2014–15, available at: https://assets.publishing.service.gov.uk/government/uploads/system/uploads/attachment_data/file/501065/EHS_Headline_report_2014-15.pdf (accessed 21/09/2018).

Energy Saving Trust. 2016. Private Rented Sector, National Archives, available at: http://webarchive.nationalarchives.gov.uk/20160105160709/http://nomisweb.co.uk/query/select/getdatasetbytheme.asp (accessed 22 August 2016).

English Housing Survey (2011-12) available at: https://www.gov.uk/government/publications/english-housing-survey-2011-to-2012-headline-report (accessed 22 August 2016).

Fernandez, R. and M. Aalbers. 2017. Housing and capital in the twenty-first century: realigning housing studies and political economy. Housing, Theory and Society 34: 151–158.

Gilmore, G. 2014. The rental revolution-examining the private rented sector 2014'. Residential Research, Knight Frank, London: p4.

Goddard, J. 2016. Am I qualified for this? Right to rent is no solution to the public's immigration concerns. RICS Property Journal 40-41. ISSN 2050-0106.

Hausman, D. and M. McPherson. 2006. Economic Analysis, Moral Philosophy and Public Policy. 2nd ed., Cambridge, Cambridge University Press.

Hills, J. 2001. Inclusion or exclusion? The role of housing subsidies and benefits. Urban Studies 38(11): 1887–1902.

Hochstenbach, C. 2017. State-led gentrification and the changing geography of market-oriented housing policie. Housing, Theory and Society 34(4): 399–419.

Hope, A. J. and A. Booth. 2014. Attitudes and behaviours of private sector landlords towards the energy efficiency of tenanted homes. Energy Policy 75: 369–378.

Horti, S. 2016. Private rented sector; can build-to-rent thrive in a Brexit-induced slowdown?

Huston, S., A. Jadevicius and N. Minaei. 2014. Talent and student rented sector bottlenecks: A preliminary UK investigation. Property Management 33(3).

Jack, G. 2012. The role of place attachments in well-being. In Wellbeing and Place, Edited by Atkinson S. and Painter, J. Abingdon, Ashgate Publishing Group.

Johnson, M. and J. Heinz. 2006. Decision Models for Affordable Housing and Sustainable Community Development'. Journal of the American Planning Association: The Future(s) of Housing, available at Research Gate (accessed 28 November 2017).

Johnson, M. 2007. Planning models for the provision of affordable housing. Environment and Planning B: Planning and Design 34: 501–523.

Kemp, P. A. and S. Kofner. 2010. Contrasting varieties of private renting: England and Germany, International Journal of Housing Policy 10(4): 385.

Maliene, V. and N. Malys. 2009. High quality housing—a key issue in delivering sustainable communities. Building and Environment 44: 426–430.

Manchester City Council. 2017. Providing for Housing Choice—Planning Guidance, available at http://www.manchester.gov.uk/planning/(accessed 29 November 2017).

Middlemiss, L. and R. Gillard. 2015. Fuel poverty from the bottom-up: Characterising household energy vulnerability through the lived experience of the fuel poor. Energy Research & Social Science 6: 146–154.

Morano, P., F. Tajani and M. Locurcio. 2015. Land use, economic welfare and property values: an analysis of the interdependencies of the real-estate market with zonal and socio-economic variables in the municipalities of Apulia region (Italy). International Journal of Agricultural and Environmental Information Systems 6(4): 16–39.

MSCI, Strutt and Parker and BPF. 2015. UK Lease Events Review-Nov 2015.

Mulliner, E., K. Smallbone and V. Maliene. 2013. An assessment of sustainable affordability using a multi-criteria decision making method. Omega 41(3023): 270–279.

Niemietz, K. 2015. Reducing poverty through policies to cut the cost of living. York; Joseph Rowntree Foundation: 11–12. available at: https://www.jrf.org.uk/sites/default/files/jrf/reducing-poverty-cost-living-summary.pdf.

New Economy. 2014. Mapping the private rented sector for young professionals and mid incomes families in Greater Manchester-New Economy (online) available at: http://neweconomymanchester.com/publications/private-rented-sector-in-greater-manchester.

New Economics Foundation. 2017. How to fix the housing crisis: take land seriously. Martin, A., Ryan-Collins, J. and Macfarlane, L. Accessed at http://neweconomics.org/ (Nov 2107).

ONS. 2013. Households, families and people-General Lifestyle Survey Overview—a report on the 2011 General Lifestyle Survey-(online) available from: http://www.ons.gov.uk/peoplepopulationandcommunity/personalandhouseholdfinances/incomeandwealth/compendium/generallifestylesurvey/2013-03-07/chapter 3 house holds familes and people generallife style survey overview are port on the 2011generallifestylesurvey (accessed 20 August 2016).

ONS. 2014. Young adults living with parents, 2013. London: Office for National Statistics.

Owens, S. and R. Cowell. 2011. Land and limits: Interpreting sustainability in the planning process. Routledge.

Oxoby, R. J. 2004. Cognitive dissonance, status and growth of the underclass. The Economic Journal 114: 727–749.

RICS. 2014. Global Affordable Housing Report, BRICS Plus Mortar, St Andrews Centre for Housing Research.

RICS. 2015. Private rented sector, Code of practice, RICS, available at: http://www.rics.org/Global/Private_Rented_Sector_code_PGguidance_amended_July_2015.pdf (accessed 7 August 2016).

RICS. 2017. Has mandatory regulation of the Private Rented Sector finally arrived?' available at http://www.rics.org/uk/news/news-insight/news/has-mandatory-regulation-of-the-private-rented-sector-finally-arrived/(accessed 30/11/17).

Romero-Lankao, P., T. McPhearson and D. Davidson. 2017. The food-energy-water nexus and urban complexity. Nature Climate Change 7: 233–235.

Scanlon, K., M. Fernández and C. Whitehead. 2014. A lifestyle choice for families? Private renting in London, New York, Berlin and the Randstad. Get Living London, London, UK.

Scanlon, K., C. Whitehead and P. Williams. 2015. Taking stock buy to let, available at http://www.lse.ac.uk/business-and-consultancy/consulting/assets/documents/Taking-stock.-Buy-to-let.pdf (accessed 30 November 2017).

Sen, A. 1999. Development as Freedom. Oxford, Oxford University Press.

Smales, C. 2016. Housebuilders down tools until referendum storm has passed, Home Truths, Professional, Property Week, 19/08/16: p53.

Smith Institute. 2014. The growth of private rented sector: What do local authorities think, available at: https://smithinstitutethinktank.files.wordpress.com/2014/09/the-growth-of-the-prs-what-do-local-authorities-think.pdf (accessed 27 February 2017).

Smith, P. 2018. Affordable housing and private rented sector reform. Ch. 15 in Smart Urban Regeneration Visions, Institutions and Mechanisms for Real Estate. Editor: Simon Huston. Abingdon, Routledge.

Souter, J. and A. Kafkaris. 2015. The politics of private rent, EGi, Legal, Estates Gazette, 24 April 2014, available at: http://www.egi.co.uk/legal/the-politics-of-private-rents/?keyword=private%20rented%20 sector (accessed 28 August 2016).

Storper, M. 2013. Keys to the City: How Economics, Institutions, Social Interaction, and Politics Shape Development. NJ, Princeton University Press.

Tiesdell, S. 2004. Integrating affordable housing within market-rate developments: the design dimension. Environment and Planning B: Urban Analytics and City Science 31(2): 195–212.

Tostensen, A., I. Tvedten and M. Vaa. 2001. The urban crisis, governance and associational life. In Associational life in African cities: popular responses to the urban crisis. Editors: Tostensen, Tvedten and Vaa. Stockholm. Nordiska Africainstitutet.

UNFPA. 2007. State of the world population 2007: unleashing the potential of urban growth, available at: www.unfpa. org/publications/state-world-population-2007 (accessed 30 March 2017).

UN Habitat. 2016. World Cities Report: Urbanization and Development: Emerging Futures http://wcr.unhabitat.org/main-report/(accessed 30 March 2017).
Wadley, D. 2010. Exploring a quality of life, self-determined. Architectural Science Review 53(1): 12–20.
Wetzstein, S. 2016. The global urban housing crisis and private rental in the Anglophone world: future-proofing a critical sector and tenure. Housing Finance International. Autumn 2016: 31–34.
Wilkinson, R. and K. Pickett. 2009. The Spirit Level. Why Equality is Better For. London, Penguin Books.

CHAPTER 3

The Implementation of a Sustainable Management System for the Delivery of Affordable Housing

Soo Cheen Khor

Introduction

The concept of sustainable housing has been developing in line with the goals and principles of the Habitat Agenda, which is a blueprint for sustainable development in the 21st century adopted by 179 nations including Malaysia, in Rio de Janeiro in June 1992. The concept of housing sustainability can be understood as a sustainability loop. Housing development is undergoing a three phase life cycle process: namely at emergent, rational and constrained levels (Kotler et al., 2003). Nevertheless, housing development now faces numerous constraints. The development of housing at various levels of performance, namely: design, construction, operation and demolition, can have a significant impact on the environment (Huby, 1998).

The development coincides with Tosics's (2004) statement that housing constraints can have implications on urban development. The imbalance between environmental, social and economic elements has resulted in a vicious circle for human lives. However, Chougill (1994) stated that sustainable housing may be understood in terms of ecological sustainability, economic sustainability, technological sustainability, cultural sustainability and social sustainability.

The concept of sustainable housing development can be defined as a joint venture between internal (housing) and external (housing neighbourhoods) aspects of housing interactions along with their performance-based characteristics. The sustainability of housing developments comprise of a multidisciplinary approach with extensive

Universiti Tunku Abdul Rahman Malaysia.

and sophisticated consequences to understand (Clayton and Radcliffe, 1996). The implementation of sustainability therefore incorporates an open system approach in order to solve complicated problems in a systematic way.

Compliance practices of management principles in a housing sustainability system can lead to pragmatic and systematic performance practices in an organization. A sustainable management system comprises of 10 steps in a management process to achieve sustainability in housing developments. In order to achieve the objective of housing sustainability, the authors of the present study grapple with questions of how integrated economic, social and environmental sustainability principles can be pragmatically implemented in order to deliver affordable housing and sustainable development in the housing sector.

Sustainability

The initial definition of sustainable development was defined by Brundtland Commission in 1987 as "development that meets the needs of the present without compromising the ability of future generations to meet their own needs". According to Blackburn (2007), the concept of sustainability has been perceived as a long-term development for countries. Sustainability practices embrace the rigorous use of scarce natural resources through practising good economy management principles, without compromising the environment for the sake of society. The incorporation of economic, environmental and social aspects into a sustainable management system should be emphasized, as well as the philosophy of sustainability.

There are three phases in a lifecycle of development which include the "Emergent" phase at a primary stage, the "Rational" phase in the middle stage, and the "Constrained" phase in the final stage of the cycle. According to the views of the economist Weintraub (1985), in that day and age, people maximized utilities while companies maximized profits. A rapid development progress stimulated the exploitation of natural resources and a lack of accountability for the consequences compromised the environment. The leaders, e.g., the developers in housing developments who played important roles in leading performance, continued to manipulate the market and pushed business production during the emergent phase of the life cycle.

Nevertheless, the emergent phase gradually evolved to a rational stage after business began to be completed. A number of stakeholders tried to maintain the initial strategy of development mechanisms, however, a number of mishaps and disasters began to occur. The growing population worldwide but disparity in the development of various countries resulted in excessive exploitation of the world's natural resources. Human activities which have a huge impact on global ecology are now threatening the life of every living creature on the earth. Proven evidence can be seen through symptoms such as: soil erosion, deforestation, desertification, species extinctions and natural disasters such as tsunamis, hurricanes, storms, floods and earthquakes. These outbreaks are presumed to be a consequence of the overwhelming development of human activities.

The constant human greed for natural resources such as fossil fuel and mineral elements has impacted the natural resource market. Indeed there is an imbalance in the market where demand exceeds supply. Moreover, the significant production of waste has exceeded the ability of the ecology system to digest and suppress waste generation. These adverse scenarios might have direct or indirect vicious effects on the development of the economy, the environment and society as a whole. As a consequence, countries will eventually fall into a phase of quandary, which might compromise the ability of future generations to meet their needs.

Thus, the transformation of a development cycle into sustainable mechanisms of development is essential. Leaders have to take the initiative to transform the cycle by adopting creative destruction of capitalism revolution. It is the time to move away from the constrained phase and pass into the rational stage of the sustainability loop. Renewal policies which incorporate sustainability key areas into new development guidelines are considered to be a good way to manipulate a sustainable way of development. In order to sustain the balanced development of a country, it is essential that all stakeholders have to work strenuously to lead change, have the support of their followers and work together to implement the guidelines of renewal policies.

Moving through a continuous loop of a cycle enables a country to undergo a renewal stage of evolution. Adaptation of change can occur by incorporating three strategic spectrums namely: political, technical and cultural changes (Kotler el al., 2003). Economists have stated that political issues go hand in hand with authorization for mandatory change, which is needed especially during quandaries faced during works of transformation. However, technical changes are used to implement new strategies which can substitute obsolete technology. A number of good practices include the incorporation of a green techno-economic paradigm (Daniels, 2005). An example of this could be to replace non-renewable resources of materials with renewable resources. Last but not least, cultural changes have a significant effect on the evolution of the Sustainability Loop. Indeed, the country's followers have the most powerful influence on making changes in a country's development. The dynamic support of people can assist leaders to drive change of a country's development in this modern day and age and implement sustainability trends.

Sustainability Management System

The incorporation of management principles in a sustainability system can help to achieve the task of sustainability. The task is comprised of two different stages (Blackburn, 2007).

1) The first stage is the establishment of policies which will lead to the goal of sustainability. Generally, this is a preliminary stage where a theoretical and conceptual framework is outlined. The feasibility of the policies will be taken into account by incorporating techniques, methodologies and theories of the framework.
2) The second stage involves implementing established policies by applying pragmatic approaches, management principles, mechanisms, strategies, and diversified options of technical techniques. The second stage is considered to be a critical path of work that involves actual decision-making processes. The

second stage will require the involvement of political authorities, particularly when mandatory work needs to be done.

A sustainability operated system is defined as "a process of proactive, holistic organizational management purpose of achieving sustainability for both the organization and society" (Blackburn, 2007). The purpose of employing a sustainability operated system is to provide guidelines, especially where there is decision-making on important issues, so that all pertinent factors to a sustainability policy is considered. Primary issues should incorporate the basic structure of sustainability in the process of drivers, efficient enablers, pathways and evaluators (Blackburn, 2007).

A management approach of a sustainability operated system which is adopted in an organization practice to accomplish a sustainability policy has been proven to drive performance-based experiences (Blackburn, 2007). The integration of teamwork into an organization is required to practice every step of an efficient management system in order to achieve sustainability goals. Synchronously, the employment of a sustainability operating system in a project organization is a strategic way to adapt change in a transformation cycle. Nonetheless, the use of measurable indicators allows an organization to illustrate the level of achievement in sustainability goals during an operational system.

Achieving Sustainability in Housing Development

Housing is a basic need of everyone. In conjunction with this argument, the Malaysian government has set a policy which aims to enable every family to have the chance to own a house. Since the introduction of the New Economy Policy (NEP) in 1971, housing programmes have been undertaken by the private and public sector in order to meet the needs of the population. The government prosperously developed housing programmes with the perception that the development of housing industries will stimulate the growth of the economic sector (Agus, 2002).

Approximately half of the world's populations live in cities (United Nations, 2003). In Malaysia, a majority of housing developments are established in cities. Land in Malaysia is comprised of West and Peninsular Malaysia. Most cities have a high population density with a high number of housing schemes developed. The reason why many people live in cities is due to an urbanization process and a push factor of rural migration to urban areas. The population growth in urban cities was six times higher after the implementation of the New Economy Policy (Agus, 2002; Evers, 1979; Agus, 1981).

The demand for houses continues to increase in congruence with the growth of the population. The allocation within the Malaysia Plan budget for housing programmes every five years has also increased. Up until the Ninth Malaysia Plan (2006–2010), the housing units required was 709,400 units, which is an increment of 15.35% compared to the 615,000 units in the Eighth Malaysia Plan (2001–2005). Thus, many housing schemes have been carried out with a particular focus on urban cities. The authors of the present study believe the evolvement of housing neighbourhood developments occurred during the Fifth Malaysia Plan (1986–1995).

This can be proven when the government introduced the human settlement concept in housing scheme projects. The idea of the concept can be interpreted to mean that housing programme development had to provide facilities and infrastructures such as schools, clinics, community halls and promote economic opportunities (Agus, 2002). The concept of a neighbourhood unit in a residential area was introduced since 1923 by Clarence Arthur Perry through his monograph "The Neighborhood Unit: a Scheme of Arrangement for the Family Life Community" (Lawhon, 2009). The definition of Neighbourhood Unit is consistent with the concept of a human settlement, which has been defined above.

In order to fulfil a neighbourhood unit, providing physical standards that encourage interaction activities, and also planning the provision of infrastructure and facilities, namely, schools, open spaces, institutional and commercial uses that provide opportunities for economic promotion, is required. The prevailing point of a Neighbourhood Unit is that it emphasizes residents security and safety. The design of the areas is required to limit residents walking distances. Walking distances must not exceed a quarter-mile distance, which is equal to approximately 1.6 km. The areas must connect facilities and infrastructure features to avoid residents crossing over main arterial streets (Lawhon, 2009).

Sustainable development is unattainable without sustainable housing and housing neighbourhoods. The development of a sustainable housing neighbourhood can help achieve urban sustainability at a macro level (Kennedy et al., 2005). It has significant potential to contribute to sustainable development. However, due to the growth of populations in cities, many unsustainable features in contemporary cities can be seen. This might be a consequence of poor planning by the Town and Country Planning Department (TCPD) at a micro or neighbourhood level (Codoban and Kennedy, 2008).

The term neighbourhood sustainable development in a residential area means the adoption of elements that embrace economic, environment and social pillars. Moore and Scott (2005) said that sustainable development is essential for human settlement development when it has full consideration of the needs of achieving economic growth, social development and environment protection. It is increasingly linked with the concept of quality of life, well-being and livability. The achievement of sustainability in a neighbourhood can lead to sustainable livelihoods (Kato, 1994).

Sustainable livelihood is an approach to social and economic activities for all societies to be compatible with preservation of the environment. In line with this, Lawhon (2009) explained that social and economic coherency is significant to social interaction and stability of a neighbourhood. Codoban and Kennedy (2008) stated sustainable neighbourhoods in residential areas require strategies that promote green buildings, integrated water systems, cycling, pedestrians, transit friendly designs, urban forestry, local energy production; and neighbourhood management.

Several system tools to evaluate the environmental performance of buildings is being established. The growth and use of buildings' environmental performance assessment systems can be considered as a great contribution to sustainable practices in the building industry. The assessment tools developed worldwide are built upon differing evaluation criteria which are based on countries' conditions to suit each countries' own characteristics. The differences between the system tools are based on criteria such as: geography, culture, technical availability in local contexts and differing scales of development (Kyvelou and Filho, 2006).

The incorporation of sustainable elements is essential to establish the variables of sustainable housing development. However, the economic, social and environmental sustainability framework must be integrated and interlinked and be mutually interdependent and influence one another (Basiago, 1999; Montmollin and Scheller, 2007).

Correlation of Economic Sustainability with Affordable Housing

Economic sustainability in a city development means "the potential to reach qualitatively a new level of socio-economic, demographic and technological output which in the long run reinforces the foundations of the urban system" (Ewers and Nijkamp, 1990 cited by Basiago, 1999). Kahn (1995) stated that the element of economic sustainability included the criteria of growth, development and productivity which had significant positive influences. Meanwhile, affordable housing is defined as a term to describe dwelling units where total housing costs are deemed to be "affordable" to a particular median income group of society and housing costs do not exceed 30% of a household's gross income (U.S. Department of Housing and Urban Development). Housing productivity which is an outcome of technical and engineering productivity, is a measure of output from a production process, per unit of input, which will influence economic trends (Pineda, 1990; Saari, 2006). Nonetheless, eco-efficiency practices during the construction of housing will provide big savings for project stakeholders. Additionally, waste prevention practices and energy efficiency during productivity will reduce operation and capitalization costs.

However, the growth of the economy has a negative impact on social and environmental aspects of countries. Significant evidence of this has been recorded in the 1980s where the destruction of the environment was caused by huge construction projects in the past. There has been global unanimous agreement that development has increased the social problem of disparities and accelerated the exploitation of biodiversity (Basiago, 1999). Thus, a country should consider the pros and cons of economic development. It is important to note that economic development should be confined to resolving social and environmental issues.

On the contrary, Ellison and Sayce (2006) justified that anticipating sustainability will have an impact on a property's worth by a number of parameter variables such as rental growth, depreciation, cash flow, letting duration, duration of sale and economic value. The examples of the variables to be assessed for economic sustainability should incorporate property transaction elements to ensure development in a designated area is available for business sectors to suit the needs of people.

Significance of Economic Sustainability in Housing Development

The fact of the matter is, the economy is a critical factor among sustainability criteria. The factor became predominant after the emergence of the Bruntland Report

(Bruntland, 1987) which emphasized the importance of sustainable development (Stern, 2006). In addition to this, the doctrine of 'sustainable development' was acquired from an economics discipline, and started to evolve two centuries ago (Basiago, 1999). Eventually, economic growth is the main concern of people but there is a relationship between environmental sustainability and economic development, as environmental sustainability will have a direct impact on economic development.

As mentioned by Basiago (1999), long term economic growth may be restrained by environmental destabilization and environmental problems. The practice of economic sustainability development or the alternative term 'steady-state economic' growth, attempts to achieve an equilibrium between the cost and benefit of environmental and social aspects. Neither is concerned with the exploitation of the environment, reproductive exponential growth nor pollution matters beyond its capacity of absorption (Daly, 1973; Alexander, 1994).

The Perspective of Economic Value and Worth in Property Sustainability Development

The pillar of 'economic sustainability' was generated by Hicks in his classic second edition work 'Value and Capital', during the years 1939 and 1946 (Basiago, 1999). The commensuration of economic sustainability implies the value and worth of an economy, which has an implication on a demand and supply system of consumption without compromising future needs (Bruntland, 1987). Figure 3.1 represents the relationships between environmental, social and other values to build performance financially.

Figure 3.1 demonstrates that economic values cannot stand alone to represent a property valuation in the niche of sustainability implementation. A property's economic value is influenced by a building's performance and a building's neighbourhood taking into account environment conservation issues and social and cultural values. Economic value is constituted by two distinguishable definitions, which are market value (i.e., exchange value) and worth (i.e., use value), as shown in the center of Fig. 3.1. The term 'worth' is defined as property value to stakeholders (house buyers including owner-occupiers) involved in the speculation of property stocks (RICS, 2004). In conjunction with investment activities, certain values of worth can be perceived by mathematical calculations. The gain of worth can be reflected through either a maximum or minimum capital sum that stakeholders are willing to pay or accept for their property.

However, the perspectives of worth differ for different investment activities. Stakeholders buying a property mainly for investment will refer to worth as a so called discounted value through the generation of a property cash flow. Meanwhile, a house owner wishing to occupy a property may perceive the worth of their asset through a beneficiary-pays principle. Therefore, no matter what the circumstances are, the calculation of worth in terms of the advantages of a sustainable property should incorporate indicators of speculative risks and resultant expenditures in incremental costs for the use of environmental conservation purposes, plus the costs for the pragmatic impetus of Socially Responsible Investment activities (Lowe and Ponce, 2009).

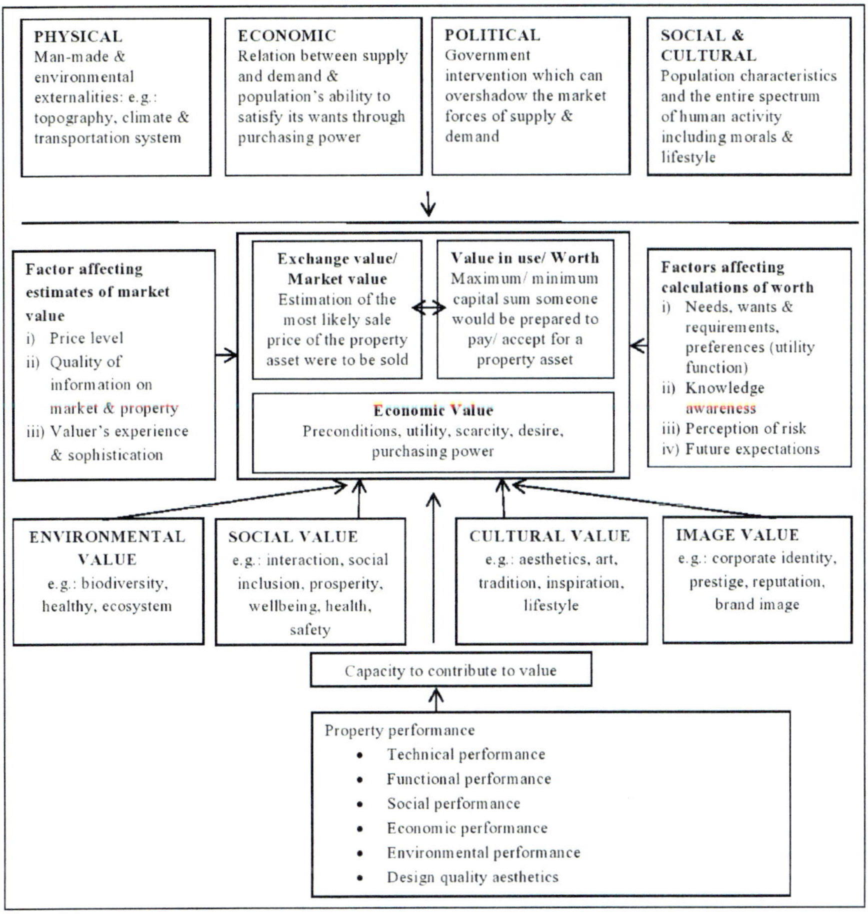

Figure 3.1. The relationships between environmental, social and other values for property performance (Source: Lowe and Ponce (2009)).

The term market value has been given a few definitions by the International Valuation Standards Council (IVSC) which are listed as follows:

1) *"Market Value is the estimated amount for which a property should be exchange for on the date of valuation between a willing buyer, and a willing seller in an arm's length transaction after proper marketing wherein the parties had each acted knowledgeably, prudently and without compulsion"* (IVSC, 2005).
2) *"The definition of market value connect closely to the concept of highest as well as best use and is a 'fundamental and integral part of Market Value estimates'"* (IVSC, 2005).
3) *"Highest and best use defined in international standards as: 'The most probable use of a property which is physically possible, appropriately justified, legally permissible, financially feasible, and which results in the highest value of the property valued"* (IVSC, 2005).

The term market valuation defined by the IVSC is an estimation of the highest and best use of a property hypothetical price analysis. The highest and best use value of a property is formed through competitive market forces. The objective of analyzing market forces are to assist in the set-up of "the foundation for a thorough investigation of the competitive position of the property in the minds of the market participants" (Sheila et al., 2014). Essentially, market valuation of a property trend should incorporate the transformation of market participants' perspectives of the benefits gained for the ownership of their property assets.

Sustainability practices have a significant impact on property market forces and will thus influence competitive trends in a property market. Lowe and Ponce (2009) suggested that the establishment of a property's rating system should incorporate valuation indicators in order to reflect how attainable a sustainability achievement is. Figure 3.2 illustrates the theoretical calculation of Market Value mechanisms and incorporates the valuation of sustainability issues with other relevant market valuation parameters. Ellision and Sayce (2006) explained five sustainability indicators which could affect the mechanisms of property valuation of worth, namely rental growth, depreciation, cash flow, duration of rental and duration of sale.

However, it seems unachievable in the real world, due to a lack of information of operational data on performance-based building descriptions in property transaction databases and the quantification of linking relationships in a property market. More critically, there is a lack of understanding of market behaviours that may underpin market forces and drivers. Thus, the establishment of an economic valuation will take years to accumulate informative data to enable the distinguishing of valuation bonuses for a sustainable building (Lorenz and Lutzkendorf, 2008).

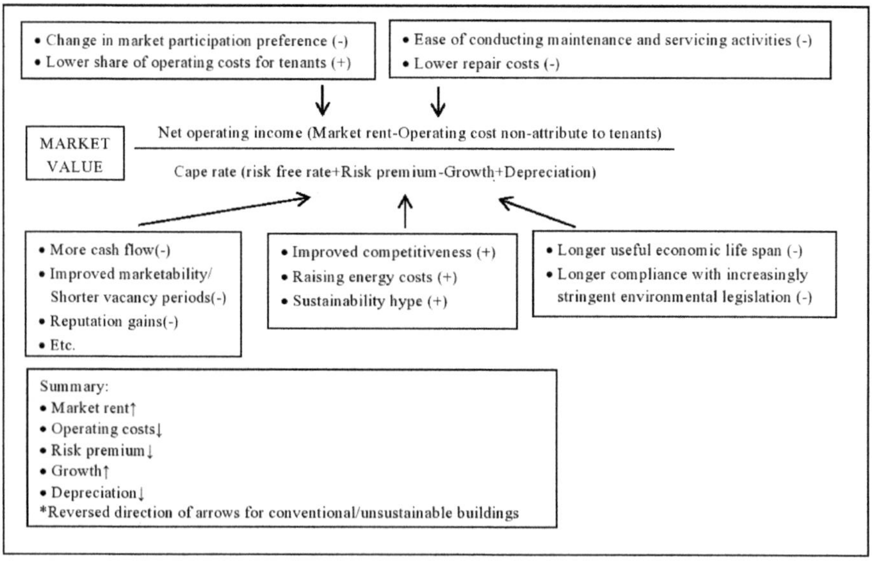

Figure 3.2. Theoretical mechanisms for Market Value calculation with the incorporation of sustainability issues (Source: Lowe and Ponce (2009)).

As cited earlier in the economic valuation section, the calculation of a market value seems unattainable in the real world due to missing data on the operation of sustainability approaches and linkages of relationships in the real-world. However, practising sustainability approaches in property development will have an impact on economic valuation. The advantages of a sustainable property cannot be perceived through direct financial gains using a market value calculation. Alternatively, the worth of a sustainable property can be evaluated through its indirect gains or benefits. The benefits of economic evaluation will be reflected through indicators such as increase in property marketability, housing occupancy and rental prices, which will reduce unnecessary costs. Moreover, sustainable property can assist with obtaining loans on favourable terms for building projects. Many have unanimously agreed that sustainable property projects can have many advantages and link the relationship between the attributes of sustainable property with the minimization of a number of property risks (Wilson et al., 1998; Heerwagen, 2002; Mills, 2003a,b; Kats et al., 2003; Royal Institution of Chartered Surveyors, 2004; 2005). The influences of sustainable property are shown in Table 3.1.

Recently, especially in the demand of sustainable housing in Malaysia, these scenarios can be seen in growing markets and the rapid sale of a building once it is completed. The concept of sustainable housing is becoming well known and has

Table 3.1. Sustainable design features that affect the minimization of property risks.

Symbolization of sustainable property	**The likelihood of risk minimization**
Flexibility and adaptability	Risks reduction through market stakeholders' transition of strict preferences imposed by third parties
Energy efficacy and water conservation	Risk minimization can be attained through energy and water usage mechanisms that may reduce expenditure on water and energy; minimized risk for business activities which may be caused by power outages which are channelled from on-site electricity power supplies with energy efficiency characteristics
Use of building materials and products that have environmental safety certificates	This may protect construction workers and building occupants from harmful particles and chemicals and therefore may assist in the reduction of denouncement risk which may subsequently reduce the risk of compensation
Relevance with socially responsible investments such as the beneficial features of convenience location that fulfils comfort and health ambiance	Reduce the risks of vacancy and tenancy unavailability
The buildings are distinguishably of better quality, reliable market acceptance with ease of maintenance, systematic management approaches	Property values are more stable with long lasting worth
Many of sustainable policies are tied to the predefined legal compliance of environmental and health requirements	Minimizes the risks of spending tremendous renovation fees and subsequently reduces losses in property valuations

Source: Adopted from Lorenz and Lutzkendorf (2008).

become a current trend to move towards a sustainable way of development, particularly in the real estate sector. End consumers nowadays are prone to stay in real-estate built with sustainable housing concepts, this increases the market value of sustainable designs (Tan, 2011). The practice of economic efficiency in sustainable development considers the efforts to fulfil the demands of the greatest goods. Initiated action plans should take into consideration all the expenses of external costs but also the prices of social and environmental elements (Blum and Grant, 2004).

Example of Benefits of Sustainability in Built Environment Features Brought to the Economic Sector

A piece of real estate property constitutes of a 'neighbourhood' and 'housing' under an urban built environment in order to form a unit of residential neighbourhoods, where there is a multitude of human mass and energy flows (Charlot-Valdieu and Outrequin, 2003). An urban scale project is defined as a project where multiple entities have an influence on a project's development and operations, where interdependencies and interactions exist between the different effects of a project, while each building functions independently. In other words, improvements of an environmental performance for a designated area has to consider entities composed of neighbourhoods and housing (including multiple buildings, sites and public environments such as infrastructure and facilities which exist in between the sites).

Empirical work has shown that open green spaces raises prices by 7.1% (Jim and Chen, 2006), while accessible open and green spaces near homes can raise house prices by 5 to 6% (Tyrvainen and Miettinen, 2000). House buyers are willing to pay for houses that promote the sustainable use of resources, energy efficiency, and have healthy indoor environments (Buys et al., 2005; Sitar and Krajnc, 2008). Another advantage of sustainable neighbourhoods can be perceived through the reduction of transportation costs in terms of economic sustainability (Choguill, 2008). The way to reduce expenditures in transportation is by improving accessibility between residential areas and destinations such as working locations and facilities. Nonetheless, for the sake of reducing transportation costs, the best practice is to improve infrastructure services by providing cycling lanes and pedestrian precincts. Tan (2011) agreed on the need to integrate infrastructure services in Malaysia's housing and sustainable neighbourhoods and improve linkages between housing areas so that movement routes of automobiles can be reduced. Accessibility within neighbourhoods for cycling or walking can reduce emissions of carbon.

In addition to this, it is also time to reduce the number of private vehicles by reducing linkages between housing provinces. A further effort can be made to enhance the services of public transportation to reduce the use of private vehicles. Choguill (2008) suggested an approach which reduces the flow of traffic within neighbourhoods by diminishing the permeability of neighbourhoods through internal roads that cross through designated areas.

Arguments have also been made that accessibility within a neighbourhood can profoundly benefit economic sustainability by the inflation of housing prices (Chin et al., 2004; Hui et al., 2007; Jim and Chen, 2006; Jim and Chen, 2007; Redfearn, 2009; Jim and Chen, 2009; Poudyal et al., 2009). However, economic sustainability is not equivalent in principle and a paradox issue was raised, where a decrease in transportation costs did not decrease expenditures. Thus, no benefit was provided for house buyers because they had to spend more than expected, especially on transportation cost.

Based on property market trends in Malaysia, a majority of housing developers are concerned with costs and benefits of developing sustainable housing neighbourhoods especially since some believe there is minimal to no enthusiasm on the practice of sustainability in built environments. Most housing developers unanimously agreed that sustainability practices are expensive, this information is based on their feedback from surveys. Nonetheless, despite the fact there are other barriers to the realization of sustainable development, it was found that half of the households from Kuala Lumpur and Selangor in Malaysia were generally willing to pay more to live in a sustainable neighbourhood when interviewed by Tan (2011). This result means that housing developers should build sustainable neighbourhoods. Moreover, housing developers realizing the benefits will result in the increase of sustainable development and sustainable built environments when there are collaborative efforts to offset the problems in property development.

Evidence from the CoStar Commercial Real Estate Information (Choi, 2009) reported that buildings with LEED and ENERGY STAR certificates gained more benefits compared to conventional buildings. Furthermore, studies from the countries which are pioneers in sustainability practices have demonstrated the benefits of sustainable development in the form of investments, design, construction, implementation, maintenance and operation. However, the benefits of sustainability can only be proven in the long term. There are interaction effects between different stages of development. Therefore, the bottom line is, if properties' operate sustainable practices from inception, it will have an impact on the overall process providing a competitive value and worth in the property market (Romm and Browning, 1998).

Incorporating a Sustainability Management System into Sustainable Housing Development Progress

The establishment of a management system approach is an assembly tool to facilitate the implementation of an agenda based on experiences gained by companies who have practised management system approaches, this was recorded in the sustainability handbook of Blackburn (2007). The tool had been concisely prescribed to enable an organization to practice sustainability policies using a disciplined approached which obligates organizations with a regulation of laws to conform to. The explicit benefits for a corporation are they can effectively manage risks, reduce and prevent waste and demonstrate a company's uniqueness to other stakeholders through their sustainability report.

A standard set of terminology for a management system approach to practice sustainability in an organization of company has been established. The methodology has been outlined in a Sustainability Management System standard model. In line with this, the authors adapted the standard model of the Sustainability Management System to incorporate it into the entire process of a housing development. Each process phase is listed below.

Sustainability Management System

Identify the Scope and Implementation Purpose

The scope of a Sustainability Management System in a housing development project is administered by stakeholders who are involved in the organization of a housing construction project. Every stakeholder is obliged to be responsible for the management process which is relevant to their duties to implement sustainable housing development. The system enables communication and encourages the integration of organization in the scheme.

Define Stakeholders' Responsibility in an Organization Structure

There are four definitions, namely a key group, an organization, sustainable development and a sustainable operating system. The key group refers to a group of people who are involved in a project who have influence in a sustainable housing development progress. The authors categorized the group into explicit and implicit stakeholders. The explicit stakeholders are the actors involved in a building project such as governments, bankers, contractors, consultants, developers, etc. On the other hand, implicit stakeholders may include governments, occupiers, investors and landowners in a housing development who are not involved in a building project but will affect housing development trends (Fisher, 2005).

The definition of an organization means the integration of a key group of people working together. The concept of sustainable development is defined as the wise use of natural resources, the management of economics, respect for people and other living things (Blackburn, 2007). A sustainable operating system is the operation of sustainability which is incorporated into management aspects and has pragmatic and systematic procedures.

Identify the Policy in the Description of a Management Process

Identifying the policy for a housing project will enable an organization to work towards sustainable objectives. A systematic management approach will record every task that has been done and will provide reports of it. This enables stakeholders to share ideas and learn together to improve their performance.

Analytical Methods and Periodic Reporting

Planning before executing a housing project is essential. Initially, an analysis needs to be made on a project execution's required tasks. This enables project stakeholders identify duties which are concurrent with organizational sustainability policies, are responsible and behave in conformity with the law. For example, the Energy Efficiency in Non-Residential building legislation (MS1527: 2007) is mandatory to follow during a construction project.

There are also prioritizations that need to be considered when handling a sustainable housing project, such as the risks and opportunities that lead to successful sustainable practices in housing development, the involvement of communities and major forces of change that can affect sustainability trends such as population, technology, transport, politics, environment and economic (Fisher, 2005).

Talent management reviews enable a leader to evaluate the strengths, needs and capabilities of their project executors' performance in an organization. Analysis results enable a leader to identify and compare gaps, which, if found, a leader can figure out an action plan to close the gaps to achieve conformity of every project staff with an organization's sustainability policies.

During a pre-planning stage, a leader will review and discuss reports with every planning team periodically. Previous audit reports, regulatory enforcement data, indemnities, penalty clauses and levies will demonstrate the capabilities of an organization to manage a project in a sustainable way. A report analysis enables leaders to determine factors that deter progress. Another benefit is it enables leaders and stakeholders to identify dynamic changes of management strategies and policies in an operation.

After reviewing and analyzing pre-planning procedures, every stakeholder has to identify measurable sustainable pragmatic management approaches. They also need to set short and long-term objectives which conform to a feasible strategic plan. Finally, policy makers have to review lists of sustainable risks, issues and objectives which have been prepared. The preparation of a tactical plan in a housing sustainable management approach is to develop, deploy and identify stakeholders' responsibilities and prepare a working schedule.

Implementation and Operation

The structure of a project organization will be categorized into multiple levels. The actors in an organization are namely, leaders, executive sponsors, group leaders and team members. A sustainability leader sits at the top of a hierarchy in an organization. The leader (the developers in a housing development) is the chief executive officer (CEO) in a project organization who has the authority to monitor and control the process of a project performance. The responsibility of a leader is to coordinate the cooperation of executors to practice and implement management tasks in sustainability approaches.

An executive sponsor sits at the second level of an organizational structure. An executive sponsor is a senior-executive advisor who works directly under a leader. An executive sponsor represents a leader in an organization by promoting housing sustainability management initiatives. Group leaders sit at the third level of an organization. Each group leader will represent a senior-executive, which has to guide a group and ensure employees practice a Sustainability Management System.

A team consists of project executors who are responsible for working out objectives. They are delegated to function and implement strategies and tactical plans. In order to practice a sustainability policy in an organization and ensure a project conforms to sustainable principles, a leader is responsible to provide effective training, communication, tools, guidance and other measurements to employees, contractors and other relevant parties who work on a mission.

It is necessary to identify necessary resources such as humans, materials, technology, financial and other resources which are necessary for a project and use the resources prudently and deliberately. Decision making considerations should take into account changes of facilities, services, programmes, processes and products. It is also necessary also to prepare a contingency plan for emergency and crises responses. The plan should try to ensure processes and procedures are consistently applicable to the sustainable policy.

There is a need to maintain document policies, regulation requirements, plans, goal missions, progress and performance reports, assessment results and audits. Moreover, procedures designed to conserve and protect the community, people, property, and the environment should be published for people to read.

Monitoring, Corrective and Prevention Actions

The organization should monitor issues, namely, sustainable goals, progress of strategic and tactical plans, the practices of sustainable housing performance, results of audits, records of regulating compliance and issues which concern the public. The organization shall also be responsible for recording, prevention and correcting the occurrence of incidents such as accidents, spills, legal, ethical violations and other problems.

Analysis and Reporting

There are internal and external reports. An internal report should include numerate issues such as the risks of practicing sustainable housing, the factors for success or failure of a performance, the benefits obtained from changing to sustainable trends, the impediments to achieving sustainability and the effectiveness of a sustainable management system. An external report will include sustainability performance results and the risks of housing developments harmful to community users.

Recognition and Accountability

The achievement of stakeholders in their performance toward sustainability will be rewarded with incentives such as compensation, promotions or bonuses. These rewards will motivate the stakeholders to keep their efforts accountable. The recognition of achievement can be a prototype of a company by encouraging the compliance of other organizations.

Management and Stakeholder Feedback

An organization can survey and communicate with various groups to obtain feedback on sustainability risks and opportunities, strengths and weaknesses of performances and reports of results. The feedback can be used as a reference for future planning for better improvement of results.

Continuation of a Process Cycle

The cycle of a sustainable management process needs to start from an initial stage. A sustainability leader and teams should reevaluate their Sustainability Management System and decide whether the system or the organization's policies need to be improved before reapplying the cycle of the management system.

The Sustainability Management System is applicable to housing management projects, as a housing managerial process has a similar methodology to a company management system. A building team will manage a project in a sustainable design approach during a housing production. This is a soft system perceived as a roadmap to understand the interlock of complex systems. The standard of system has been established by the diverse principles of various models, analytical tools and constructs. The purpose of the soft system is to integrate differing perspectives and technical approaches (Clayton and Radcliffe, 1996).

Conclusion

The impact of housing developments have implications on economic, social and environmental problems. The implication of environmental pollution in housing developments is already undergoing a constrained stage of the sustainability loop cycle. Attention needs to be paid to the crisis occurring before the problems become detrimental to the whole of society, the environment and creatures living on earth. Retaining peacefulness is to secure the balance of the world. It is time to change housing development methods and pave way for sustainable ways of development.

The evolvement of a sustainability management system is the result of the pragmatic practice of sustainability. A sustainability management system is an open system approach which has the initiative to solve multi-dimensional, dynamic and evolving matters by systematic, practical and pragmatic measures.

Inevitably, the application of a sustainability management system in a housing development project is strategic to achieving the objective of affordable housing. The transition to sustainability requires the cooperation and integration efforts of all housing stakeholders. The need for sustainability indicators, particularly in housing developments, has been identified and the variables should be incorporated into a housing sustainability management system approach and convert them into a measurable performance. This in conjunction with the anticipation that sustainability has an impact on a property's worth with parameter variables such as rental growth, depreciation, cash flow, letting duration, duration of sale and economic value.

The current practices of sustainable housing development in other countries cannot be directly adopted in Malaysia due to constraints such as differing cultural patterns, a lack of green technology facilitators, a lack of government incentives and a lack of information on values mentioned by TISSUE (2007). In order to overcome these obstacles, what is required is a knowledge of information transfer into a sustainable management approach which is incorporated into housing development practices to undergo transitions of sustainability performances. Lorenz et al. (2007) stated economy changes are majorly influenced by socio-cultural, technological and political transformations around the world.

Although housing projects are tentative and temporary during a construction, nevertheless. the effects of a housing project will have a long-term impact on social, economy and environment elements. The practices of a sustainable housing development will have implications on the construction sector and have perceived symptoms of change. The observed difference between conventional housing project types in the housing sector with sustainable housing projects are that the latter requires the advice of consultants and experts in sustainable housing project schemes in order to incorporate sustainability practices and implement a sustainability management system.

The principles of social and environmental sustainability are closely interrelated. However, the practice of both sustainability policies in sustainable housing development will enable the feasibility of economic sustainability and vice versa. Basiago (1999) mentioned that an economy is fundamental to society. Thus, it can be perceived that the synergy between economic, social and environmental implementation will assist in the achievement of sustainability in housing development. Lastly, full cooperation of an organization is essential to implement sustainable housing development practices. This will inevitably have an implication on the rise of sustainability awareness in the whole of society and will have an effect on the transition from conventional types of housing development towards more sustainable housing designs.

References

Agus, M. R. 1981. Problems of Squatters in Malaysia, M. Soc., Sc. Dissertation. Unpublished Sc. Dissertation.

Agus, M.R. 2002. Housing Policy System in South and East Countries (Malaysia): Palgrave Macmillan.

Albrecht, W. P. 1983. Economics. Englewood Cliffs: New Jersey: Prentice-Hall.

Alexander, W. M. 1994. Aug. 27–29. Humans sharing the bounty of the Earth: hopeful lessons from Kerala. Paper presented at the Proceedings of the International Congress on Kerala Studies, Kerala, India.

Basiago, A. D. 1999. Economic, social, and environmental sustainability in development theory and urban planning practice. The Environmentalist 19: 145–161.

BCA Green Mark-RB. 2008. Building and Construction Authority Green Mark for Residential Building Version 3.0 ed., the Building and Construction Authority Retrieved 8 June 2009 from: http://www.bca.gov.sg/GreenMark/others/ gm_resiv3.pdf.

Berke, R., P. Conroy and M. Manta. 2000. Are We Planning for Sustainable Development? Journal of the American Planning Association, Winter 66(1): 21–33.

Blackburn, W. R. 2007. Sustainability Handbook: The Complete Management Guide to Achieving Social, Economic and Environmental Responsibility. (Vol. 1). London: Earthscan Publication.

Blum, A. and M. Grant. 2004. Sustainable Neighbourhoods: Assessment Tools for Renovation and Development, HQE2R Deliverable 16: IOER-UWE.

BREEAM (Building Research Establishment of Environmental Assessment Method) for Eco-Homes. 2006. Building Research Establishment Ltd (BRE), retrieved on 26 May 2009, from: http://www.breeam.org/filelibrary/EcoHomes_ 2006_Guidance_v1.2_-_April_2006.pdf.

Brundtland, C. 1987. World Commission on Environment and Development, in Our Common Future: Oxford University Press.

Buys, L., K. Barnett, E. Miller and C. Bailey. 2005. Smart housing and social sustainability: Learning from the residents of queensland's research house. Australian Journal of Emerging Technology and Society 3(1): 43–57.

CASBEE (Comprehensive Assessment System for Building Environmental Efficiency) for Urban Development. 2007. Tokyo: Institute for Building Environment and Energy Conservation (IBEC), JSBC (Japan Sustainable Building Consortium). Retrieved on 19 December 2008, from: http://www.ibec.or.jp/CASBEE/english/download/CASBEE_UDe_2007manual.pdf.

CASBEE (Comprehensive Assessment System for Building Environmental Efficiency) for CASBEE for Building (Detached House). 2007. JSBC (Japan Sustainable Building Consortium), Institute for

Building Environment and Energy Conservation (IBEC), Tokyo, retrieved on 3 March 2009, from: http://www.ibec.or.jp/CASBEE/english/download/CASBEE-H(DH)e_2007manual.pdf.

Charlot-Valdieu, C. and P. Outrequin. 2003. HQE²R: Towards a methodology for sustainable neighbourhood regeneration: HQE²R Brochure 1. Retrieved 2 Dec 2010, from www.eukn.org/dsresource?objectid=157014.

Chin, T. L., K. W. Chau and F. F. Ng. 2004. The impact of the asian financial crisis on the pricing of condominiums in Malaysia. Journal of Real Estate Literature 12(33): 33–50.

Choi, C. 2009. Removing market barriers to green development: Principles and action projects to promote widespread adoption of green development practices. JOSRE: CoStar Commercial Real Estate Information 1(1): 107–138.

Choguill, C. L. 1994. Sustainable Housing Programme in a World of Adjustment. Habitat International 18(2): 1–11.

Clayton, A. M. H. and N. J. Radcliffe. 1996. Sustainability a System Approach (Vol. 1). UK: Earthscan Publication Limited.

Codoban, N. and C. A. Kennedy. 2008. Metabolism of neighborhoods. Journal of Urban Planning and Development 134(1): 21–31.

Daly, H. E. 1973. Towards a Steady State Economy: San Francisco: Freeman.

Daniels, P. L. 2005. Technology revolutions and social development: Prospects for a green technoeconomic paradigm in lower income countries. International Journal of Social Economics 32(5): 454–482.

DDC Method, prepared for NYC Department of Design & Construction Office of Sustainable Design by Gruzen Samton Architects LLP with Mathews Nielsen Landscape Architects P, published on June 2008, retrieved on June 2009, from: http://www.nyc.gov/html/ddc/html/pubs/publications.shtml.

Ellision, L. and S. Sayce. 2006. The Sustainable Property Appraisal Project.: Kingston University, London.

Evers, H.-D. 1979. The Structure of Malaysian Urban Society. Paper presented at the International Conference on Malay Studies on Language, Arts and Malay Culture.

FGBC-GD. 2009. Florida Green Building Coalition for Green Development Design Standard Reference Guide. Version 6 ed., Florida Green Building Coalition Retrieved 16 June 2009 from: http://www.floridagreenbuilding.org/-db/standards/devs/default.htm.

FGBC-GHS. 2009. Florida Green Building Coalition for Green Home Standard Version 6 ed., Florida Green Building Coalition. Retrieved 10 August 2010 from: http://floridagreenbuilding.org/files/1/file/HomeStandard5.pdf.

Fisher, P. 2005. The property development process. Property management 23: 158–175.

Foster-Fishman, P. G., D. Cantillon, S. J. Pierce and L. A. V. Egeren. 2007. Building an active citizenry: The role of neighbourhood problems, readiness, and capacity for change. Am. J. Community Psychol. 39: 91–106.

Green Building Index (GBI). 2009. Malaysia system developed by Pertubuhan Akitek Malaysia (PAM) and the Association of Consulting Engineers Malaysia (ACEM). Retrieved on 6 May 2009, from http://www.greenbuildingindex.org/.

Heerwagen, J. 2002. Sustainable design can be an asset to the bottom line. Retrieved 10 April, 2009, from http://www.edcmag.com/CDA/ArticleInformation/-features/BNP__Features__Item/0,4120,80724,00.html.

HQE2R. 2004. Sustainable Revolution of Buildings for Sustainable Neighbourhoods. CSTB Catherine CHARLOT-VALDIEU, La Calade Philippe OUTREQUIN and Celia ROBBINS UWE, Energy, Environment and Sustainable Development programme. Retrieved 12 February 2009 from http://hqe2r.cstb.fr/.

Huby, M. 1998. Social policy and the environment. Buckingham: Open University Press.

Hui, E. C. M., C.K.P.L. Chau and M. Y. Law. 2007. Measuring the neighboring and environmental effects on residential property value: Using spatial weighting matrix. Building and Environment 42: 2333–2343.

International Valuation Standards (IVS) 1-Market value basis of valuation. Seventh Edition. Retrieved 2 February 2010, from http://www.ivsc.org/.

IVSC. 2005. International Valuation Standards 2005: International Valuation Standards Committee, London.

Jim, C. Y. and W. Y. Chen. 2006. Impacts of urban environmental elements on residential housing prices in Guangzhou (China). Landscape and Urban Planning 78: 422–434.

Jim, C. Y. and W. Y. Chen. 2007. Consumption preferences and environmental externalities: A hedonic analysis of the housing market in Guangzhou. Geoforum 38: 414–431.

Jim, C. Y. and W. Y. Chen. 2009. Value of scenic views: Hedonic assessment of private housing in Hong Kong. Landscape Urban Planning 91(4): 226–234.

John, H.H. 1975. Adaptation in Natural and Artificial Systems: University of Michigan Press (Ann Arbor).

John, E. 1997. Cannibals with Forks: Triple Bottom Line of 21st Century Business: Capstone Ltd.

Kahn, M. 1995. Concepts, definitions, and key issues in sustainable development: the outlook for the future. Paper presented at the Proceedings of the 1995 International Sustainable Development Research Conference Mar. 27–28 Keynote Paper, 2–13, Manchester, England.

Kats, G., L. Alevantis, A. Berman, E. Mills and J. Perlman. 2003. The costs and financial benefits of green buildings—a report to California's sustainable building task force. Retrieved 22 Jan, 2009, from www. usgbc.org/ Docs/News/News477.pdf.

Kato, S. September 3–10, 1994. Salzburg Seminar on Environment and Diplomacy. Salzburg Austria: Working Group on Sustainable Development Manuscript on file at Salzburg Seminar.

Kennedy, C. A., E. Miller, A. Shalaby, H. L. MacLean and J. Coleman. 2005. The four pillar of sustainable urban transportation. Transport Rev. 25(4): 393–414.

Kotler, P., H. Kartajaya, D. H. Hooi and S. Liu. 2003. Rethinking Marketing-Sustainable Marketing Enterprise in Asia (Vol. 1): Prentice Hall Pearson Education Asia Pte Ltd.

Kyvelou, S. and W. L. Filho. 2006. Sustainable management and urban space quality in the Mediterranean-Challenges and perspectives. Management of Environmental Quality: An International Journal 17: 611–624.

Lawhon, L. L. 2009. The Neighborhood unit: Physical design or physical determinism? Journal of Planning History 8(2): 111–132.

Lowe, C. and A. Ponce. 2009. UNEP-FI/SBCI's Financial & Sustainability Metrics Report. Retrieved 3 Mar, 2011, from http://www.unepfi.org/fileadmin/-documents/metrics_report_01.pdf.

Lee, G. K. L. and E. H. W. Chan. 2008. SURPAM (Sustainable Urban Renewal Project Assessment Model)-A sustainability evaluation of government-led urban renewal projects. Journal of Facilities 26(13/14): 526–541.

LEED for Homes Rating System. 2008. Retrieved 12 January 2010, from http://www.usgbc.org/ShowFile.aspx?DocumentID=3638.

LEED. 2009. Leadership in Energy and Environmental Design for Neighborhood Development Rating System. From http://www.usgbc.org/-ShowFile.aspx? DocumentID=5275.

Lorenz, D. P. and T. Lutkendorf. 2008. Sustainability in property valuation: Theory and practice. Journal of Property Investment and Finance 26(6): 485–521.

Lorenz. David, P., S. Truck and T. Lutkendorf. 2007. Exploring the relationship between the sustainability of construction and market value-Theoretical basics and initial empirical results from the residential property sector. Property Management 25(2): 119–149.

Mills, E. 2003a. The insurance and risk management industries: New players in the delivery of energy efficient and renewable energy products and services. Energy Policy 31(1): 1257–1272.

Mills, E. 2003b. Climate change, insurance and the buildings sector: Technological synergisms between adaptation and mitigation. Building Research & Information 31(3/4): 257–277.

Montmollin, A. d. and A. Scheller. 2007. MONET indicator system: the Swiss road to measuring sustainable development. International Journal of Sustainable Development 10(1/2): 61–72.

Moore, N. and M. Scott. 2005. Renewing urban communities: Environment, citizenship and sustainability in Ireland. Aldershot: Ashgate.

NAHB-GHB. 2006. National Association of Home Builders for Green Home Building National Association of Home Builders. Retrieved 7 May 2009 from: http://www.nahb.org/generic.aspx?genericContentID=56077.

Ninth Malaysia Plan. 2006–2010. Economic Planning Unit (EPU) Retrieved 6 January 2010. From http://www.epu.gov.my/html/themes/epu/html/-rm9/ html/english. htm.

Poudyal, N. C., D. G. Hodges and C. D. Merrett. 2009. A Hedonic Analysis of the Demand and Benefits of Urban Recreation Parks: Land Use Policy.

Pineda, A. 1990. A Multiple Case Study Research to Determine and respond to Management Information Need Using Total-Factor Productivity Measurement (TFPM). Virginia Polytechnic Institute and State University.

Redfearn, C. L. 2009. How informative are average effects? Hedonic regression and amenity capitalization in complex urban housing market. Regional Science and Urban Economics 39: 297–306.

Romm, J. J. and W. D. Browning. 1998. Greening the Building and the Bottom Line: Rocky Mountain Institute.

Royal Institution of Chartered Surveyors (RICS). 2004. Sustainability and the built environment-an agenda for action. Retrieved 10 Nov, 2011, from www.rics.org/NR/rdonlyres/DE2FC8A1-9600-46F4-9673-D13D6B686023/0/Sustainability_and_built_environment.pdf.

Royal Institution of Chartered Surveyors. 2005. Green value-green buildings, growing assets. Retrieved 6 May, 2011, from http://www.rics.org/NR/rdonlyres/ 93B20864-E89E-4641-AB11-028387737058/0/GreenValueReport.pdf.

Saari, S. 2006. Productivity. Theory and Measurement in Business (PDF). Espoo, Finland: European Productivity Conference.

Samton, G. and M. Nielsen. 2008. DDC-Sustainable Urban Site Design.

Saw, S. H. 1972. Patterns of Urbanization in West Malaysia, 1911–1970: The Malaysian Economic Review.

Sheila, A. Murphy and Tax Court (U.S.). 2014. Reports of the United States Tax Court, Volume 139, July 1, 2012, to December 31, 2012, Tax Court; First, First in hardcover; Semi-annual edition (May 28, 2014).

Sinou, M. and S. Kyvelou. 2006. Present and future of building performance assessment tools. Management of Environmental Quality: An International Journal 17(5): 570–586.

Sitar, M. and K. Krajnc. 2008. Sustainable housing renewal. American Journal of Applied Sciences 5(10): 61–66.

Stern, N. 2006. Stern review on the economics of climate change. Retrieved 12 May, 2010, from: www.hmtreasury.gov.uk/independent_reviews/stern_review _economics_climate_change/sternreview_index.cfm.

SURPAM (Sustainable Urban Renewal Project Assessment Model). 2008. Journal of Facilities 26(13/14): 526–541.

Tan, T. H. 2011. Measuring the willingness to pay for houses in a sustainable neighbourhood. International Journal of Environmental, Cultural, Economic and Social Sustainability (forthcoming): Munich Personal RePEc Archive (MPRA) paper. Retrieved 15 May 2012, from http://mpra.ub.uni-muenchen.de/30446/1/MPRA_paper_30446.pdf.

TISSUE browser. 2007. System of the current level. Specific Targeted Research for Programme: Integrating and Strengthening the European Research Area. retrieved on June 2009, from: http://ce.vtt.fi/tissuebrowser_public/index.jsp.

Tosics, I. 2004. European urban development: Sustainability and the role of housing. Journal of Housing and the Built Environment 19: 67–90.

Tyrvainen, L. and A. Mitettinen. 2000. Property prices and urban forest amenities. Journal of Environmental Economics Management 39: 205–223.

U.S. Department of Housing and Urban Development. Retrieved 15 September 2009, from http://portal.hud.gov/portal/page/portal/HUD/topics/buying_a_home.

United Nations. 2003. World urbanization prospects: The 2003 revision. Economic and social affairs. Retrieved 12 January 2010 from: http://www.un.org/-esa/population/publications/wup2003/WUP2003Report.pdf.

Weintraub, E. R. 1985. Neoclassical Economics (Vol. 1): Library of Economics and Liberty.

Wilson, A., J. Uncapher, L. McManigal, L. Hunter Lovins, M. Cureton and W. D. Browning. 1998. Green Development: Integrating Ecology and Real Estate: Wiley: New York.

CHAPTER 4

Sustainable Urban Design

A Tool for Creating Green Residential Neighbourhoods

Marek Kozlowski

Introduction

Over the past decades, there has been an on-going discourse on the issue of housing affordability and how to develop sustainable housing accessible to the medium and lower income groups. Property-led development triggered by the rise and expansion of neoliberalism has been one of the main drivers of national economies throughout the world. As a result, a substantial amount of new residential stock is market driven and with basic sustainable gadgets creating urban utopias that are mainly targeting the higher income brackets (Weaver, 2016; Drodz, 2014; Houston et al., 2017). The aim of this study is to demonstrate how sustainable urban design can contribute to creating a green residential neighbourhood affordable to a wider spectrum of the population. The corresponding objectives of this study are; to identify the basic principles and practical application of sustainable urban design, and to list the major challenges in creating affordable, green residential neighbourhoods.

The research methodology is based on an identification of the problem, the major aim, and the objectives. In order to address the objectives, this study concentrated mainly on the use of qualitative research methods. The major qualitative research methods include literature review, qualitative analysis, and observations. Orestad in Copenhagen, Vasta Hamnen in Malmo and Punggol in Singapore are presented as the case study areas. This case study approach is partially based on Yin (2013) including the definition of the problem and main objective, data collection and qualitative data analysis. A portion of this study is based on secondary data sources, such as planning documents and policies, and on information from professional literature and journals.

University Putra Malaysia.

The Concept of Sustainable Urban Design

The term urban design had its inception at a conference that took place at the Harvard Graduate School of Design in 1956. The meeting's participants, a group of leading scholars in planning and architecture, were debating the future of cities (Krieger, 2004; Lang, 2005). The post World War II era witnessed a growing gap between architecture and planning. The difference resulted because architecture focused more on the design of single buildings and town planning increased its concentration on the broad scale problems and implications of urbanization (Caves, 2005; Krieger, 2004). The discipline of the civic design was not sufficient to fill this gap, as it focused only on the design and site planning of civic buildings including government and institutional precincts. The conference participants were determined to introduce a new discipline that would halt the split between architecture and planning and also directly address the physical form of the cities.

Carmona et al., 2010 claimed that two different traditions of urban design thought, regarding interpreting design and appreciating the products of the design process, emerged in the late 1950s. The two approaches were the visual-artistic tradition and the social usage tradition. The 'visual-artistic' tradition, greatly influenced by the works of Camillo Sitte, centred on visual qualities and aesthetic experience of the urban environment. Cultural, social, economic and political factors were considered of least importance. This tradition is very dominant in Gordon Cullen's townscape approach developed in the early 1960s (Cullen, 1961). The second approach, the 'social usage' tradition, focused on how people use urban spaces. The prominent supporters of the 'social usage' tradition were Kevin Lynch, Jane Jacobs and Christopher Alexander. The social usage tradition emphasizes the appreciation of the urban environment as commonplace for everybody and the importance of identifying people's perceptions and mental images of urban form (Carmona et al., 2010; Alexander et al., 1977; Jacobs, 1964; Cullen, 1961; Lynch, 1961).

However, as a result of the political transformation, globalization and European integration made not only the problems became more common but also the differences in approach to problem-solving became less apparent. The traditional streamlines in the last two decades had undergone a process of mutual re-adaptation developing an integrated version which can be defined as contemporary urban design. The contemporary urban design is a universal notion aimed at creating places for people. It is concerned with producing a functional, high-quality urban environment that supports the diversity of activities and uses (Carmona et al., 2010). About the place making tradition, the notion of urban design is '... The design and management of the public realm' defined as public face of buildings, spaces between frontages, the activities taking place in and between these spaces, and the managing of these activities, all of which are affected by the uses of buildings themselves or otherwise known as the private realm...' (Carmona et al., 2010).

A meaningful task of portraying the wide scope of urban design has been carried out by Carmona et al. (2010), who identified its six overlapping and interrelated dimensions the morphological, perceptual, social, visual, functional and temporal. The six dimensions focus on the urban form and layout including the size and configuration of blocks and street patterns, environmental perception of the city and construction of a place, the functioning of the public realm and its dependence on social interaction and inter-communication between various groups of users, the visual-aesthetic aspects of the urban environment and the implications of time on buildings. The sustainable urban design adapts all major principles of contemporary urban design mentioned above but provides additional emphasis on creating fully sustainable urban places, neighbourhoods, districts and cities.

According to Clarke (2009) sustainable urban design should be addressed at the urban structure level as well as at the neighbourhood community level. At the urban structure level, it is imperative to achieve an integrated and holistic interdependence between the city and the entire surrounding region. At the community-neighbourhood level, the important sustainable design and planning principles include creating walkable communities where basic community facilities such as schools, health centres and recreational space, retail centres and transport stations are all within walkable distance from all the residences. Shops and retail services should be situated along the main street of a particular neighbourhood or district and also around key facilities such as major transport hubs. The residential neighbourhood should provide a range of different housing opportunities in dwelling size and building type as well as in terms of affordability. Housing densities should be the highest at the district centres and along the major transport routes. A strong accent is placed on the orientation of blocks with the consideration of the solar potential. To maximize solar potential an east-west alignment of streets is the most appropriate solution. There should be a high integration between public transport and land uses with the provision of intensified mixed development around major transport hubs. Clarke also addresses the basic social principles by promoting socially mixed and inclusive communities. To avoid social polarization a wider mix and choice of housing opportunities must be provided. Another important issue is community engagement. Cotemporary communities should fully participate in planning, design and decision making regarding the future of their residential environment (Clarke, 2009; Kozlowski, 2010).

One of the strong prerequisites for sustainable urban design and ecological planning is the bio integration of the natural and built environments (Yeang, 2009). The provision of natural environments in the city is elaborated by Hagan (2015) who introduces the term ecological urbanism. According to the author ecological urbanism overlaps urban ecology when it addresses the urban environment that is a hybrid of natural and human made elements. However ecological urbanism focuses on human activities. Ecological urbanism demands that the built form is characterized by an irreducible and undeniable relationship between the human culture and its biophysical environments.

The role of nature and landscape design on the development of urban areas is further elaborated by Van Brocke, 2009. The author stresses that both nature and landscape in the city are critical in improving the quality of life and making the urban areas more sustainable from the ecological, social and economic points of view. The provision of natural landscapes within the dense city limits is one of the ways of combating the urban exodus to suburban areas. One of the most successful design measures is planting canopy trees along major and residential streets. Tree planting along streets creates a form of natural urban enclosure, deflects attention from parked cars, and in warmer climatic zones creates a pleasant specific type of microclimate. The author also emphasizes on the importance of Water Sensitive Urban Design (WSUD) in future planning. WSUD is based on collecting all storm-water and rainwater from the area, the provision of water retention systems including ponds and canals. WSUD is also a land planning and engineering design approach which integrates the urban water cycle, including storm-water, groundwater and wastewater management and water supply, into an urban design to minimize environmental degradation and improve aesthetic and recreational appeal. WSUD uses better urban planning and design to reuse stormwater, stopping it from reaching our waterways by mimicking the natural water cycle as closely as possible (Melbourne Water, 2017).

Thorne, Filmer-Sankey and Alexander (2009) argue that sustainable urban design has a vital role to play in making places suitable for pedestrians and cyclists. Determining appropriate street width, the treatment of different types of streets and building frontages can all contribute to a variety of experience by a cyclist and pedestrian. The promotion of public transport with quality design of buses, trams, trains as well as transit stops, stations and interchanges are imperative as the best way to counteract the widespread impression that public transport is only for those who cannot afford a car.

The New Urbanism principles introduced in the 1990s placed a strong emphasis on residential building design promoting design measures such as close proximity of the building to the street, provision of entry porches, relocation of garages to the rear, and creating articulated building facades that add visual interest to the surrounding streetscape (Charter of New Urbanism, 1999). Sustainable urban design goes one step further by emphasizing on achieving energy efficiency, thermal comfort and water conservation through passive design. Thomas and Ritchie (2009) address the importance of sustainable building design. One of the most important steps is to reduce energy demand by introducing the passive design that reduces the required energy for heating, cooling and lighting. The important issue is to design to achieve thermal comfort, optimize natural ventilation and obtain acoustic comfort. The authors stress that it is important to design buildings in such a way that they can be altered within their lifetimes. Providing quality façade treatment by incorporating sun shading devices, extended eaves, balconies and terraces, and cantilevers improves the aesthetic qualities of the buildings and significantly reduces its heat impact. Green roofs with soils and plants are also an important step in achieving quality sustainable building design. The design of landscapes vertically on the high/medium rise building façade connecting to the roof-top, allows for sufficient links between the building complex and the natural habitats (Yeang, 2009; Brisbane City Council, 2016).

Other important elements of sustainable urban design include water harvesting and recycling of waste including the provision of energy from waste (Thomas and Ritchie, 2009; Ritchie, 2009). One of the most important steps is to introduce sustainable water urban management and aim at reducing waste and maximizing the recycling of waste. The re-use of existing waste can generate revenue and employment and as such contribute to the local economy. A proportion of housing waste can be recycled in the form of materials that can burn without additional fuel thus creating a potential energy source (Ritchie, 2009). The proper selection of materials for the buildings and the hard and soft landscape areas is also vital. All materials should be chosen based on local availability and their responsiveness to the local climate. One of the key objectives of sustainable design is to reduce the waste of existing materials by re-using them for construction purposes (Royse, 2009).

Green Residential Neighbourhoods

Planning and designing of ideal residential neighbourhoods have a long history and can be traced back to the 1920s when Clarence Perry developed a neighbourhood unit diagram which was published as part of the Regional Plan of New York and the Environs. Perry's concept was based on a maximum 800 m distance between the residences and the centre of the neighbourhood. The centre of the neighbourhood contained the civic buildings. The neighbourhood also had a network of small walk-to-parks. The main idea is that the centre of the neighbourhood should be within a convenient walking distance from any of the residences. The ideal size of the neighbourhood was estimated at 160 acres (65 hectares), and the entire area was surrounded by highways. Perry's neighbourhood concept influenced a generation of plans for residential neighbourhoods around the world (Douglas Farr, 2009).

The New Urbanist neighbourhood concept developed in the 1990s by Duany Platter Zyberk (DPZ) was based on Perry's principle with a few amendments. The major amendments included the conversion of highways to tree lined boulevards and the expansion of the neighbourhood centre by providing retail outlets and a bus stop. The New Urbanist concept located a school and additional retail at the edge of the neighbourhood which could also service the adjacent neighbourhood (Douglas Farr, 2008).

Farr Douglas (2008) makes a brief comparative analysis of a classical neighbourhood introduced by Charles Perry in 1929, the New Urbanist neighbourhood and the sustainable urbanist neighbourhood. The author argues that sustainable neighbourhood principles are built on the previous models adapting them to the current sustainable requirements and cotemporary community needs and aspirations. A sustainable neighbourhood must be located along a transit corridor. The main transport hub of the neighbourhood is located within its centre together with other civic and retail facilities. The neighbourhood must be fitted out with innovative infrastructure and have quality streetscape design. It must provide a pedestrian friendly environment, mixed uses, a variety of residential densities and be connected to the natural habitat by a network of greenways. The important element of a sustainable residential neighbourhood is the provision of a pedestrian and cycling network and quality housing design. The houses have to be designed to increase the aesthetic qualities of the surrounding street environment.

Churchill and Baetz (1999) developed a set of design guidelines for sustainable communities, which address a broad range of design factors, including population density, alternative modes of transportation, community agriculture, water re-use, and green building techniques. The guidelines emphasize the need to take advantage of infrastructure system interactions. Provided examples include implementation of natural wastewater treatment at the local scale and reduction in water use through low-water-use landscaping and greywater system installation.

An innovative and cutting edge approach to sustainable planning/design and developing a green built environment is identified by renowned Malaysian architect Ken Yeang. The author develops a new terminology called ecological master planning based on the environmental concept of urban planning and design as environmentally benign and seamless bio-integration of four infrastructures. The four infrastructures comprise, the green infrastructure including all the green natural areas, the blue infrastructure including sustainable drainage and water conservation system and the overall hydrological management, the grey infrastructure including roads, drains and sewerage utilities and the red infrastructure including the built environment, the enclosures, and hardscapes as well as the basic human activities. The integration of these four infrastructures provides the basis for eco-master planning and for designing urban ecological environments. The model identified by Ken Yeang can be easily applied to a residential neighbourhood or complex (Yeang, 2009; Yeang, 2010).

A new type of residential community which addresses all principles of sustainable planning and design is an urban village. In urban planning and design, an urban village is an urban development typically characterized by the medium-density housing, mixed use zoning, good public transit and an emphasis on pedestrianization. Unlike traditional residential neighbourhoods, urban villages are usually designated within an existing urban footprint. The basic premise of an urban village is to create new workplaces within walking distance from home. Urban villages have been successfully established in inner city areas of major cities including Seattle, Sydney and Amsterdam (Urban Villages, 2017).

Existing challenges to residential neighbourhood design are identified by Yigitcanlar et al., 2015. The major findings derived from the literature review of residential development areas reveal that rapid urbanization has brought environmentally, socially, and economically great challenges to cities and societies. To build a sustainable neighbourhood, these challenges need to be faced efficiently and successfully. In this regard the authors recommend that the first step is to determine the sustainability levels of neighbourhoods. The findings from this research clearly identify that master-planned communities provide more option for sustainable living than traditional sub-division and piecemeal development. Another important principle of sustainable neighbourhood communities is its management. A sustainable management framework must be established to coordinate all services and operations related to the neighbourhood and ensure its low impact on the surrounding environment (Franklin, 2006). Nicole (2012) further argues that good management supported by on-going cooperation between all actors of the housing stock is the key to achieve a well-designed and well-functioning residential neighbourhood.

Case Studies: Vastra Hamnen, Malmo, Orestad Copenhagen, Punggol, Singapore

As of today, there are still a few urban residential projects that fully address all principles of sustainable urban design. However, through the introduction of sustainable policies and growing community awareness, their number is on the increase. As part of this study three fully sustainable residential communities are discussed. They include Orestad in Copenhagen, Vasta Hamnen in Malmo and Punggol in Singapore.

Vastra Hamnen, Malmo

Sweden has been for years the frontrunner of green urban development. The city of Malmo with a population of 328,000 is located in the southwest corner of the country. Malmö is a compact city, facilitating the provision of collective services, including transport and bicycle pathways while simultaneously incorporating mixed use planning and green space; creating favourable conditions for sustainable urban development. The main aim of the current master plan is to create a robust and long term sustainable urban structure (Malmo Stad, 2017).

Vastra Hamnen (western harbour) is an entirely new ecological district located along the Baltic seafront next to the central part of Malmo. Previously the area was a typical redundant industrial land located at the fringe of the city centre. Västra Hamnen was primarily used as a port and industrial area and was home to the Kockums shipyard from the 1890s until late 1970 when the shipping industry began to decline. In the early 1990s, the site became a typical derelict post-industrial area, and the city of Malmo recognized the enormous potential of the land for future living and education. In 1998 the Vastra Hamnen branch of Malmo University was opened on the site. Three years later, in 2001 a Swedish Housing Expo Bo01 was organized. These two milestones marked the beginning of a new urban district coming to life in Malmö. The success of the Bo01 exhibition was followed by further sustainable planning and development creating one of the most sustainable and ecological, friendly residential neighbourhoods in the world. The newer developments in Västra Hamnen include the residential neighbourhoods of Dockan and Flagghusen and the University District of Universitetsholmen (Nicolle Foletta, IDTP Europe, 2017).

Today Vestra Hamnen is home to 4,500 inhabitants and covers an area of 75 hectares with its neighbourhoods offering a diverse variety of houses, activities and land uses. The Turning Torso Tower by world's renowned architect Santiago Calatrava is the dominant iconic landmark of the area. The final projected population is estimated at 10,000 covering an area of 175 hectares (Nicolle Foletta, IDTP Europe, 2017; Raboff, 2015). To achieve desired ecological outcomes, the City of Malmo established a Quality Program to define the standards that must be met by all stakeholders participating in the development process in Vastra Hamnen (Dalman and Von Scheele, 2009). Almost a 100% of districts energy comes from renewable resources. The renewable resources include solar, and wind energy, as well as the energy form, refuse and sewerage. All buildings have to be designed to reduce the demand for heat and electricity. A large percentage of the heating energy is extracted from the sea aquifers. Biogas is extracted from organic refuse and sewerage and after purifying is returned to the district via the city's natural gas system. Fifty three percent of the waste from the district is treated to produce biogas. Eco-cyclic adaptation of water and sewerage system is based on the coordination of activities in the community and the rest of city. Also, refuse chutes are installed in the area to deal with organic waste and other household rubbish. The structure of the green space has a special meaning in the district. The main aim is to create a network of natural environments within a medium to a high-density residential area. A rich variety of habitats is created into parks, gardens and courtyards. Each courtyard contains a traditional cottage garden. The planners want citizens to be able to walk from Västra Hamnen to the city centre of Malmö through parks and green spaces (Nicolle Foletta, IDTP Europe, 2017).

All rainwater is collected and dealt by open channels in the landscape areas that are further connected to discharge ponds, canals and the sea. A sophisticated and innovative IT system is used to improve the overall environmental performance of the area (Dalman and Von Sheele, 2009). The entire district is covered by a network of pedestrian walkways and cycling lanes making walking and cycling the preferred movement mode for shorter trips. The public transport fleet is made up of electric and gas powered and hybrid vehicles. The maintenance vehicles used in the area are electrical powered (Dalman and Von Scheele, 2009). Vasta Hamnen offers a variety of housing types targeting all social groups. A special cap for developing a percentage of affordable low-cost housing for rent purposes was imposed on the private developers. Vastra Hamnen is one of the best examples of transforming an industrial wasteland into a cutting sustainable district. Current urban landscapes of Vastra Hamnen are shown in Fig. 4.1.

Figure 4.1. Vastra Hamnen, Malmo–provision of outdoor recreational areas. Source; M. Kozlowski.

Orestad, Copenhagen, Copenhagen

Copenhagen, covering an area of 74.4 km^2 with a population of 542,000 in 2011, is the capital and largest city of Denmark and regarded as the world's greenest city. The European Commission has considered that Copenhagen is a highly successful role model for the green economy, with an effective citizen engagement strategy. It is a good model for urban planning and design. It is also a transport pioneer, aiming to become the world's most practicable city for cyclists. Public-private partnerships have been placed at the core of the Copenhagen's approach to eco-innovation and sustainable employment. The city works with companies, universities and organizations to develop and implement green growth. A typical example, could be the North Harbour project. This model of green economic development tackling environmental, economic and social concerns has the high potential for replication in other cities of the world (European Green Capital, 2017). The Danish capital has been named the 'World's most livable city' repeatedly, and Danes have been ranked the 'World's Happiest People' over and over again. This is partly due to a favourable work-life-balance, high levels of tolerance and a peaceful democracy (Copenhagen Convention Bureau, 2017).

Orested is the largest consolidated urban development project in the history of Copenhagen. The area, located on Amager Island south of Copenhagen was reclaimed for urban development as a result of population growth and plans to expand the capital city. For decades Amager was something of a backwater, a place where Copenhagen's garbage and human waste were disposed of. In the 1990s, the Danish Parliament came up with a plan to revitalize Copenhagen. As part of the revitalization, it was decided to develop a long strip of land in Amager that was 600 meters by 5 km, totalling 310 hectares. Amager is also home to the Copenhagen International Airport (Leonardesen, 2015; New in Sustainability, 2017). It was also decided to connect Amager not only to Copenhagen Metropolitan Area but also to other EU countries. As a result in 1991, Denmark and Sweden agreed to build the Øresund Bridge. Construction took place between 1995 and 1999, and the combined bridge/tunnel opened to the public in July 2000. The bridge is the longest road/rail bridge in Europe. It's about 8 km long (New in Sustainability, 2017).

According to Leonardsen 2015, the northern part of Orestad was developed for mixed purposes with a new science centre for culture, media and IT. The southern part of the district was developed for medium to high-density residential with the 8 Tallet (8 House), VM Mountain and VM House as the iconic spearheads of innovative and sustainable residential development. All areas with Orestad are within walking distances of the metro stations connecting the district with central Copenhagen, the Copenhagen International Airport and other urban centres located within its urban conurbation. Today Orestad is home to a diverse urban population. The current residential population of Orestad is 8,000 with 12,000 people working there and almost 20,000 studying there. The majority of the residential population living in Orestad also work there, labelling the district as one of the most sustainable urban villages. The Fields Shopping Centre situated along one of the metro stations is the main retail/commercial centre for the entire district.

Orestad is earmarked by a few examples of sustainable and cutting edge residential complexes. The 8 Tallet complex is in itself a self-sustaining village with the main aim of promoting urban life inside the building sphere. Shops, businesses and day-care centres are situated on the ground level. On the next level, the housing units meander upwards creating a 8 figure. The whole complex offers a wide range of apartments targeting different social and income groups. The entire complex is connected by a path that takes one right to the top of the building. At the ground level, there are two courtyards which serve as the recreation area for the residents. Both courtyards and the building path are open to the public. The 8 Tallet building is a typical example of contemporary inclusive architecture (Boll, 2016). The VM Mountain building resembles a gigantic high tech machine with large metal and glass surfaces. The building dwellings have been constructed on plots stacked on top of each other. The building also contains business and retail uses on the ground level. Together with the adjacent elevated metro line, The VW Mountain Complex resembles a futuristic urban scene (Boll, 2016). Next to the VW Mountain is the VM Houses Complex which represents another futuristic and cutting edge design. When viewed from the air the buildings are shaped like letters 'V' and 'M'. The aim of the architects is to provide maximum daylight to the interiors thus reducing electricity demand. Another example of a fully sustainable building in Orestad is the Crown Plaza Hotel. Located next to the Fields Shopping Centre and one of Orestad's Metro stations, The Crowne Plaza Hotel is regarded as one of the most sustainable hotels in the world. Sustainability is incorporated into all aspects of the hotel, and it uses 65% less energy than other comparable hotels. The lobby of the hotel features a natural woodland area (BC Hospitality Group, 2017).

The Orestad district contains the large amount of recreational space including the Orestad commons and the three Orestad islands, the latter situated on the canal running parallel to the elevated metro line. Each of the islands contains its form and function, however, all three islands share the same basic concept of creating an integrated space for recreation and cultural activities (Stubbegaard, 2016).

Cotemporary Orestad with its excellent connectivity to the Copenhagen Metropolitan Area, Copenhagen City Centre, Copenhagen International Airport and also to Malmo Sweden, with its pedestrian and cycling friendly environment, green building design, mixed uses and mixed income constitutes one of the most sustainable urban districts in the world. Current images of Orestad are shown in Fig. 4.2.

Punggol, Singapore

Singapore is recognized in the global marketplace as one of the prominent and leading cities of Asia with an innovative sustainable development approach (Rimmer and Dick, 2009). Singapore was established by the British as a colonial trading post in 1819, and for the next 150 years, it became the dominant city of British Malaya. In 1965, Singapore shifted away from hinterland Malaya and became an independent city-state. As a result of flexible economic policies and openness to foreign investments, Singapore's economic development tripled per capita income in the 1980–1995 periods (Rimmer and Dick, 2009; Yuen, 2011). Within half a century, the low-rise former British colonial trading post became a high-rise post-industrial garden city of 5.4 million covering an urban area of 700 km^2 (Department of Statistics, Singapore, 2014). The transformation to a garden city together with

Figure 4.2. Residential Complexes at Orestad, Copenhagen (The 8 Tallet Complex–top and the VW Mountain Building–bottom. Source; M. Kozlowski).

innovative and sustainable urban planning policies attracted the attention of many urban scholars and professionals (Yuen, 2011).

Punggol is a residential planning area and new town situated on the Tanjong Punggol peninsula in the North-East Region of Singapore. Punggol has been selected by HDB (Housing Development Board) to be developed into its first eco-town in Singapore, for the tropics. Construction of the Punggol 21 project commenced in 1998 but was halted as a result of the slump in the construction industry in 2003. The Punggol 21 Plus project was launched by Singapore's Prime Minister Lee Hsien Loong as a revamp of the former Punggol blueprint introduced in 1998. This residential project focuses on multiple initiatives for sustainable living through urban mobility, effective energy, water, and waste management. Punggol residential district is a new additional of innovative housing developments in the city managed and coordinated by Singapore's Housing Development Board (HDB) (Koon Hean, 2017; History SG, 2017).

The 'Punggol 21-plus' project involves 18,000 new HDB and private flats, and about 3,000 new units will be built in Punggol New Town every year. Punggol New Town is expected to have 96,000 units when fully developed in the long term. The waterfront housing features stepped courtyards which complements the local tropical climate. An integrated waterfront commercial and residential development is developed at the town centre on both banks of the waterway. There are also other facilities within the town centre, which includes a community club, regional library and a hawker centre.

Punggol is connected by MRT public transport system to the centre of Singapore. Apart from providing excellent public transport connectivity policies have been introduced encouraging residents to share cars, there are plans to promote the use of electric vehicles through the car-sharing scheme. Charging points for the electric vehicles will also be strategically located for greater convenience to residents using an electric vehicle (Housing Development Board, 2017).

Solutions that adopt cleaner energy are deployed. These include solar photovoltaic system, elevator energy regeneration system, energy-efficient lighting in common areas, and smart grid/meters (Housing Development Board, 2017). To meet a 3-times increase in recyclables, recycling points are provided at every level in the residential blocks. To achieve this, a second centralized refuse chute that is dedicated for recyclables is being developed. This will not only enhance convenience for residents to dispose of their recyclable waste but also improve collection efficiency for waste collectors (Housing Development Board, 2017). To date, Punggol Waterway and Treelodge@Punggol are two successful iconic developments that set Punggol as a sustainable waterfront town of the 21st century (Housing Development Board, 2017). Punggol Waterway is Singapore's first human-made waterway by HDB. With a length of 4.2 km, it meanders through the entire Punggol Eco-Town. The design of the water navigation system and the landscaped promenade embraces the area's rich coastal heritage, providing more opportunities for water-based recreational activities and sports right next to the heartlands. The waterway and the new communal spaces along the promenade provide a vibrant living environment, transforming Punggol into a sustainable waterfront town of the 21st century. Treelodge@Punggol completed in 2010 is the first HDBs eco-precinct. The precinct has various green technologies and solutions to enable residents to have an eco-friendly lifestyle. This includes tropical greening to reduce the heat impact, LED lighting and Elevator Energy Regenerative system. It has also introduced PV solar panels to reduce dependency on grid electricity (Housing Development Board, 2017). Punggol is one of the best examples of a sustainable, eco-friendly township dominated by affordable housing. Current images of Punggol are shown in Fig. 4.3.

Figure 4.3. Punggol, Singapore – Penetration of the green areas and waterways into the residential neighbourhoods. Source: M. Kozlowski.

Key Findings

The review of three residential neighbourhoods in Copenhagen, Malmo and Singapore revealed that comprehensive sustainable design and planning followed by systematic and coordinated project management could significantly contribute to the creation of entirely green/sustainable residential community. The development of the three selected neighbourhoods is preceded by a comprehensive and sustainable urban planning and design process. All the three case studies have excellent physical and visual connectivity with the surrounding areas and are linked to other parts of the city by an innovative rapid transit system. The three case studies feature quality sustainable building design, friendly pedestrian and cycling environments and provide natural green habitats that penetrate right into the heartlands of the neighbourhoods. All three areas score high in the categories of energy efficiency, thermal comfort and water conservation. Regarding social sustainability all three provide a variety of housing choice for all income groups and Punggol is labelled as an affordable housing district of Singapore. The three neighbourhoods contribute to the local economy as they provide new employment opportunities and attract local retail and commercial businesses. For example, in Orestad the population engaged in working and studying in the district is higher than the local residential population. The area together with the rest of Copenhagen successfully promotes green economy aimed at reducing the environmental risks.

In all three case studies one can observe the dominance of passive design strategies including design for additional day lighting, provision of semi-outdoor spaces, introducing cross ventilation as an alternative to air conditioning (the latter applicable mainly in Punggol located in a tropical climate zone), provision of solar gain for moderate climates and sun-shading for warmer climates, and the inclusion of building insulation (Atkins, 2017). The sustainable urban design principles are addressed not only at the building/site scale level but also at the urban scale level targeting issues such as the provision of public/recreational spaces, schools, childcare centres, medical clinics and shops and ensuring good connectivity and accessibility to surrounding areas and the public transport modes.

It should also be noted that the three examples are located in countries that have a strong track record in promoting and enforcing sustainable development. In a majority of developing or even developed countries, there are still a lot of challenges facing the creation of fully sustainable residential neighbourhoods that would address the three precepts including physical, social and economic sustainability.

One of the biggest obstacles in achieving sustainable urban outcomes is the neoliberal policies practised by all levels of government in a majority of countries around the world. Neoliberalism[1] has strongly influenced urban policies, especially in the major global cities Instead of developing residential communities for all social groups, urban local authorities prefer to depend on the private sector and allow in creating semi-privatized and revenue producing residential enclaves that boost the image of the city to the outside world. Policies aimed at disadvantaged neighbourhoods and developments of affordable housing are overshadowed by the development of master planned residential communities, large scale urban transformation and gentrification (Purcell, 2011; Weaver, 2016).

One of the outcomes deriving from this neoliberal approach in urban planning and design is property led development.[2] In the last decades in the main cities around the globe, urban policy has increasingly relied on private-sector property developers to provide the driving economic force (Turok, 1992). As a result, property led development is dictated by the market and targets specific income groups. According to Davies and Monk (2009), architecture and urban design outcomes guided by neoliberal principles and property led development policies result in the development of fortified enclaves in the form of luxury themed environments, super exclusive shopping malls, and

[1] The notion of neoliberalism emerged strongly in the late 1970s era of stagnation and economic recession. Its driving argument was that a liberal free market is much more efficient in allocating resources than state interventions. Since then the neoliberal doctrine has become a major driver of globalization and dominant on the broad political spectrum (Purcell, 2011).

[2] Property led Development can be defined as "the assembly of finance, land, building materials and labour to produce or improve buildings for occupation and investment purposes" (Turok, 1992).

artificial island/utopian suburbs and gentrified downtown districts. As a consequence, urban design is gradually shifting from a universal discipline aimed at developing inclusive urban places and areas for all people to a master planning tool delivering new exclusive and often utopian urban enclaves. In the master planned precincts, the sustainability outcomes are often achieved through the dominance of active design measures. In contrast to passive design, the active design includes measures such as forced-air HVAC (High-Velocity Air Conditioning) systems, photovoltaic panels, heat pumps, radiant panels or chilled beams, and electric lights (Autodesk Sustainability Workshop, 2017; Atkins, 2017). Many master planned communities also label themselves as 'sustainable' by providing sustainable gadgets such as solar panels, wind turbines, automatic window shutters for energy efficiency and thermal comfort and swales, water tanks and reticulation ponds for water conservation. The provision of these sustainable gadgets often increases the property prices making the new residential neighbourhoods unaffordable for the average wage earner. For example, average median house price in the Currumbin Eco-Village a sustainable community located in the City of Gold Coast Australia is around 800,000 AUD which is 30% more than the average median house price for the City of Gold Coast (Gold Coast Bulletin, 2017; Currumbin Eco-Village, 2017). The average house prices at eco labelled new residential neighbourhoods in the Greater Kuala Lumpur area in Malaysia (Setia ECO-Glades and Symphony Hills, Cyberjaya) are well above 1,000,000 Malaysian Ringgit (MYR) (US$ 240,000) which is over three times the national average house price of MYR 334,736 (US$ 75,390) and almost 40% higher than the already spiked average house price for Malaysia's capital Kuala Lumpur (Global Property Guide, 2017; Setia Eco-Glades, 2017). There are many other similar examples from around the world. It seems that, similar to the idyllic rural residence escapees of the 18th and 19th century, the notion of green, sustainable living, is reserved for the upper middle classes. The master plans focus primarily on physical sustainable design. The whole idea of social sustainability characterized by total inclusivity of the urban areas and equal opportunities for all social groups is being missed out.

Combating the negative influences of neoliberalism and property led development is one of the biggest challenges facing urban planners, designers and decision makers. In many countries property led development has become one of the main driving forces of the economy and the term 'sustainable or green living' is another selling jargon to attract more customers. Therefore it is imperative to promote and achieve universal sustainable/green residential building design through passive design measures and sustainable urban design outcomes for the entire residential neighbourhoods.

For example, Australia has been heavily influenced by neoliberal policies with property led developments a common phenomenon in every major city. The average median house price in the Sydney Metropolitan area has reached a staggering figure of 905,000 AUD (US$ 751,000) (news.com.au, 2017). Nevertheless, every state in Australia is developing regional, municipal and local planning policies addressing sustainable planning and design for new residential development. Local authorities in major cities in Australia for the past years have been incorporating sustainable passive design measures derived from the Green Building Council of Australia into the local statutory plans as mandatory development requirements. These development requirements include a provision of sun shading devices, extended eaves, and necessary minimum floor to ceiling heights to achieve satisfactory cross-ventilation, provision of semi-outdoor spaces and pedestrian plazas at ground level (Brisbane City Council, Gold Coast City Council, Sydney City Council 2017, Green Building Council of Australia).

The Government of New South Wales went one step ahead by introducing the Building Sustainability Index and ensuring that a sustainable state-wide policy is implemented at a local street and site levels. The Building Sustainability Index provides single and multiple dwelling units are designed to use less potable water and be responsible for fewer greenhouse gas emissions. This is enforced by setting energy and water reduction targets for houses and residential apartments. BASIX is an online program that is accessible to all users. The building designer or builder enters data relating to the house or unit design—such as location, size, building materials, etc.—into the BASIX tool. BASIX analyzes this data and determines how it scores against the energy and water targets. The design must pass specific goals (which vary according to location and building type) before the user can print the BASIX Certificate. One cannot commence with constructing a house or residential apartment building without a BASIX certificate (New South Wales Government, 2017).

Conclusions

Addressing sustainable residential development must be conducted at a regional, city wide, neighbourhood-district and site level. Regional and city-wide municipal plans must address sustainable residential development identifying overarching sustainable urban design principles. Local statutory and neighbourhood master plans must respond to regional and municipal plans by listing implementation actions aimed at low, medium and high-density residential development. All actions should address the building-site scale level as well as the overall urban context level. At the building-site level, the actions should include smart form based codes for development addressing issues such as thermal comfort, energy efficiency, water conservation and the provision of landscaping and community areas. At the urban context level actions should target good visual and physical connectivity, provision of walking distance to public transport stations, shops, community services, recreational areas, schools and child care centres. To achieve social sustainability a mandatory cap for the provision of affordable housing must be imposed on the developers. Introducing such caps ensures the creation of mixed income and multi-ethnic green residential neighbourhoods. Efforts must be made on promoting the green local economy that reduces adverse impacts on the environment. Imposing all sustainable urban design measures on building owners, construction companies and property developers and other stakeholders involved in the development process is the only way to achieve green residential neighbourhoods that are physical, economically and socially sustainable.

References

Alexander Christopher. 1977. Notes on the Synthesis of Form, Harvard University Press: Cambridge.

Atkins. 2017. Passive & Active Design, CIBSE Building Simulations Group http://www.cibse.org/getmedia/22be4d3b-2b5f-410d-bda4-79ce098baf66/Peter-Brown-(Atkins)-Passive-and-Active-Design(1).pdf.aspx.

Autodesk Sustainability Workshop. 2017. Passive Design Strategies https://sustainabilityworkshop.autodesk.com/buildings/passive-design-strategies.

BC Hospitality Group. 2017. Cowne Plaza Hotel, Orestad http://bchg.dk/en-GB/Brands-Business-areas/Crowne-Plaza-Copenhagen-Towers.aspx.

Boll, M. 2016. The 8 House' in A Guide to New Architecture in Copenhagen Sarensen, A. (editor) Danish Architecture Centre, pp. 64.

Brisbane City Council. 2017. Brisbane City Plan https://www.brisbane.qld.gov.au/planning-building/planning-guidelines-tools/brisbane-city-plan-2014.

Brisbane City Council. 2016. Buildings that Breathe https://www.brisbane.qld.gov.au/planning-building/planning-guidelines-tools/neighbourhood-planning/neighbourhood-plans-other-local-planning-projects/city-centre-neighbourhood-plan/new-world-city-design-guide-buildings-breathe.

Carmona, Mathew, Tim Heath, Taner Oc and Steven Tiesdell. 2010. Public Spaces-Urban Spaces, Architectural Press: Oxford.

Churchill, C. J. and B. W. Baetz. 1999. Development of decision support system for sustainable community design. ASCE Journal of Urban Planning and Development 125: 17–35.

City of Gold Coast. 2017. City Plan http://www.goldcoast.qld.gov.au/planning-and-building/city-plan-2015-19859.html

Clarke, P. 2009. Urban planning and design. pp. 12–21. *In*: Ritchie, A. and T. Randall (eds.). Sustainable Urban Design: An Environmental Approach. London and New York: Taylor and Francis.

Commission for Architecture and Built Environment (CABE) (2001). Urban Design in Planning System: Towards a Better Practice, Department for Environment, Transport and Regions, United Kingdom' www.cabe.com.org accessed 12.05. 2008.

Congress for New Urbanism. 1999. Charter for New Urbanism (New York: McGraw Hill).

Copenhagen Convention Bureau. 2017. This is Copenhagen. (http://www.copenhagencvb.com/copenhagen/copenhagen-32017.)

Cullen, G. 1961. The Concise Townscape. London: Architectural Press.

Currumbin Eco-Village. 2017. www.theecovillage.com.au.

Dalman, E. and C. Von Scheele. 2009. Bo01 and flagghusen: Ecological city districts in malmo, Sweden. pp. 161–169. *In*: Ritchie, A. and T. Randall (eds.). Sustainable Urban Design: An Environmental Approach (London and New York: Taylor and Francis).

Davis, M. and D. B. Monk. 2009. (eds.). Evil Paradises: Dream World of Neoliberalism. (New York: New Press.)

Department of Statistics, Singapore. 2014. www. singstat.gov.sg.

Douglas, F. 2008. Sustainable Urbanism: Urban Design with Nature' Wiley and Sons; London.

Drodz, M. 2014. Spatial Inequalities. Neoliberal Urban Policy and the Geography of Injustice in London in Justice Spatiale|Spatial Justice. https://www.jssj.org/wp-content/uploads/2014/05/Drozdz-Eng-n%C2%B06-jssj.pdf.

European Green Capital. 2017. Copenhagen 2014 Cities Fir For Life http://ec.europa.eu/environment/europeangreencapital/winning-cities/2014-copenhagen/.

Franklin, B. 2006. Housing Transformations: Shaping the Space of Twenty First Century Living (London: Routledge).

Global Property Guide. 2017. Subdued house price rises in Malaysia https://www.globalpropertyguide.com/Asia/malaysia/Price-History.

Gold Coast Bulletin. 2017. CoreLogic data reveals Gold Coast house prices up 6.2 per cent over the 12 months to $600,000. https://www.realestate.com.au/news/corelogic-data-reveals-gold-coast-house-prices-up-62-per-cent-over-the-12-months-to-600000/.

Green Building Council of Australia. 2017. Green Star Rating http://new.gbca.org.au/Hagan, S. 2015. Ecological Urbanism: Nature of the City (London and New York: Routledge).

History, S.G. 2017. Punggol 21 is Announced http://eresources.nlb.gov.sg/history/events/c31a9cc5-3c2a-4fd5-9f4a-3054f95dffa3.

Housing Development Board. 2017. About us http://www.hdb.gov.sg/cs/infoweb/about-us.

Housing Development Board. 2017. http://www.hdb.gov.sg/cs/infoweb/about-us/our-role/smart-and-sustainable-living/punggol-eco-town.

Housing Development Board. 2017. Punggol Waterway http://www.hdb.gov.sg/cs/infoweb/about-us/our-role/smart-and-sustainable-living/punggol-waterway.

Houston, S., A. Jadevicius and Zafer A. Sahin. 2017. Smart Urban Planning in Smart Urban Regeneration: Visions, Institutions and Mechanisms for Real Estate (Houson, S. editor) London; Routledge.

Jacobs, J. 1961. The Death and Life of Great American Cities. Penguin Books: New York.
Koon Hean, C. 2017. Evolution of HDB Towns in 50 Years of Urban Planning in Singapore Chye Kiang, H. (ed.). pp. 101–127.
Kozlowski, M. 2010. Urban Design: Shaping Attractiveness of the Urban Environment. Lambert Academic Publishing; Saarbrucken, Germany.
Krieger, A. 2004. Territories of Urban Design. www.chankrieger.com/cka/essays/territoriesud.pdf.
Lang, J. 2005. Urban Design : A Typology of Procedures and Products (Oxford: Architectural Press).
Lehmann, S. 2010. The Principles of Green Urbanism: Transforming the City for Sustainability. Earthscan; Washington.
Leonardsen, L. 2015. Master Plans: Orestad in A Guide to New Architecture in Copenhagen Sarensen, A. (ed.). Danish Architecture Centre, pp. 157.
Lynch, K. 1960. The Image of the City. MIT Press: Cambridge USA, London.
Malmo Stad. 2017. Sustainable Urban Planning in Malmo. http://malmo.se/Nice-to-know-about-Malmo/Sustainable-Malmo-/Sustainable-Urban-Development/Sustainable-Urban-Planning.html.
Melbourne Water. 2017. Introduction to Water Sensitive Urban Design www.melbournewater.com.au/wsud
News.com.au. 2017. Home prices jump 3.7 per cent since start of year. http://www.news.com.au/finance/real-estate/buying/home-prices-jump-37-per-cent-since-start-of-year/news-story/3b7368464e6989a25b02b214ed1d1a48.
New in Sustainability. 2017. Sustainable Cities: Orestad, Copenhagen. http://www.newinsustainability.com/orestad/.
New South Wales Government. 2017. Building and Sustainability Index (BASIX). https://www.basix.nsw.gov.au/information/about.jsp.
Nicol, L. A. 2012. Sustainable Collective Housing: Policy and Practice for Multi-family Dwellings (New York: Routledge).
Nicole Foletta, ITDP Europe. 2017. Vastra Hamnen Site and Facts. https://www.itdp.org/wp-content/uploads/2014/07/25.-092211_ITDP_NED_Vastra.pdf.
Purcell, M. 2011. Neoliberalisation and democracy. pp. 42–55. *In*: Fainstein, S. and S. Campbell (eds.). Reading in Urban Theory. London: John Wiley and Sons.
Raboff, J. L. 2015. Vastra Hamnen 2015: The Western Harbour in Malmo, Sweden Raboff AB/LLC.
Rimmer, P. J. and H. Dick. 2009. The City in Southeast Asia: Patterns, Processes and Policy (Singapore: NIUS Press).
Ritchie, A. 2009. Waste and resource in Thomas, R and Ritchie, A 'Building Design'. pp. 87–97. *In*: Ritchie, A. and T. Randall (eds.). Sustainable Urban Design: An Environmental Approach. London and New York: Taylor and Francis.
Roger, C. 2005. Encyclopedia of the City. Routledge: London, New York.
Royse, S. 2009. 'Materials' in 'Sustainable Urban Design: An Environmental Approach' Ritchie, A and Randall, T (eds.). pp. 74–79. (London and New York: Taylor and Francis).
Setia Eco-Glades. 2017. http://www.propwall.my/cyberjaya/setia_eco_glades/2552?tab=classifieds&lang=en.
Stubbegaard, D. 2016. Orestad Islands in A Guide to New Architecture in Copenhagen Sarensen, A. (editor) Danish Architecture Centre, pp. 52.
Sydney City Council. 2017. Sydney Development Control Plan (2012). http://www.cityofsydney.nsw.gov.au/development/planning-controls/development-control-plans.
The Eco-Village at Currumbin. 2017. http://www.propertyartists.com.au/7-coolamon/.
Thomas, R. and A. Ritchie. 2013. Building Design. pp. 43–55. *In*: Ritchie, A. and T. Randall (eds.). Sustainable Urban Design: An Environmental Approach. London and New York: Taylor and Francis.
Thomas, R. and A. Ritchie. 2013. Water in Sustainable Urban Design: An Environmental Approach. A. Ritchie and T. Randall (eds.). pp. 81–87. (London and New York: Taylor and Francis).
Thorne, R., W. Filmer-Sankey and A. Alexander. 2009. In Sustainable Urban Design : An Environmental Approach. Ritchie, A. and T. Randall (eds.). pp. 31–42 (London and New York; Taylor and Francis).
Turok, I. 1992. Property-Led Urban Regeneration: Panacea or Placebo in Environment and Planning A, 24(3): 361–379.
Urban Villages. 2017. http://www.urbanvillages.com/.
Von Brocke, C. 2009. Landscape and Nature in the City. pp. 31–41. *In*: Ritchie, A. and T. Randall (eds.). Sustainable Urban Design: An Environmental Approach. London and New York: Taylor and Francis.
Weaver, T. P. R. 2016. Blazing the Neo-liberal Trail: Urban Development in the United States and United Kingdom (Philadelphia: University of Pennsylvania Press).

Yeang, K. 2009. Eco-Master Planning, John Wiley and Sons, London.

Yeang, K. 2010. Ecodesign: A Manual for Ecological Design (London: John Wiley and Sons Ltd.).

Yigitcanlar, T., Md. Kamruzzaman and S. Teriman. 2015. Neighborhood 'Sustainability Assessment: Evaluating Residential Development Sustainability in a Developing Country Context' Sustainability 2015 nr 7 pp. 2570–2602.

Yin, R. K. 2013. Case Study Research: Design and Methods. (New York: Thousand Oaks).

Yuen, B. 2011. Singapore planning for more with less. pp. 201–220. *In*: Hamnett, S. and D. Forbes (eds.). Planning Asian Cities: Risks and Resilience. London and New York: Routledge.

CHAPTER 5

Toward Sustainable Building

Adopting the Life Cycle Assessment Approach

Sumiani Yusoff and *Ahmad Faiz Abd Rashid**

Introduction to the Life Cycle Assessment

In line with global awareness on climate change and other global environmental issues, a more concerted and holistic approach toward environmental improvement and management is pertinent to address and manage effectively environmental impacts and burdens from human activities and interventions.

Life Cycle Assessment (LCA) is an environmental management tool that enables quantification of environmental burdens and their potential impacts over the whole life cycle of a product, process or activity (Azapagic, 1999). Initially LCA was introduced primarily for use in product manufacturing for the purpose of tracing the direct impacts as well as impacts associated with a product throughout the entire life cycle from cradle to grave in order to get a holistic overview of the environmental burden associated with the products. It is also one of the essential environmental management tools for developing and implementing strategies towards overall process optimization, preservation of environment, environmental labels and design for environment (Yusoff, 2006).

Life cycle assessment is an important tool to address potential environmental problems when a new technology, process or material is incorporated within a system. In an attempt to divert solid waste from ending up in the landfill, LCA plays an important role to evaluate the alternative method or comparison of better a method whether it will reduce or aggravate the environmental impacts. LCA is a methodology that can be used to estimate environmental impacts of a product, process or service from production of the raw materials to ultimate disposal of wastes.

Sustainability Science Research Cluster, University of Malaya.

* Corresponding author

LCA will be a very important methodology in the near future due to the general public and industry's awareness on the sustainability issues related to the product that they consume and with the introduction of more stringent environmental laws worldwide. Cook (2010) mentioned that in Europe, LCA is a cornerstone of the European Integrated Product Policy, the Thematic Strategies on Waste Prevention and Recycling and on the Sustainable Use of Natural Resources.

Life cycle assessment is carried out based on the ISO 14040 series of standards, which consists of goal and scope definition, inventory analysis, impact assessment and result interpretation. The details of the elements of the LCA framework are as below:

a) Goal and scope definition consist of the goal of the study which states the reasons for carrying out the LCA, description of the functional unit to which the inputs and outputs of the system are related, defining the scope of the study such as selection of impact categories or substances that the study will take into account, followed by scope which include a damage assessment, and also whether aggregation is employed in carrying out the impact assessment, and construction of system boundary of a product systems, like a flowchart of process or stages involved in the system.
b) Life Cycle Inventory (LCI) consists of quantity of input and output inserted into the software such as quantity of feedstock used, energy consumption and emission to environment in specified units such as kg of CO_2. Life Cycle Impact Assessment (LCIA) will convert these flows into simpler indicators. The inventory results are a list of the total inputs and outputs from the product system. They are calculated by iteratively totalling the resource consumption and emissions from each stage, and the totals from each process referenced by the stages, in the system.
c) Life cycle impact assessment will evaluate the degree of the input and output of the system. There are many assessment methods employed in the life cycle impact assessment. This includes assessment methods such as the Eco-indicator 99, Ecopoint and EPS methods. Weighting is an aggregation method to obtain a single integrated value which represents the overall impact of the product system. Eco-indicator 99 method uses classification, characterization, normalization, and aggregation, to obtain a single combined index used to assess the overall impact of the system.
d) LCA Interpretation, which is the explanation and analysis of the results in the study also assess the outcome from all phases to reach the conclusions and recommendations. This final stage is a systematic procedure to identify, qualify, check and assessed the results outcomes of the LCI and/or LCIA product system. It defines if the study results could meet the requirements in the goal and scope of the study. It includes communication to comprehend easily and is useful for decision makers to give the results credibility of the other LCI and LCIA phase. By identifying the most significant environmental impacts category, it expands the eco-efficiency opportunities in achieve improvements to environment and economic performance of product/process/service life cycle. Figure 5.1 below shows a typical LCA study stage in ISO 14040: Life Cycle Assessment System.

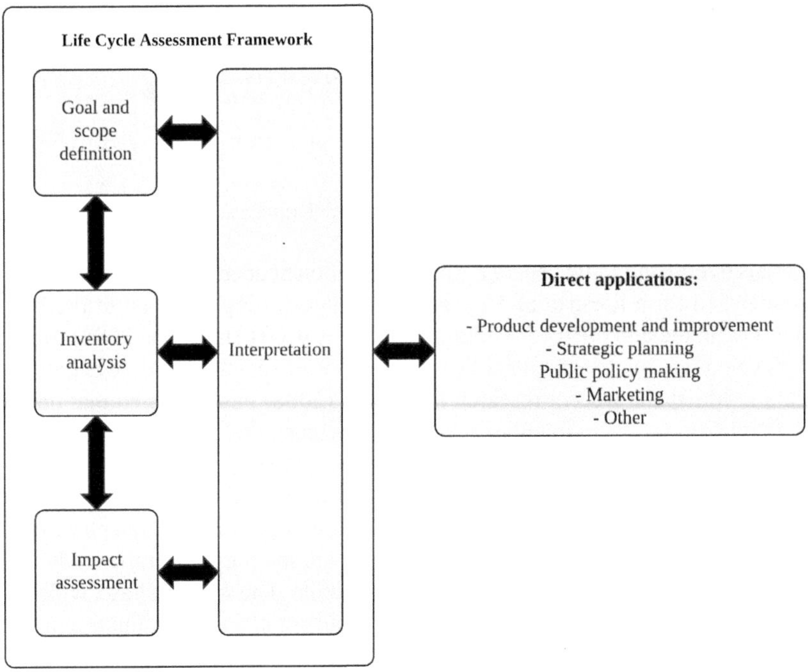

Figure 5.1. LCA Study Stages (ISO 14040, 2006).

Implementation of Life Cycle Assessment in the Building Industry

Research on the implementation of LCA has been increasing steadily since the last decade in relation to the manufacturing of building materials and also construction processes. In general, the development of buildings are more complex compared to other products because of their unique characteristics such as its size, diverse materials and inconsistent production method (Scheuer et al., 2003). The other significant limitation is that there is limited data with reference to the environmental impact of the production and manufacturing of construction materials or the actual process of construction and demolition (Scheuer et al., 2003).

The LCA methodology applied in the building industry, however, is still in a fragmented state due to a variety of case study buildings with diversity in materials selection, locations, construction process, building design and usage that will produce a different definition of goal and scope and will bind to certain limitations (Abd. Rashid and Yusoff, 2012). A whole life cycle of a building from cradle-to-grave which consists of the pre-use phase, construction phase, use phase, and End-Of-Life (EOL) phase as shown in Fig. 5.2.

EOL phase was rarely incorporated previously, but recent research identifies that the ability of recycling potential of building materials will contribute to the reduction in life cycle impact (Blengini and Di Carlo, 2010). During end-of-life phase after the demolition of the building, the building waste and debris were either sent to the landfill or to the recycling plant. Recycling of building materials can reduce building

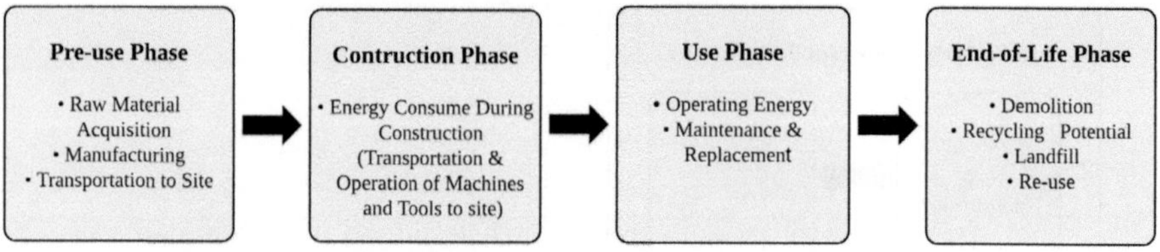

Figure 5.2. Life cycle phase of a building (Abd. Rashid and Yusoff, 2015).

materials overall embodied energy. An energy efficient apartment building in Sweden was analyzed for a lifespan of 50 years, and the recycling potential of the building materials can reclaim up to 15% of the total energy used (Thormark, 2002). However, as a developing country, the demolition process for residential buildings hardly ever occurs in Malaysia. Recent research by Arham (2008) has identified that only steel and aluminium are being regularly recycled whereas other materials are transported to the landfill.

Most of LCA research related to the building industry followed the LCA framework from the ISO 14040 which consists of four phases. The first phase of LCA, which is defining goals and scopes, will determine the purpose of the study, system boundaries, and selection of suitable functional units. The second phase, which is life cycle inventory (LCI) is the data collection process of all relevant inputs and outputs of a product life cycle. The third phase, the life cycle impact assessment (LCIA) will use data from LCI and subsequently evaluates potential environmental impacts and estimate resource used in the study. The last phase is the interpretation that identifies significant issues, assesses results to reach conclusions, explain the limitations and provide recommendations. A cradle-to-grave LCA research will respond to a predetermined system boundary, functional unit, building lifespan. Currently, the international standard for LCA framework for building is still being developed however Fig. 5.3 is an example of an LCA framework for building which was adapted based on various published LCA research aligned with the ISO 14040 standard.

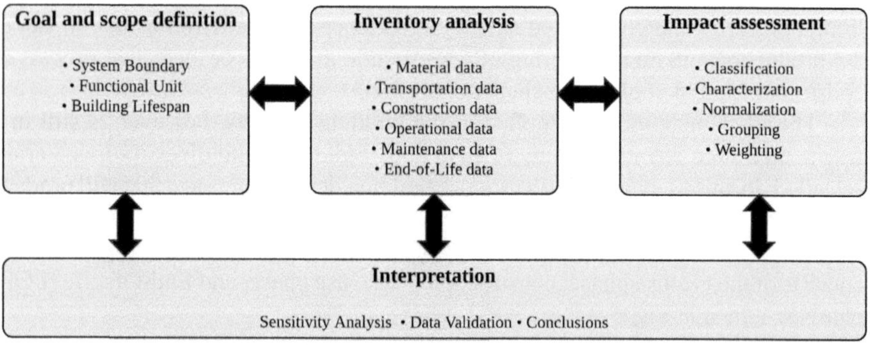

Figure 5.3. LCA framework for the building industry. Adapted from (Blengini and Di Carlo, 2010; ISO, 2006; Ochsendorf et al., 2011; Ortiz-Rodríguez et al., 2010; Ove Arup and Partners Hong Kong Ltd., 2007).

Impact of Building Materials in Life Cycle Assessment Perspective

The selection of building materials will affect the total embodied energy. It also influences the total energy consumption and the environmental impact in the use phase and the recycling potential during the end-of-life phase. Initially, LCA research were conducted to evaluate the impact of various building materials only and eventually expanded to the whole building life cycle.

A research conducted by Asif et al. (2007) for a residential building in Scotland suggested that concrete, timber, and ceramic tiles have the highest initial embodied energy compared to other materials. Concrete has been identified as the largest initial embodied energy. The research suggests that the subtantial amount of concrete used contribute to significant total embodied energy. Similar findings on concrete were also reported in other research (Blengini and Di Carlo, 2010; Cole, 1998). However, this research only considered the pre-use phase of the building only. Other researchers suggested that building material with low initial embodied energy may not typically have low life cycle energy (Utama and Gheewala, 2008; 2009).

Other concrete related research were conducted Mithraratne and Vale (2004) to compare three identical design residential buildings by using LCA. The buildings have different core materials namely light construction (timber frame), concrete construction and light construction with superinsulated construction. Concrete and superinsulated buildings were identified to produce higher initial embodied energy compared to light construction by 8 and 14% but have lower life cycle energy by 5 and 31% respectively. The additions of insulating materials were identified to reduce the consumption of total life cycle energy of a building. Similar findings in relation to insulation were also identified in a low energy house in Italy (Blengini and Di Carlo, 2010), a green home in Australia (Fay et al., 2000) and a residential building in Netherland (Huijbregts et al., 2003) that produced similar results. Recent research also suggested that residential buildings using Insulated Concrete Form (ICF) in the USA are more efficient during its life cycle compared to a light frame timber house with a similar design (Ochsendorf et al., 2011).

Most case studies described earlier were conducted in a temperate climate. In a tropical climate, clay-based products performed better compared to cement-based products. Utama and Gheewala (2008) suggested that a landed residential building in Indonesia using clay bricks and clay roof tiles have better life cycle energy in comparison to cement based bricks and roof tiles due to lower thermal transfer thus preserving the cooling effect of air conditioning. The researchers also suggested that high rise residential apartments using a sandwich wall of external clay brick, internal gypsum plasterboard, and air gap in between have lower life cycle energy compare to single clay brick wall by up to 59% (Utama and Gheewala, 2009).

López-Mesa et al. (2009) conducted an LCA research for two seven-story residential buildings with similar concrete based products by using *in situ* cast concrete floors and precast concrete floors. The advantage of the precast concrete floor system is the ability to have a longer span between beams thus minimizes columns and footings hence have lower concrete volume and subsequently reduce the environmental impact by 12.2%.

Impact of Different Building Phases in Life Cycle Assessment Perspective

The phases in building can be divided into four different phases as in Fig. 5.2. The use phase has been identified as the highest environmental impact contributor due to the extensive duration compared to the other phases. The emissions produced during the use phase is associated with the fossil fuel combustion in electrical generation and space heating (Scheuer et al., 2003). Other researchers also have concluded the same findings. Kofoworola and Gheewala (2009) identified that the use phase of a 60,000 m^2 office building in Thailand for a period of 50 years accounted for 81% of total energy consumption. Mithraratne and Vale (2004) compared three construction type of residential building in New Zealand with different construction materials specifically the light construction, the concrete construction and the insulated construction. The use phase of the buildings for the lifespan of 100 years contributed 74, 71, and 57% respectively. A comparative research was conducted in Italy on low energy and standard house by Blengini and Di Carlo (2010). The research suggested that the use phase in a standard house contributed more than 80% total energy consumption compared to lower than 50% in a low energy house.

In Michigan, USA, Scheuer et al. (2003) identified that building operation in a new university building representing 94.4% of total energy consumption, while Ding (2007) conducted a life cycle energy analysis of 20 secondary schools in Australia for a 60 years lifespan and operational energy in building use phase represent 62% of energy used. Ooteghem and Xu (2012) conducted LCA research to five single storey retail buildings in Canada for 50 years lifespan and identified that space heating consume the highest energy during building use phase (42%), followed by lighting (37%), ventilation fans (7%), space cooling (6%) and miscellaneous equipment (6%). There are also other studies that produce similar results in which heating is the biggest energy consumer (Blengini and Di Carlo, 2010; Iyer-Raniga and Wong, 2012; Mithraratne and Vale, 2004; Zabalza et al., 2009).

However, buildings in a different location will produce a different energy use pattern. A building in the tropical region has a different outcome from a building in a cold climate region as space heating is unnecessary. A semi-detached house in Malaysia was analyzed and air-conditioning was identified to consume the highest electricity compared to lighting and other electrical equipment (Abd. Rashid et al., 2017). Similar to other research, the use phase was identified as responsible for most of the environmental impact assessed especially on Global Warming Potential (GWP) due to fossil fuel usage in the electricity generation mix in Malaysia.

Life Cycle Assessment and the Green Building Rating System

The building industry consists of numerous phases starting from mining, manufacturing, construction, usage and demolition. In each phase, a substantial amount of energy is consumed, and at the same time, a significant emission is released. Energy is consumed directly during building construction, use and demolition while indirectly through the production of materials (embodied energy) used in the building (Sartori and Hestnes, 2007).

Most research has identified that the use phase is the largest energy consumer. Between 1971 and 2004, the carbon dioxide emissions had grown at a rate of 2.5% per year for commercial buildings and 1.7% per year for residential buildings (Levine et al., 2007). However, the building industry has the largest potential for significantly reducing GHG compared to other industry for both developed and developing countries by an estimated 30 to 80% throughout the building lifespan (UNEP-SBCI, 2009). The introduction of rating tools to the building industry helps to push the idea of sustainability to the industry players including the building users.

Green Building Rating System in the Building Industry

Due to the significant impact of the buildings on the environment, various sustainable indicators have been introduced in recent years. The main purpose of the indicators is to promote the reduction of energy and GHG emission, by improving the building design, incorporating energy efficiency equipment, incorporating recycled building materials, and promote better site planning and management (Green Building Index, 2012).

The first building sustainable was developed by Building Research Establishment Environment Assessment Methodology (BREEAM) in the 1990s in the UK and later with Leadership in Energy & Environmental Design (LEED) that become the generic guidelines in other rating tools available worldwide (Green Building Index, 2009). Figure 5.4 shows the building rating tools available worldwide.

Currently, there are more than 600 sustainable rating systems offered throughout the world, and the numbers keep growing as new systems are being introduced and regularly updated (Berardi, 2012). Although numerous rating system are available, the most advanced and leading rating tools available are BREEAM (Building Research Establishment's Environmental Assessment Method) in the UK, LEED (Leadership in

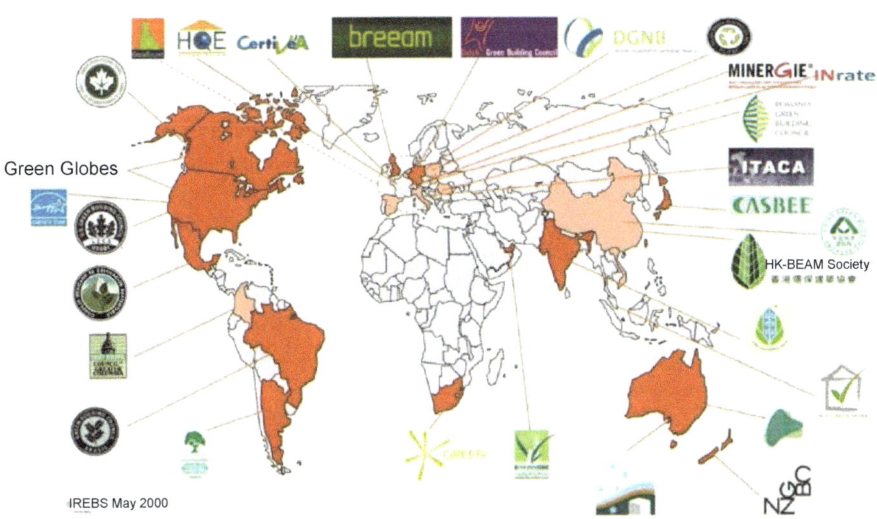

Figure 5.4. Building Rating Tools Available Worldwide (Reed et al., 2011).

Energy and Environmental Design) in the US, CASBEE (Comprehensive Assessment System for Building Environmental Efficiency) in Japan, Green Star in Australia and HK-BEAM in Hong Kong (Nguyen, 2011).

The rating methods developed with these tools are built upon various principles and different evaluation items, data, and criteria, based on the original condition of the buildings without taking into consideration a lifetime parameter such as a modification of the building elements (MD Darus and Hashim, 2012). The weighting systems are being formulated according to different environmental categories; then the points will be summed into a single final score that represents overall ratings. Although most rating systems have almost similar environmental categories, the value of each weighting is different.

Malaysia's Green Building Index (GBI)

The Green Building Index (GBI) is a voluntary scheme, co-developed by Pertubuhan Akitek Malaysia (PAM)—Malaysian Institute of Architects—and Association of Consulting Engineers Malaysia (ACEM) officially launched on 21st May 2009 (PAM, 2009). The GBI was derived from existing rating tools, which include the Green Mark from Singapore and Green Star from Australia, but is extensively modified for Malaysian tropical weather, environmental context, cultural and social needs (Green Building Index, 2009).

The GBI system evaluates six main criteria including energy efficiency, indoor environment quality, sustainable site planning and management, material and resources, water efficiency and innovation as shown in Table 5.1. The system is being created to promote sustainable development in the building industry. The final result of the assessment will be rated with Platinum (86+ points), Gold (76 to 85 points), Silver (66 to 75 points) and Certified (50 to 65 points).

The application of GBI is not limited to residential buildings but spans to non-residential buildings, industrial building, retail building, and township. The buildings are divided into two categories, namely new and existing construction except for residential and township that only focus on new construction. Each category has different weighting points allocation set in the pre-determined six criteria. Since its launch in 2009 to July 2013, a total of 146 projects have been certified, while the majority of the project is for non-residential new construction (72 projects), followed by residential new construction (61 projects) (Green Building Index, 2013). Among notable projects awarded with Platinum awards are the Energy Commission building (Diamond building) in Putrajaya, SP Setia Berhad Corporate Headquarters in Shah Alam, Kompleks Kerja Raya 2 (KKR 2) in Kuala Lumpur, Bangunan Perdana Putra in Putrajaya, S11 House in Petaling Jaya and Tun Razak Exchange township in Kuala Lumpur.

Overall, the introduction of GBI promotes the idea of sustainable buildings, although most of the projects were concentrated in the urban areas in Kuala Lumpur, Selangor, and Penang. It is estimated that the GBI certified buildings capable of reducing the CO_2 emission by 243,789 tonne CO_2 per annum (Green Building Index, 2013). However, recent research has identified that there are barriers to the

Table 5.1. GBI rating system criteria (Green Building Index, 2016).

GBI Criteria	Scope
Energy Efficiency (EE)	Improve energy consumption by optimizing building orientation, minimizing solar heat gain through the building envelope, harvesting natural lighting, adopting the best practices in building services including use of renewable energy, and ensuring proper testing, commissioning, and regular maintenance.
Indoor Environment Quality (EQ)	Achieve good quality performance in indoor air quality, acoustics, visual and thermal comfort. These will involve the use of low volatile organic compound materials, application of quality air filtration, proper control of air temperature, movement, and humidity.
Materials & Resources (MR)	Promote the use of environment-friendly materials sourced from sustainable sources and recycling. Implement proper construction waste management with storage, collection, and re-use of recyclables and construction formwork and waste.
Sustainable Site Planning & Management (SM)	Selecting appropriate sites with planned access to public transportation, community services, open spaces, and landscaping. Avoiding and conserving environmentally sensitive areas through the redevelopment of existing sites and brownfields. Implementing proper construction management, storm water management and reducing the strain on existing infrastructure capacity.
Water Efficiency (WE)	Rainwater harvesting, water recycling, and water-saving fittings.
Innovation (IN)	Innovative design and initiatives that meet the objectives of the GBI.

implementation of GBI that need to be overcome such as lack of awareness and technical understanding, the perception of higher cost, insufficient supply of green products, and lack confidence in the sustainable options (Algburi et al., 2016).

Green Building Rating System and LCA

The integration of LCA to the Green Building Rating System (GBRS) is the next step in creating a more comprehensive GBRS in relation to the environmental impact assessment. At the moment, some GBRS have integrated LCA such as in Leadership in Energy and Environmental Design (LEED) in the US, Building Research Establishment Environmental Assessment Methodology (BREEAM) in the UK, Green Globes in Canada, California Green Building Standards (CALGreen) in California, German Sustainable Building Certification (DGNB) in Germany, Comprehensive Assessment System for Building Environmental Efficiency (CASBEE) in Japan, and Green Star in Australia (Al-Ghamdi and Bilec, 2016; Anand and Amor, 2017; Roh et al., 2016; Zuo et al., 2017).

The implementations of LCA in each GBRS were varied in scope, qualification standard, and LCA evaluation method. Roh et al. (2016) have discussed the different LCA methodology applied in three major GBRS which are BREEAM, LEED and CASBEE. The BREEAM and LEED assess the impact on building materials according to the building elements, while in CASBEE only assess the total building materials used. Each GBRS uses different evaluation method. BREEAM uses materials database from Green Guide and IMPACT LCA software, LEED uses Building Energy Efficiency Standard (BEES) database and Athena Eco Calculator, and CASBEE uses its own Building Materials spreadsheet (Roh et al., 2016).

Moreover, the usage of building materials with certified Environmental Product Declarations (EPD) will be given credits when the whole building is assessed under the BREEAM and LEED (Zuo et al., 2017). The integration of LCA in Green Star was made available in 2014 which embedded it as a core credit in the rating tools and reward credit for the usage of EPD certified building materials (Green Building Council Australia, 2015). EPD is a verified detail transparent document of life cycle environmental impact of a product that follows guidelines in the ISO 14025 which may be used for many different applications, including green public procurement (GPP) and building assessment schemes (EPD International AB, 2016).

There are also other initiatives by researchers trying to integrate LCA into GBRS such as in CALENER VYP and CALENER GT in Spain (Zabalza et al., 2009) and Building Life Cycle Carbon Emission Assessment Program (BEGAS 2.0) to complement Korea's Green Building Index certification system (Roh et al., 2016). The integration of LCA in GBRS is a dynamic process since the LCA methodology in the building industry is still being developed and thriving. The standardization of the method is important to ensure the environmental impact assessed is comparable and thus establishing a robust database for future references.

Conclusion

Life cycle assessment is a reasonable tool for analyzing comparable aspects of quantifiable systems. For a LCA study, there are a variety of limitations and obstacles faced by LCA researchers and practitioners which may include data gap: Lack or accessibility of data in some parts requires assumption and estimation and some aspects are excluded from the study due to non-availability of data or minimal contribution. Others would include choice of measurement tools for the assessment, uncertain fate and transport of the substances and lack of national database. In Malaysia, the lack of access to such a national database can cause the results obtained to be not truly representative of the actual impacts in a Malaysian context figuratively.

Hence it is important for the LCA study to be meaningful and accurate, the strategy is to make sure that by 2020 there is the availability of national average data on LCA for the whole sectors of the building industry and the data can be used as a guide in formulating strategies to reduce the environmental impacts from the whole sectors of the building industry.

LCA results can help to pinpoint many aspects for improvement of the building sector. Therefore, it is a very powerful tool for assessing the potential environmental impacts of building materials, methods, etc., by identifying specific contribution of every process in the building to the overall environmental performance of the system.

LCA is a tool used to quantify environmental performance of products and processes. It is recommended for building industry as it can assist to understand and improve the overall environmental performance. It is a reliable instrument for environmental assessment because it can provide quantification information and evaluation systematically and discretely. Other sectors that attempt to start or move forward in sustainable green initiatives can adopt the life cycle assessment approach.

References

Abd. Rashid, A. F., J. Idris and S. Yusoff. 2017. Environmental impact analysis on residential building in Malaysia using life cycle assessment. Sustainability 9(3): 329. http://doi.org/10.3390/su9030329.

Abd. Rashid, A. F. and S. Yusoff. 2012. Sustainability in the building industry: A review on the implementation of life cycle assessment. *In*: Wijesuriya, K. (ed.). 2nd International Conference on Climate Change & Social Issues. Kuala Lumpur: International Center for Research and Development (ICRD).

Abd. Rashid, A. F. and S. Yusoff. 2015. A review of life cycle assessment method for building industry. Renewable and Sustainable Energy Reviews 45: 244–248. http://doi.org/10.1016/j.rser.2015.01.043.

Al-Ghamdi, S. and M. Bilec. 2016. Green building rating systems and whole-building life cycle assessment: Comparative study of the existing assessment tools. Journal of Architectural Engineering 23(1): 1–9. http://doi.org/10.1061/(ASCE)AE.1943-5568.0000222.

Algburi, S. M., A. A. Faieza and B. T. H. T. Baharudin. 2016. Review of green building index in Malaysia; existing work and challenges. International Journal of Applied Engineering Research 11(5): 3160–3167.

Anand, C. K. and B. Amor. 2017. Recent developments, future challenges and new research directions in LCA of buildings: A critical review. Renewable and Sustainable Energy Reviews 67: 408–416. http://doi.org/10.1016/j.rser.2016.09.058.

Arham, A. 2008. A methodological analysis of demolition works in Malaysia.

Asif, M., T. Muneer and R. Kelley. 2007. Life cycle assessment: A case study of a dwelling home in Scotland. Building and Environment 42(3): 1391–1394. http://doi.org/10.1016/j.buildenv.2005.11.023.

Azapagic, A. 1999. Life cycle assessment and its application to process selection, design and optimisation. Chemical Engineering Journal 73(1): 1–21. doi: http://dx.doi.org/10.1016/S1385-8947(99)00042-X.

Berardi, U. 2012. Sustainability assessment in the construction sector: Rating systems and rated buildings. Sustainable Development 20(September 2011): 411–424. http://doi.org/10.1002/sd.

Blengini, G. A. and T. Di Carlo. 2010. The changing role of life cycle phases, subsystems and materials in the LCA of low energy buildings. Energy and Buildings 42(6): 869–880. http://doi.org/10.1016/j.enbuild.2009.12.009.

Blengini, G. and T. Di Carlo. 2010. Energy-saving policies and low-energy residential buildings: an LCA case study to support decision makers in Piedmont (Italy). The International Journal of Life Cycle Assessment 15(7): 652–665. http://doi.org/10.1007/s11367-010-0190-5.

Bong, C. P. C., L. Y. Lim, W. S. Ho, J. S. Lim, J. J. Klemeš, S. Towprayoon and C. T. Lee. 2012. A review on the global warming potential of cleaner composting and mitigation strategies. Journal of Cleaner Production. doi:http://dx.doi.org/10.1016/j.jclepro.2016.07.066.

Cole, R. J. 1998. Energy and greenhouse gas emissions associated with the construction of alternative structural systems. Building and Environment 34(3): 335–348. Retrieved from http://www.sciencedirect.com/science/article/pii/S0360132398000201.

Cook, S. 2010. Sustainability as a Global Issue. MRB Rubb. Technol. Dev. 10(2): 12–15.

Ding, G. 2007. Life cycle energy assessment of Australian secondary schools. Building Research & Information, [online] 35(5), pp.487-500. Available at: https://doi.org/10.1080/09613210601116408 [Accessed 14 Sep. 2018].

EPD International AB. 2016. What is an EPD - Environmental Product Declarations. Retrieved July 22, 2017, from http://www.environdec.com/en/What-is-an-EPD/.

Fay, R., G. Treloar and U. Iyer-Raniga. 2000. Life-cycle energy analysis of buildings: Acase study. Building Research and Information 28(1): 31–41.

Green Building Council Australia. 2015. Life Cycle Assessment in Green Star - Materials Category - Green Building Council Australia (GBCA). Retrieved July 18, 2017, from https://www.gbca.org.au/green-star/materials-category/life-cycle-assessment-in-green-star/.

Green Building Index. 2009. The development of green building index malaysia. Architecture Malaysia 14–17.

Green Building Index. 2012. What is a green building? Retrieved October 12, 2012, from http://www.greenbuildingindex.org/why-green-buildings.html.

Green Building Index. 2013. Greenbuildingindex.org - GBI Certified Buildings. Retrieved July 31, 2013, from http://www.greenbuildingindex.org/organisation-certified-buildings-Summary.html.

Green Building Index. 2016. Green Building Index Rating System. Retrieved February 26, 2016, from http://new.greenbuildingindex.org/how/system.

Huijbregts, M. A. J., W. Gilijamse, A. M. J. Ragas and L. Reijnders. 2003. Evaluating uncertainty in environmental life-cycle assessment. A case study comparing two insulation options for a dutch one-family dwelling. Environmental Science & Technology 37(11): 2600–2608. http://doi.org/10.1021/es020971+

ISO. 2006. ISO 14040:2006 Environmental management—Life cycle assessment—Principles and framework. Geneva, Switzerland: International Organization for Standardization.

Iyer-Raniga, U. and J. P. C. Wong. 2012. Evaluation of whole life cycle assessment for heritage buildings in Australia. Building and Environment 47(0): 138–149. Retrieved from http://www.sciencedirect.com/science/article/pii/S0360132311002447.

Kofoworola, O. F. and S. H. Gheewala. 2009. Life cycle energy assessment of a typical office building in Thailand. Energy and Buildings 41(10): 1076–1083. Retrieved from http://www.sciencedirect.com/science/article/B6V2V-4WHFD60-2/2/1a1c171a94496071d1e58f2fa0e9b1ff.

Levine, M., D. Ürge-Vorsatz, K. Blok, L. Geng, D. Harvey, S. Lang and H. Oshino. 2007. Residential and commercial buildings. *In*: Metz, B., O. R. Davidson, P. R. Bosch, L. Dave and A. Meyer (eds.). Climate Change 2007 - Mitigation of Climate Change: Working Group III Contribution to the Fourth Assessment Report of the IPCC. Cambridge University Press.

López-Mesa, B., Á. Pitarch, Tomás, A. and T. Gallego. 2009. Comparison of environmental impacts of building structures with in situ cast floors and with precast concrete floors. Building and Environment, [online] 44(4), pp.699-712. Available at: https://doi.org/10.1016/j.buildenv.2008.05.017 [Accessed 14 Sep. 2018].

MD Darus, Z. and N. A. Hashim. 2012. Sustainable building in Malaysia: The development of sustainable building rating system. pp. 113–144. *In*: Ghenai, C. (ed.). Sustainable Development - Education, Business and Management—Architecture and Building COnstruction - Agriculture and Food Security. InTech. Retrieved from http://cdn.intechopen.com/pdfs/29275/InTech-Sustainable_building_in_malaysia_the_development_of_sustainable_building_rating_system.pdf.

Mithraratne, N. and B. Vale. 2004. Life cycle analysis model for New Zealand houses. Building and Environment 39(4): 483–492. Retrieved from http://www.sciencedirect.com/science/article/B6V23-4B1X4GJ-1/2/8a345afe66113d09f02363e056c19da5.

Nguyen, B. 2011. Tall Building Projects Sustainability Indicator. University of Sheffield.

Ochsendorf, J., L. K. Norford, D. Brown, H. Durchlag, S. L. Hsu, A. Love and M. Wildnauer. 2011. Methods, Impacts, and Opportunities in the Concrete Building Life Cycle, (August). Retrieved from http://www.greenconcrete.info/downloads/MITBuildingsLCAreport.pdf.

Ortiz-Rodríguez, O., F. Castells and G. Sonnemann. 2010. Life cycle assessment of two dwellings: One in Spain, a developed country, and one in Colombia, a country under development. Science of The Total Environment 408(12): 2435–2443. Retrieved from http://www.sciencedirect.com/science/article/B6V78-4YJCTJF-3/2/9b3775e33fce59f6334c6022f487c694.

Ove Arup and Partners Hong Kong Ltd. 2007. An Introduction to Life Cycle Energy Assessment (LCEA) of Building Developments. Hong Kong.

PAM. 2009. The Development of Green Building Index Malaysia. Architecture Malaysia 14–17.

Reed, R., S. Wilkinson, A. Bilos and K. W. Schulte. 2011. A Comparison of International Sustainable Building Tools—An Update. In The 17 th Annual Pacific Rim Real Estate Society Conference. Pacific Rim Real Estate Society. Retrieved from http://www.academia.edu/897809/A_Comparison_of_International_Sustainable_Building_Tools_An_Update.

Roh, S., S. Tae, S. J. Suk, G. Ford and S. Shin. 2016. Development of a building life cycle carbon emissions assessment program (BEGAS 2.0) for Korea's green building index certification system. Renewable and Sustainable Energy Reviews 53: 954–965. http://doi.org/10.1016/j.rser.2015.09.048

Sartori, I. and A. G. Hestnes. 2007. Energy use in the life cycle of conventional and low-energy buildings: A review article. Energy and Buildings 39(3): 249–257. Retrieved from http://www.sciencedirect.com/science/article/B6V2V-4KST3K4-1/2/a4990b7e5be60e7f824fbcb2688ec02f.

Scheuer, C., G. A. Keoleian and P. Reppe. 2003. Life cycle energy and environmental performance of a new university building: modeling challenges and design implications. Energy and Buildings 35(10): 1049–1064. Retrieved from http://www.sciencedirect.com/science/article/pii/S0378778803000665.

Thormark, C. 2002. A low energy building in a life cycle--its embodied energy, energy need for operation and recycling potential. Building and Environment 37(4): 429–435. Retrieved from http://www.sciencedirect.com/science/article/B6V23-452V6BK-D/2/e8fa0870b8d25163863d447518f384bf.

UNEP-SBCI. 2009. Buildings and Climate Change: Summary for Decision-Makers. Paris. Retrieved from http://www.unep.org/SBCI/pdfs/SBCI-BCCSummary.pdf.

Utama, A. and S. H. Gheewala. 2008. Life cycle energy of single landed houses in Indonesia. Energy and Buildings 40(10): 1911–1916. Retrieved from http://www.sciencedirect.com/science/article/B6V2V-4SG4HKT-1/2/c56c6bacbbc6b383c5ed4ea0313a1867.

Utama, A. and S. H. Gheewala. 2009. Indonesian residential high rise buildings: A life cycle energy assessment. Energy and Buildings 41(11): 1263–1268. Retrieved from http://www.sciencedirect.com/science/article/B6V2V-4WWG345-4/2/2a58c0df30e50f4c1fe77f6b568088ad.

Van Ooteghem, K. and L. Xu. 2012. The life-cycle assessment of a single-storey retail building in Canada. Building and Environment 49(0): 212–226. Retrieved from http://www.sciencedirect.com/science/article/pii/S0360132311003131.

Yusoff, S. 2006. Renewable energy from palm oil – innovation on effective utilization of waste. Journal of Cleaner Production, [online] 14(1), pp. 87–93. Available at: https://doi.org/10.1016/j.jclepro.2004.07.005 [Accessed 14 Sep. 2018].

Zabalza, I., A. Aranda, S. Scarpellini, S. Diaz, C. Koroneos and A. Dompros. 2009. Life cycle assessment in building sector: State of the art and assessment of environmental impact for building materials, pp. 1–8. *In*: 1st International Exergy, Life Cycle Assessment, and Sustainable Workshop & Symposium. Greece.

Zabalza Bribián, I., A. Aranda Usón and S. Scarpellini. 2009. Life cycle assessment in buildings: State-of-the-art and simplified LCA methodology as a complement for building certification. Building and Environment 44(12): 2510–2520. Retrieved from http://www.sciencedirect.com/science/article/B6V23-4W9XB8N-1/2/cbe2a47f936298e5c06080b0408bf830.

Zuo, J., S. Pullen, R. Rameezdeen, H. Bennetts, Y. Wang, G. Mao and H. Duan. 2017. Green building evaluation from a life-cycle perspective in Australia: A critical review. Renewable and Sustainable Energy Reviews 70(September 2016): 358–368. http://doi.org/10.1016/j.rser.2016.11.251.

CHAPTER 6

Green Building Policies in Hong Kong and Singapore

Yajing Liu[1] and *Zhonghua Gou*[2,*]

Introduction

Since the 1990s, a green building revolution swept across the world. What the chapter discusses is not about these green certification systems, such as U.S. LEED, U.K. BEAM, and Singapore Green Mark. Green Building is like a revolution, springing up all over the world. Most of the countries, in facing the challenge of energy crisis and environmental threatens, resort to green building. The benefits of green building have been widely discussed, however, after so many years of promotion, the practice of green building has not launched on a large scale (Koeppel and Ürge-Vorsatz, 2007). The private investments are still reluctant to voluntarily enter the green building market, due to the challenges and barriers in its market penetration. There are many reasons behind the reluctance to adopt green building: administrative burden of certification (Athens, 2012), doubt of the green building performance (Gou et al., 2013), cost and payback (Bond and Perrett, 2012) and so on. Policy is a strong and inevitable support to green building practice.

The chapter investigates the green building related policy tools, aiming to gain experience and find opportunities to bring affordability issues into the current green building policy. Following with the work of Pearce et al. (2007), policies tools in this chapter are categorized into four aspects for the investigation.

- Policy options which are a formal execution, applied as an official procedure to mandate or encourage green building action;
- Programs which support green building practice with monetary or information;

[1] Kerry Properties Limited, Hong Kong.
[2] Griffith University, Australia.
* Corresponding author: gouzhonghua@gmail.com, z.gou@griffith.edu.au

- Implementation which offers guidance and technical support on the project level when developers or designers would like to achieve certain goals;
- Evaluation which serves to measure the actual performance of green building.

The chapter selects Hong Kong and Singapore as case studies for the investigation. Singapore and Hong Kong are the leaders in Asian green building revolution.

The Hong Kong government keeps a fundamental principle in minimizing interference in economic activities. The government stays outside of the market tide and only gets involved when general interests are impacted. This principle has influenced Hong Kong for decades. In the field of building industry, the Hong Kong government often uses voluntary approaches in improving buildings' sustainability. Regulations of sustainability in the building sector are relatively light-handed. A large number of green building related policies are on a voluntary basis. To further encourage the action of sustainability from private sectors, the government takes the lead and applies green buildings in public buildings, which provides a model and guidance in technologies and practices for private sectors.

Singapore is a city-state with limited land size and scarcity of natural resources. It has to entirely depend on import on energy supply, which pushes its development to be extremely vulnerable to global energy trading (Wen and Clifford, 2007). The Singapore government has long been aware of this risk. Since 1970, the Singapore government has put energy security as the top concern in national policy. The energy concern in Singapore is also closely integrated with environmental sustainability. Singapore has been committed to reduce greenhouse gas emissions in several international declarations. It emphasized sustainability through promotion on clean energy, energy efficiency and environmental conservation.

Green Building in Hong Kong and Singapore

The number of green building projects is the most intuitive reflection of green building penetration and development in the society. The popularity of green building in both Hong Kong and Singapore is a product from the combined power of policy and market force. The extent of its booming in the building industry shows the validity of related policy strategies and the acceptance of the market.

Currently in Hong Kong, three green building certifications are popular. HKBEAM is a local brand developed on the BREEAM in UK. Green Building Labelling (GBL) system is a revised version from the green building certification system in China Green Building Council, managed by its Hong Kong chapter. LEED, established in US, but gained an international reputation in all around the world. GBL certification in Hong Kong commenced in 2010. Until 2014, only eight projects have been successfully certified (Gou and Lau, 2014). The wide influence of LEED in Hong Kong attracted a big group of followers. Until 2014, around 90 projects has been registered in LEED certification, 11 ranks on platinum, 32 on gold, 5 on silver and 4 on certified (USGBC, 2014).

The local certification system HKBEAM has the largest group of audiences (Table 6.1). During the almost 20 years' development, HKBEAM has evolved through several versions. Statistics of the project number in different versions are calculated separately.

Table 6.1. Quantity statistics of beam plus certified projects in Hong Kong from 2009–2014 (Source: HKGBC, 2014).

	Final platinum	**Final gold**	**Provisional platinum**	**Provisional gold**	**Provisional silver**	**Provisional bronze**	**Provisional unclassified**	**Registered**	**In total**
No. of public Projects	2	1	22	11	4	3	3	33	79
No. of Private Projects	3	0	8	32	25	27	44	359	498
No. in total	5	1	30	43	29	30	47	392	577

It is mainly counted in two phases. Based on the information on Beam Society (BEAM Society, 2014), as of Oct 2009, BEAM had certified nearly 199 landmark properties in Hong Kong, including around 75% private projects and 25% public ones. According to the information released on HKGBC website (HKGBC, 2014), until 2014, totally 577 projects had been registered or certified on BEAM Plus, including 79 government projects, making up one seventh of the total project number. However, only five projects achieved the final Platinum level, including two public projects. Since the launch of BEAM Plus in 2009, only six projects received their final certifications and half were public projects. Many projects were still on the provisional certification, which means they had not passed the final check after construction. Understandably, some projects with provisional certification were due to their unfinished construction. But undeniably, many other projects only got the provisional certification, which is useful enough for project marketing. The final certification was after project construction, which had no help to project sale. Thus, many developers ignored the final certification.

The suddenly jumped number of private projects on the registered level is probably attracted by the incentives on GFA (Gross Floor Area) concession, launched in 2011. Projects registered on the BEAM Plus certification are eligible for the incentive of GFA concession with a 10% cap. However, due to no requirement on the certified level in BEAM Plus, most projects just stay on the entry level without any certification levels. Unclassified level is for projects which cannot reach the Bronze level but fulfil the prerequisites. It is easy to estimate the number of projects on the unclassified level would continuously increase if the requirements remain no certification level for the incentives. Combining number of projects under both BEAM 4/04 & 5/04 and BEAM Plus, until November 2014, Hong Kong totally had nearly 776 BEAM Plus projects.

The green building certification market in Singapore is dominated by its local green building brand-Green Mark, developed and managed by Building & Construction Authority (BCA). The International brand LEED only takes up a small proportion. Until 2014, the total number of LEED projects was 58, including six on platinum, 21 on gold, five on silver, six on certified, others on uncertified (USGBC, 2014). Since Green Mark was launched in Singapore in 2005, the number of certified buildings has increased dramatically with the mandatory implementation of Green Mark on public buildings and phased promotions of Green Building Masterplans for private buildings. The official figures demonstrated that until 2013, Singapore had more than 1600 Green Mark certified buildings. However, data available on Green Mark website only covered 1060 projects (see Table 6.2) (BCA, 2014). From the distribution of green building in different levels on the available information, green building level in Singapore is more than that in Hong Kong. Government/public projects makes up of one fifth of the total number.

Even though the consciousness of green building commenced in Hong Kong 10 years earlier than Singapore, Singapore has caught up with more than double the number of green building in Hong Kong under respective local certification standards. If considering projects with high-level green building certification (Gold and Platinum), Singapore performs even better. The slow pace in Hong Kong resulted from the special context and legislation framework. Hong Kong always pursues the free economy with less government intervention. Green building development mainly relies on market force; while Singapore is a government-dominated legal system and

Table 6.2. Quantity statistics of green building projects in singapore until Nov, 2014.

	Platinum	Gold plus	Gold	Certified	In total
No. of public projects	58	31	55	41	185
No. of private projects	209	181	361	124	875
No. in total	267	212	416	165	1060

Source: BCA, 2014.

Note: Due to the different characteristics of the search engines, public buildings in Hong Kong and Singapore are searched under different conditions. In HKBEAM, a public building is identified based on building types and developers. Building type in government, institutional and community buildings and building developed by public sectors, listed as Housing Authority, Architecture Service Department, Drainage Department, Hong Kong Housing Society, Immigration Authority, Urban Renewal Authority, West Kowloon Cultural District Authorities, hospital, university and school are included. In Singapore Green mark, public building is categorized on building types, including institutional of healthcare, school, sports and recreation facilities, community buildings, church, public house, park, infrastructure and transit system.

government maintains a strong power in directing the society development. The rapid development of green building in Singapore represents strong policy effectiveness.

Policy Tools in Hong Kong

Policy Options

The Hong Kong government had concern about building energy efficiency since a long time. With the challenge in global warming and large energy consumption, the Hong Kong government was determined to improve local environment sustainability. The first commission of a consultant report on building energy efficiency regulation was launched in 1990 by Hong Kong Government (Wen and Clifford, 2007). An Energy Efficiency Advisory Committee (EEAC) was authorized by government to consult on improving energy efficiency (Chan and Yeung, 2005). This committee recommended an Overall Thermal Transfer Value (OTTV) standard for building envelope, which was adopted in July 1995 for commercial buildings and hotels in Hong Kong, as a mandatory standard. However, an interesting point is that during the past decades, OTTV standard has only been further updated in 2000.

Besides OTTV, which applies to building envelope, there are standards for building services, including lighting, air conditioning and electrical equipment launched in 1998, as well as for elevators in 2000, initially in a voluntary basis under the Hong Kong Energy Efficiency Registration Scheme for Buildings (HKEERSB), which is to promote Building Energy Codes (BECs) by providing certification to buildings complying with BECs' requirements. These four codes were respectively updated in 2005 and 2007.

At this time, all those codes were in a prescriptive approach, which addressed requirements by making minimum standards. Prescriptive standards were simple and easy to apply. However, those isolated number in prescriptive standards never perceive buildings as a whole (Hui, 1998). Thus, the minimum requirements often turned to be the maximum due to the reluctance of projects in going beyond of code compliance. Contrarily, a performance-based regulation could be open to innovation and flexibility, as well as optimize efficiency (Xie and Gou, 2017). It overlooked the partial performance of building service but addressed the holistic energy performance. It reviewed the building in a systematic manner, agreed on trade-off among building components (Ma and Wang, 2009) and resulted in the freedom for innovation. Admittedly, performance-based approach is generally complicated and requires intensive skills.

In 2005, a performance-based option was added to all the five standards. In 2012, the four voluntary schemes were updated to mandatory. Exception for lighting, which was looser than international standards, the other three standards were similar to worldwide requirements. Since 2000, OTTV with the other four standards cover the same scope as the US ASHRAE 90.1 standard. The advancing sensitivity of the Hong Kong government in building energy efficiency and green building indicated its desire on promotion sustainability (Fig. 6.1). However, investigation showed, during the time ahead of the mandatory requirements, most buildings didn't voluntarily meet the requirements of the codes. Even after the launch of mandatory OTTV standards, without a holistic system controlling the energy consumption in building service, the "actual" energy savings from OTTV standards were limited (Hui, 1998). It was perceived that during the initial period, the educational influence of prescriptions in OTTV in raising concerns and awareness of energy efficiency issue was more significant than its actual result of control in energy saving. However, there was a

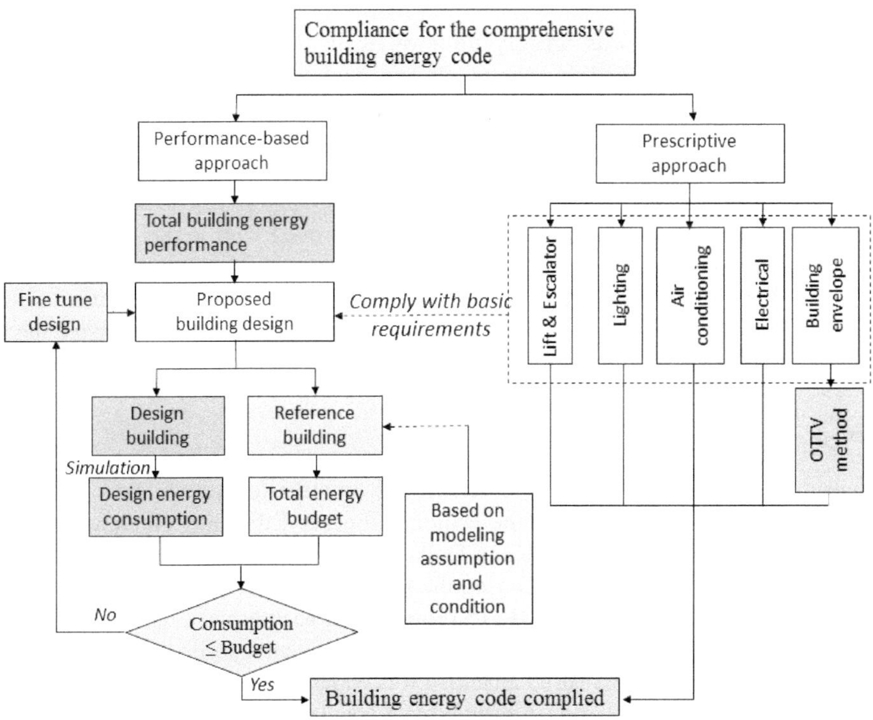

Figure 6.1. Framework of the comprehensive building energy code in Hong Kong (Source: Ma and Wang, 2009).

dramatic change in 2005, when private sectors, such as developers, building owners and operators realized the pressure from environment protection and benefits from building costs (Ma and Wang, 2009). Mandatory energy efficiency requirements in appliances currently cover five types of products, listed as air conditioners, refrigerating appliances, Compact Fluorescent Lamps (CFLs), washing machines and dehumidifiers. Under the Mandatory Energy Efficiency Labelling Scheme (MEELS), energy labels are required to show on the required selling equipment.

Hong Kong has been working on green building since 1996. As a non-government green building initiative, the Building Environmental Assessment Method (BEAM) is promoted on a voluntary basis. It was commenced by a private sector—the Real Estate developers Association in Hong Kong with the aim to "enhance the quality" as well as "reduce the environmental impact of buildings". It promotes exemplary practices in planning and design, management, operation and maintenance in the local context of Hong Kong (BEAM Society, 2014). Its directing organization HK-BEAM Society is a non-profit organization consisting of various stakeholders in the building industry, covering members from developer, engineer, contractor, designer and other building professionals. It covers two independent versions for both new and existing buildings and discusses the impact of buildings on the environment. According to BEAM Society, until 2014, BEAM had certified over 140 million square feet of green building developments and over 50,000 residential units. It is considered as "one of the most widely used voluntary green building labelling schemes" on per capita basis (BEAM Society, 2014).

BEAM Scheme was largely developed on the UK Building Research Establishment's BREEAM. It was initially launched with two versions, for new and existing office buildings. In 1999, both of them updated with an extension to high-rise residential buildings, for reducing environmental impact of new residential buildings with best techniques and reasonable costs. Later an updated version in 2003 and 2004 further defined the sustainability in buildings. In 2009, in response to climate change and global warming, the scheme was further improved with higher emphasis on greenhouse gas emission reduction. The recent update was in 2012, which introduced passive design, performance-based assessment and other minor amendments. In defining the energy performance criteria, HKBEAM made reference to Building Energy Codes (BECs) and used the minimum requirements in BECs as baseline (Fig. 6.2).

Program

Incentives were offered under several schemes. In order to encourage energy audit among private building owners, the Buildings Energy Efficiency Funding Scheme allocation of 450 million Hong Kong Dollars to support energy-cum-carbon audits and energy efficiency projects. During the three years period—from April 2009 to April 2012, the scheme supported 115 projects. For each project, up to 50% of the total actual expenditure in energy auditing and subsequent reporting was subsidized. However, this kind of cost-based incentive was widely criticized (Wen and Clifford, 2007) for its inefficiency or even resulted in a counteractive effect. Compared with performance-based incentives, which could increase competition and encourage

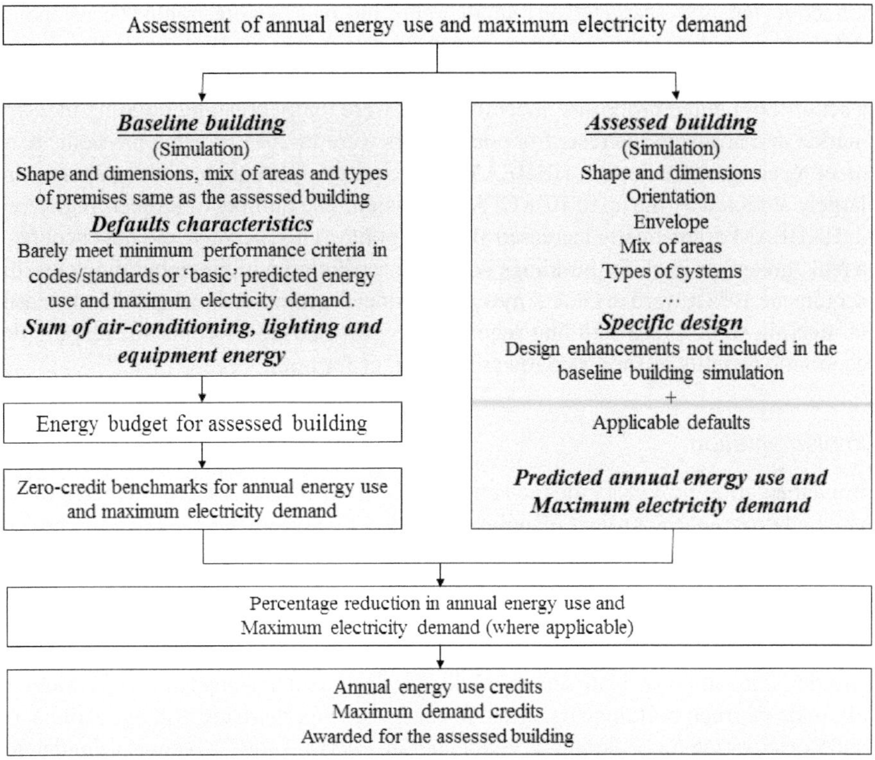

Figure 6.2. Building energy performance assessment in HK-BEAM (Source: Ma and Wang, 2009).

innovation, cost-based incentives mainly depend on expenditure that may cause cost raise.

"Hong Kong Energy Efficient Award" is held by EMSD. Through competition, it intended to encourage private sectors to efficiently use energy as well promote energy saving among the general public. According to the website information, it attracted more than 230 entries. However, it only held one session between 2004 and 2005 without any future competitions. It was easy to estimate that its influence was quite limited. Other funds targeted to a general field could be applied in building industry. One is "Innovation and Technology Fund" (ITF), which helps local companies to upgrade their technological level and introduces innovative ideas to their businesses. "Cleaner Production Partnership Program" offers a total of 143 million to encourage Hong Kong-owned factories in the Pearl River Delta region to adopt cleaner production technologies and practices in: "Minimize Air Pollutant Emissions"; "Improve Energy Efficiency"; "Reduce and Control Effluent Discharge"; "Reduce Production Cost". Even though these funds are open to the whole society, its target groups are rather small. Thus, their influence on the technology improvement in building industry is trivial.

Obviously, all the mentioned incentives and funds, in a sense, positively promote the concept of sustainability among the industry and the general public, but those funds can hardly motivate an actual change towards "green". The subsidies were not attractive; incentives targeted on a small group; and awards were granted occasionally. All these reveal that the reward system lacks a systematic plan. Rather than being integrated within a foreseeing roadmap, incentives and rewards are more likely to be a set of fragmented proposals. Incentives that were really breaking building industry market and arousing interests from developers were in 2011, when a prerequisite of meeting certain standards in HKBEAM was required. Nevertheless, developers were largely attracted by the up to 10% GFA concession. The number of projects registered in HKBEAM dramatically increased after this policy. This incentive did not encourage a real "green" project, for buildings without a certification of green building can still acquire the 10% reward in GFA. Anyway, this indeed, promotes developers' enthusiasm in meeting some green building requirements, with purposes either for reaping the economic benefits or for a step towards green, or for both.

Implementation

Implementation provides guidance on projects. It introduces the requirements of the targeted program the project planning to meet, while orientates the level the project may achieve. Those assistances, summarized as technical support, guidance documents and exemplary projects, are advised from experts in green building. Technical support and guidance focus on the confusion from professionals. Without much mandatory requirements, technical support and guidance are one of the main sources to promote sustainability among private sectors. Guidance document in Hong Kong covers almost all fields of green building strategies. In energy saving, there are "Energy Efficiency and Conservation for Buildings", "Guidelines on Energy Audit", "Technical Guidelines on Grid Connection of Small-scale Renewable Energy Power Systems" by EMSD in 2007 and "building life-cycle energy use". Besides guidance in the pamphlet, the government uses website to widely promote green, "HK EE Net" is a website for Energy Efficient Technologies. In waste management, a website "Construction Waste" is set to provide instructions on waste management, reduction guidelines and technical circulars for professionals' reference. In indoor air quality, "Indoor Air Quality (IAQ) Information Centre" is opened for the public as well as professionals. Technical advice as well as general information on IAQ is available. Noise is a serious problem in Hong Kong. Local government provides several design guidelines in noise control, such as Good Practices on different Noise Control, Noise Control Guidelines and Guidelines on Design, intending to create a noiseless environment in a persuasive manner. Information is also available on the website. Supported by EPD, "Innovative Noise Mitigation designs and measures" website provides interested parties with innovative noise control design and solid effective real examples in various housing developments, including both public and private projects. The approaches have been classified into a series of categories, easy for designers' reference. Besides, the government develops some innovative programs. Life cycle is an advanced concept. The Hong Kong government tried to bring it into industry and to the public. However, conflict of interests in life-cycle analysis harms its promotion. The two schemes ended with a narrow influence. "Life-Cycle energy Analysis (LCEA) of Building Construction" is one of these programs (EMSD, 2002). Developed by EMSD in 2002, this tool is intended to assess life-cycle costs and life-cycle performances of building material and components and suggest alternative materials and systems in improving environmental, energy and economic performance of buildings. It provides an online template with data-entry sheet. After inputting building material information, it shows an energy use report. Successful projects could be showcases of new technologies. Another benefit of exemplary projects is its introduction to innovative technologies and new ideas, which are successful in those projects providing compelling persuasion to others. Some of the exemplary projects are public buildings funded by the government as models for public education and visits, such as the zero carbon centre and The Physical Centre by EPD, with information display. In many of those projects, new technologies are implemented mainly on the purpose of education rather than their actually sustainability performance. Thus, many new technologies are exhibited at less consideration of cost. It is easy to understand that these technologies are only applied to several demonstrated showcases.

Many of the exemplary projects are for experimenting and showcases of new technologies, usually funded by the government as well. The wide used pre-casting and prefabrication of components in the Housing Department is a model of prefabrication technology. Some technologies are tested in a small district, such as the district cooling system for centralized air-conditioning testing at the Kai Tak Development. Another case is an EMSD's new headquarters building. It used the largest number of building integrated photovoltaic (BIPV) panels in Hong Kong and contributed 3–4% of the electricity consumption of the building. Showcases are also from private projects. Those advanced private projects get famous from their top ranking in green building certification and would be invited by both public and private organizations for experience sharing. This also helps their branding.

Evaluation

In Hong Kong, building certification is conducted by non-government organizations as an important evaluation tool for green building. HKBEAM is the major local brand. CGBCHK only certified several projects. Local developers chase popular international certifications as well, such as LEED. Before the mandatory requirements of the four Building Energy Codes (BECs), the Hong Kong government has launched Energy Efficiency Registration Scheme for Buildings since 1998. It was updated in 2003 with additional Performance-based Building Energy Code (BEC (PB)).

For applicants, the Hong Kong government developed a labelling scheme under a voluntary basis. Twenty-one types of household appliances and office equipment are covered under voluntary Energy Efficiency Labelling Scheme. Each product's level of energy consumption and efficiency rating is required to be shown on the label for better advice to the customers in choosing energy efficient products. According to Building Energy efficiency Ordinance, energy audit is required for commercial buildings only, on the designated four key types of central building services installation in every 10 years. There are no regular reporting requirements for residential buildings. For other building types, the government has provided various guidance documents in assisting self-energy audit.

EMSD collected energy end-use data in Hong Kong with an annual publication for open access. It covers the consumption data of various energy fuel types and its specific consumption purposes (EMAD, 2014). Since 2001, EMSD further developed benchmark tools for three-selected major types—residential, commercial and transport in Hong Kong. The major groups are detailed and explained in subgroups, which offers a possibility for the general public to benchmark their own living premises. With the help of online benchmark tools, participants can estimate the energy consumption of their own facilities. An average energy consumption indicator is available. The benchmark tool can give advice regarding specific energy consumption modes.

It would be wise for the Hong Kong government to address the significance of benchmark; however, due to its voluntary basis, it has not fully shown its potential in improving energy efficiency and saving. The systematic benchmark tool works without concrete incentives. Thus, rare parties or individuals take this tool voluntarily. The self-conduct energy consumption tool is quite possible to be imprecise and not that reliable. Hong Kong has no requirements for building performance monitoring and post-occupancy evaluation. However, in 2002, EMSD published a "Guideline on Application of Central Control and Monitoring Systems" in supplementing of energy management, which is not covered in BECs. The guideline introduced Central Control Management System (CCMS) to save energy, and also open to other means in achieving energy efficiency.

Policy Tools in Singapore

Policy Options

Singapore is one of the first countries in Asia to establish energy codes. Different from Hong Kong where building energy code started on a voluntary basis, Singapore commenced energy code on a mandatory scheme. Since the first version came out in 1979, developed on the 1975 version of the ASHRAE (American Society of Heating, Refrigerating and Air-Conditioning Engineers) energy standard in US with substantial adaption to local conditions, the standard has been revised in every decade. Further review is conducted every three years or even more frequently, depending on the requests from local professionals, according to BCA. It covered (1) the building envelope; (2) air conditioning; (3) lighting; (4) electrical system; and (5) service water heating. Singapore is also the first country introducing performance-based system in policy. In the revision of Energy Code in 1999, the new system that allowed trade-off compliance was added into energy code. The mandatory minimum energy performance standard programs established in 1998 by Singapore's National Environmental Agency and covered three products: air conditioners, refrigerators and clothes dryers.

Some of the strategies in Singapore targeted on consumption-intensive companies. Under the Energy Conservation Act (ECA), more than 165 large energy consumption private sectors implemented energy management practices and submitted energy-use reports to NEA since 2013. Similar requirements set to the intensive-consumers in water-use. Even though currently, it is a voluntary scheme, more than 370 companies submitted their reports. This scheme switched to a mandatory requirement in 2015. Without incentives and mandatory requirements, the voluntary scheme still gained such a high response, which indicates the strong influence of the Singapore government on industry sectors.

Singapore established its own green building certification brand in 2005. Launched by the Building and Construction Authority (BCA), Green Mark is committed to promote environmental awareness in both construction and real estate sectors. It incorporates a series of good practice and technologies in environmental design and construction. It covers both new and existing buildings with emphasis in five key aspects-energy efficiency, water efficiency, environment protection, indoor environmental quality and other green and innovative features that contribute to better building performance.

Green Mark is a government authorized green building certification system in Singapore, associated with government policies and programs. It has different levels corresponding to building sustainable performance. Buildings of public sectors are required to implement certain level of Green Mark as modelling; while for private sectors, the government provided phased incentives to promote and supported companies' sustainable efforts. In order to inspire the industry, the Singapore government launched a series of Green Building Masterplan schemes with phased emphasis and incentives, which were quite effective.

At its very beginning, building industry in Singapore was suspicious about green building. The first phase of Green Building Masterplan (GBMP) commencing in 2007 was to introduce green building, prepare technology and professions and establish a rudimentary green market. To spur the quiet market, the Singapore government clarified its leading role in green building implementation. It increased the demand of green building in building industry by applying green building in public sectors. Direct monetary incentives were given to developers with low requirement, which was to meet the threshold level of Green Mark. Funds were assigned to innovative technology research and development. Moreover, training for professionals was launched extensively.

Since the new building market accepted Green Mark, the second phase of GBMP policy focused on accelerating building retrofit. The Singapore government gained confidence from the success in the first phase. In 2009, it claimed its ambitious target to achieve 80% green in the building industry by 2030. Its determination in promoting green building was also revealed from its commitment of greening all the new, large and air-conditioned public buildings to the highest Green Mark rating; while for government owned existing buildings, the government set a deadline that all had to achieve Green Mark Gold Plus by 2020. Requirements on high performance of green building are mainly on public buildings, only a few added to private buildings. The required level of green building was an extra condition of purchasing land in special areas for the private sectors. Continuing the heartened response from private sectors to the direct monetary incentives in the 1st GBMP, the Singapore government further launched a new scheme in award of additional gross floor area for their excellent performance in new construction, instead of the used-up monetary incentives. A strong hit was launched for encouraging existing building retrofit with large cash incentives.

After the 2nd GBMP, green building gained a wide acceptance among professionals in the building industry. Thus, the following 3rd GBMP was for further advancing technology in existing building retrofit and engaging end users, including tenants, occupants, building owners and property managers. Singapore has built its advanced position in green building development in Asia and spread a global influence. This 3rd GBMP was to continue its leadership and kept moving on those established paths. The local government was committed to further develop on sharing and training through organizing international conferences, tightening up building industry standards, improving and inventing technical support and breeding a strong professional team. Based on the mature technologies and efficient energy performance, this 3rd phase focused on demonstrating building performance and encouraging collaboration with all stakeholders. The 3rd GBMP contained less incentive. The whole plan was dominated by cost and benefits calculation of private models. The cost of green technology and its payback time were displayed. Those vivid numbers proudly proved the matureness of green technology in Singapore. The benefits of green building, in environment, society, and especially in economy, which was the biggest concern among private developers and investors have finally demonstrated itself, after more than 10 years of advocating by the Singapore government. The whole plan was rather like a victory reward, which reflected the old saying, "facts speak louder than words".

Programs

The Singapore government offers a variety of funds and incentives in pulling sustainable practice. Funds for sustainable development are available concerted with the targets in the previously mentioned "Inter-Ministerial Committee on Sustainable Development (IMCSD)" and several sustainable development masterplans. Rather than mandatory requirements to public building and minimum standards to all new buildings, the promotion of green building among the private sectors is supported by a host of incentives, following the 3-phased Green Building Masterplans (GBMP) with different emphasis.

The incentive in the first GBMP scheme was unveiled in 2006, targeted on new construction. Considering the possible wait-and-see attitude in the immature market, financial concern could be a key issue for developers' decision in implementing green building. A direct monetary incentive—20 million Singapore Dollars incentives were offered to developers in motivating the appliance of Green Mark standards. At the very beginning, the threshold was rather low that rating above the basic certified level can get this award. Later, it was tightened to Gold rating or higher level. Another 50 million Singapore Dollars was set for R&D fund to develop advanced technologies and energy efficient solutions. A warm response from industry and research institutions were received that until 2009, 49 proposals in advanced green building technology were accepted with a distribution of 32 million Singapore Dollars of the funding.

The following incentives in 2nd GBMP were shifted to a different emphasis. Developed on the low threshold and direct monetary incentives to new construction in the first phase, incentives in the second phase tightened up requirements. Only buildings meeting Green Mark Gold Plus or platinum, whether new development, redevelopment or major A&A can get the incentive in 1–2% additional GFA according to the buildings' Green Mark rating levels. For an existing building, which is the major target in the 2nd GBMP, 100 million Singapore Dollars funds were offered to co-fund building retrofit cost. Lasting from 2009 to 2014, this scheme offered cash supplement in the cost of both equipment updating and professional services. This scheme also co-funds a "health check" of air-conditioning plants.

Compared with the previous two incentive schemes, incentives in the 3rd GBMP are less but with higher requirements. Another 52 million Singapore Dollars fund is offered for Green Building Innovation Cluster, which reveals the forcoming support from the Singapore government towards research and development in green building. Incentives for existing buildings are continued but the total fund decreased to 50 million Singapore Dollars. But a new award encouraging developers and building owners who cultivate building users to reduce energy consumption is launched, in regards to the highlighted scheme in 3rd GBMP to engage building end-users.

Monetary support to building retrofit is also offered by several other schemes. The Green Mark Incentive Scheme- Design Prototype is a support for developers and building owners to engage sustainable design consultants for integrated design and simulation with the target to achieve beyond Green Mark Platinum. The Pilot Building Retrofit Energy efficiency Financing (BREEF) Scheme finances for the cost of energy efficient or renewable energy equipment and its installation fee. Green Mark Incentives Scheme for Existing Building and Premises (GMIS-ESP) are tailored for small and medium private companies in improving energy efficiency.

Funds and incentives programs in Singapore are kept updated. Thus, many programs only last for a certain period. They are the official baits to motivate public participation. Their existence depends on government targets during the specific period of green building development. Once the objective has been achieved, government quickly launches other incentives and grants further motivation. Those incentives track the needs in the roadmap of national sustainable policy. The strict building environment performance in Singapore is sometimes accompanied by penalty. If a building fails to meet the minimum environmental standards, according to Code for Environmental sustainability of Buildings, the building owner may convict to a fine less than $10000 Singapore Dollars.

Implementation

In Singapore, the public sector dominates the exemplary projects and showcases. The Singapore government puts environmental sustainability as a priority. The public sector is committed to take the lead in sustainability. Requirements for public sectors to achieve sustainability are quite strict. They have to mandatorily meet standards in several labelling systems and certifications, addressing energy efficiency, water efficiency and recycling.

In general, all the public sectors must reach Eco-office Green Office label which is a voluntary scheme open to the applicants from the private sector. This label, besides designated eco-materials, offers a self-audit metrics determining the rate of eco-office label. Large new public buildings, fully or partly funded by the public sector must attain the Green Mark Platinum rating; while large existing public sector buildings must achieve Gold Plus rating by 2020. In energy efficiency, all the large public sector buildings have to conduct energy audit. Besides, all the public sectors are required to keep the air-conditioned premised at 24°C or higher. Training is compulsory to their facility operation managers. New appliances procured in public sectors have to fulfil the latest standards in Energy Star. Water efficiency has to follow the instructions in Water Efficient Building (WEB) label. Several materials are designed to be recycled.

Moreover, the Singapore government not only addresses efficiency in device selection, but also emphasizes the actual long-term saving of those advanced facilities. The public sector introduces a building-performance management system and implements in public projects as a showcase. Many public sector agencies adopt the Guaranteed Energy Savings Performance (GESP) contracts, which cooperate with Energy Services Company (ESCO), who is accredited by a committee of experts. ESCO is responsible for the guarantee of energy performance during the contract period. The determination of the Singapore government in sustainable development enabled public buildings of Singapore working as experimental base, testing newly invented technologies and building the benchmark of building performance. The preparation and construction of a large number of public green buildings helped establish the green building market, encouraging enough facility suppliers and necessary professionals. The actual data of these buildings' performance dispels market doubts and inspires confidence in green building performance from private investors and developers.

The Singapore government always promotes extensive publicity for outstanding green building projects, equally for public and private projects. Actual data from real projects are more persuasive. For those demonstrated private projects, publicity is a reward in branding and advertising. Real projects as showcases on booklets, those outstanding projects enjoy considerable benefits from propagation and property value enhancement, beyond the cost saving in operation. In the 3rd Green Building Masterplan and BCA Building Energy Benchmarking Report 2014, many private projects are selected as showcases presenting on the reports.

Evaluation

Green Mark as the only local government recognized, holistic certification system of green building, its rating result is associated with government requirements and incentives. This clear relationship between certification system and policy enables a gradient reward for certified green building in policy. The government could make mandatory requirements as minimum standards in building performance or offer incentives and supplement according to the rating levels in the authorized certification system. This is a big advantage compared with Hong Kong where the non-government certification system HKBEAM cannot be responsible for any government policy results.

Benchmark is perceived as a significant criterion in building performance management to identify best practice and help update to efficient standards. The Singapore government addressed benchmark from the 3rd Green Building Masterplan and required a regular reporting in building energy consumption. In 2014, BCA of Singapore published Building Energy Benchmark Report 2014, based on statistical analysis of 2013 building energy benchmarking data. This report focuses on commercial buildings, which consume most energy in the building sector. Data was collected from the previous year operation of the Building Energy submission System (BESS), which is a mandatory data submission system for commercial buildings. It covers 884 existing buildings, which is equivalent to 92% compliance.

Performance monitoring and regular reporting are a significant part in energy control in Singapore. In order to better understand building energy consumption and further improve building energy efficiency policy, the Singapore government pushed energy report in legislation. It formulated a strict requirement in regular reporting for existing buildings. Under the Building Control Act launched in 2012, commercial buildings have to mandatorily meet environmental sustainability standards and take the annual monitor of overall energy performance. The targeted commercial building stocks include offices, hotels, retail buildings and mixed developments, which are major energy consumers in building sector. Through an online Building Energy Submission System (BESS), building owners have to submit building information and energy consumption data annually, which provides the possible resource for energy monitor and support national benchmarking for commercial buildings in Singapore. Besides, many of the management programs are through contracts. Cooperating with private sectors, those contracts cover energy saving. Post-occupancy evaluation remains in their research stage with not much practice in Singapore.

Lessons from Green Building Policy

Clear Roadmap

A long-term commitment to sustainability forms a solid basis of Singapore's regulations, standards and schemes and allows broad policies and targets related to sustainability. Its strong commitment enables Singapore to set clear quantitative targets in various fields, open to public and international supervision. To make operational and practical plans, Singapore breaks down the final ambitious goal into a series of phased targets and coordinate steps in different fields and industries.

Market-Driven and Top-Down Approach

Policy is determined by legislation system and political structure. Singapore represents a typical model of the top-down approach to sustainable development. The centralized power of government ensures the harmony of private stakeholders in applying standards, directives and even governmental suggestions related to green building. The building industry complies the strict regulations on environmental performance. Government promotion of high rating in Green Mark is supported by a plenty of private sectors.

Government Project Take the Lead

In both Hong Kong and Singapore, public sectors play a significant role in making demonstrate projects. Every year, both governments would launch several public projects with outstanding performance in building environmental assessment, representing the most advanced technology development and leading the entire industry. The Singapore government demonstrated that exclusively large public buildings, no matter new or existing ones, have to comply with certain levels of Green Mark certification. This requirement applies to all government funded or co-funded projects. The enormous demand of green building in public projects moves local green market, cultivating green facilities suppliers and experienced professionals. This has paved the way of a wide development of green building among private sectors that can get green facilities and technicians much easier.

Incentives Design

Both regions have developed almost a full range of incentives for green building, including tax relief, technical support, product labelling system, professional training and grants. The phased incentives are under deliberate consideration. GFA concession is an interesting incentive working in both regions but results in contrast effects. In Singapore, GFA concession is offered as an encouragement to sky gardens. The policy effect is quite noticeable. Sky garden shows on almost every building and has become an inevitable element in Singapore buildings. Similar GFA concession was used to offer in Hong Kong to encourage design of sustainable features, including balcony, sky garden and natural ventilated corridor. However, those good intentions resulted in severe problems in Hong Kong. Developers took advantages of the loopholes in the incentive policy and made tremendous profits. Thus, good intentions of the government should be accompanied by considerable policy design.

Approaches Towards Different Stakeholders

Green building promotion is first emphasized among professionals, later to the wide building users. In the initial period, policies focused on encouraging more green building projects. Funds were offered to investors and developers to build new green building projects. Existing buildings were promoted after green building had been accepted in the new construction, since building retrofit to green costs extra efforts and capital. For existing buildings, incentives are for building owners. After green building has been approved in the market, emphasis switched to encourage actions from building end-users. Data collected from real projects reveals the cost-benefit statements, which shows the payback period and helps building owners make their choices based on the superior performance of green building. Besides, developers and building owners could get an award by encouraging more sustainable tenants.

Building Performance Information Disclosure—End Users

Among all their positive features, information disclosure cannot be ignored. Both regions grant open assess to codes, standards and various supporting documents online. A wealth of information is available for free access, including detailed code provisions, methods on compliance tools, guidance documents and application procedure for various programs. However, disclosures in Hong Kong lacks detailed explanation in building performance than Singapore. Hong Kong only reveals building's information on their green building certification level, but information in Singapore includes project features in sustainable design and construction. The information could help enhance consultants' confidence in sustainable design, support buyers and renters with wise purchasing decisions and further enhance public awareness of green building projects.

Conclusion

Green building development is becoming more and more favoured in governmental policies and building industry's response to climate change. This chapter, using Hong Kong and Singapore as two case studies, identifies a series of green building policies and related programs to encourage green building development. These policies and related programs (such as GFA concession and expedited land approval or review process) greatly help developers to reduce the capital costs and increase affordability of green building in the respect of development and construction (Gou et al., 2013). However, these policies and programs havenot addressed the affordability issue of buying green buildings. Future research is needed to look at how to reduce the cost of not only building green but also buying green to make sure that the affordability is well addressed in the green building development.

References

Athens, L. 2012. Building an Emerald City: A Guide to Creating Green Building Policies and Programs, Island Press.

BCA. 2014. BCA Green Mark Project Directory. Retrieved Nov 20, 2014, from http://www.greenmark.sg/building_directory.php.

BEAM Society. 2014. BEAM Certified Building. Retrieved Nov 30, 2014, from http://www.beamsociety.org.hk/en_beam_assessment_project_4.php.

Bond, S. and G. Perrett. 2012. The key drivers and barriers to the sustainable development of commercial property in New Zealand. The Journal of Sustainable Real Estate 4(1): 48–77.

Chan, A. T. and V. C. Yeung. 2005. Implementing building energy codes in Hong Kong: energy savings, environmental impacts and cost. Energy and Buildings 37(6): 631–642.

EMAD.2014. Hong Kong Energy End-use Data. Retrieved Nov 5, 2014, from http://www.emsd.gov.hk/emsd/eng/pee/edata.shtml.

EMSD. 2002. Life Cycle Energy Analysis (LCEA) of Building Construction.

Gou, Z. and S.S.-Y. Lau. 2014. Contextualizing green building rating systems: Case study of Hong Kong. Habitat International 44: 282–289.

Gou, Z., D. Prasad and S. Lau. 2013. Are green buildings more satisfactory and comfortable?" Habitat International 39: 156–161.

Gou, Z., S. Lau and D. Parad. 2013.Market readiness and policy implications for green buildings: case study from Hong Kong. Green Building 8(2): 162–173.

HKGBC. 2014. HKBEAM Project Directory. Retrieved Nov 20, 2014, from https://www.hkgbc.org.hk/eng/BeamPlusDirectory.aspx.

Hui, S. 1998. A review of building energy standards and implications for Hong Kong. Building Research and Information: 131–140.

Koeppel, S. and D. Ürge-Vorsatz. 2007. Assessment of policy instruments for reducing greenhouse gas emissions from buildings. Report for the United Nations Environment Programme (UNEP) Sustainable Buildings and Construction Initiative, Central European University, Budapest, September.

Ma, Z. and S. Wang. 2009. Building energy research in Hong Kong: A review. Renewable and Sustainable Energy Reviews 13(8): 1870–1883.

Pearce, A. R., J. R. DuBose and S. J. Bosch. 2007. Green building policy options for the public sector." Journal of Green Building 2(1): 156–174.

USGBC. 2014. LEED Project Directory.Retrieved Nov 20, 2014, from http://www.usgbc.org/projects.

Wen, H. and M. Clifford. 2007. Building energy efficiency: Why green buildings are key to Asia's future. Asia Business Council, Hong Kong.

Xie, X. and Z. Gou. 2017. Building performance simulation as an early intervention or late verification in architectural design: Same performance outcome but different design solutions. Journal of Green Building 12: 45–61.

CHAPTER 7

Affordable Housing and Circular Economy for Middle Income Group (M40) in Malaysia

A Case Study on the FlexZhouse Business Model

Mohd Zairul

Introduction

A housing development in Malaysia is pressured to include the sustainability agenda in its development planning. The issue of green is prioritized in the recent Malaysia Plan (2016–2020) under strategies for sustainable development (Economic Planning Unit, 2015). However, the challenges to provide steady development for a modern community and the desire to promote a healthy environment have become the main setbacks especially in developing countries like Malaysia (Zairul and Geraedts, 2015). The Construction Industry Development Board (CIDB), which is one of the country's key construction players, has identified the need to strengthen the awareness of green issues in the construction industry (CIDB, 2008). The effort towards green is supported by the Malaysian National Institution of Valuation (INSPEN), Malaysian Science and Technology Information Centre (MASTIC) and several grants to public universities to promote sustainable development in the new construction paradigm. It was reported that, in Malaysia, 10–30% of the waste disposed of in landfills is mainly from the construction and demolition activities in the building industry (Papargyropoulou, 2011). Despite all efforts made by these bodies, the delivery of a green concept in the housing industry is still relatively minimal and has become a low priority among contractors and housing players in Malaysia.

Universiti Putra Malaysia.

It is generally recognized that the construction industry, and especially the housing construction industry, must transform its modus operandi from linear production into making green agenda its overarching element in planning. With the current problems of global warming, resources depletion and the excessive destruction of the ecology and biodiversity, the sustainability issue has garnered some attention from industry players worldwide (Schaltegger et al., 2011). However, in practice, the housing industry in Malaysia still consumes an enormous amount of raw materials used for construction materials such as cement, aggregates, steel, plastic and timber (Cagamas Berhad, 2013). Furthermore, the concept of sustainability is not widely accepted by the main players in the industry, partly because there are no strict regulations from the government (Abidin et al., 2013). Nevertheless, the mass housing industry has not changed since it was introduced back in the 1970s. As a result, more resources are exploited causing pollution and climate change.

Theoretical Framework

To address this, the author is looking into how he can develop a new 'business model' (BM) for the green affordable housing in Malaysia. Currently, there is no single definition of the term 'business model'. Some describe it as a company invention, a formula to generate more money (Afuah, 2013) or a framework that explains how a company does business with clients, partners and vendors (Raphael Amit and Zott, 2012). Another author states that a BM explains how enterprises work, create, deliver and capture value (Magretta, 2002). It is also defined as a strategy and organizational theory that focuses on designing transactions (Brege et al., 2014) or an analytical tool (Kley et al., 2011). BMs are a combination of various aspects that include the notion of value, financial aspects, and aspects related to the architecture and network between partners (Amit and Zott, 2009). Further, Zott and Amit (2010) analyzed 103 BM publications and found that more than one-third of the publications do not have a clear definition of the term and that almost 20% used other scholars' definitions. Nevertheless, existing definitions only partially intersect, making possible multiple definitions.

Achieving an affordable housing price for customers depends on how the company or the housing developer creates revenues for its business. The price of the house could be reduced if the housing manufacturer could benefit from the economies of scale and from recurrent payments. In this chapter, the term "affordability" is being proposed and it will show how a company could reduce the cost of manufacturing and production by greening the lifecycle chain through extending the lifespan of the products. Therefore, for the solution, the author revisited the meaning of affordability by introducing a new BM as an alternative form of green affordable housing for the Malaysian market. But in order to convert housing into simple and affordable products, the solution has to propose an innovative BM strategy, one that would offer innovative leasing. In this strategy, the revenue of the company will rely on the housing product and stock. The author was inspired by a circular economy strategy and logistics streams that adopt industrialized production and focus on quality and defects control and maintenance of the products, thereby improving after-sales and the occupant's satisfaction.

Studies of industrialized house production in the literature are abundant, for example, pre-engineered, modular homes, factory manufactured, timber IBS, prefabricated house and many more (Nawi and Lee, 2013). The idea of flexible housing is not new; it is a paradigm that responded to the advance of technology in the construction industry. In contrast, the idea of conventional buildings or houses proves to be static, rigid and intractable. A study by Kendall (2012) highlighted that the prefabrication of housing was considered a process of mass customization and supported the previous remark by Akhtar and Tabucanon (1993), that market demands can only be addressed if the housing industry adopts the industrialized and appropriate manufacturing concept. Therefore, flexZhouse will highlight the important endeavours in the manufacturing sector and try to make an attempt to combine it with construction industry with the circular economy principles.

The circular economy principles promote the idea of innovative leasing. The idea of innovative leasing is that the manufacturers or housing developers have to agree with the potential customers on the leasing of the housing components. The number and quality or size of the housing components increases along with the financial capability of the customers during the period of the contract. In summary, the customer is free to customize the housing unit, thus providing the advantages of the 'ownership'. It allows the customer to rent the housing components, but at the same time enables the unit to be flexible according to certain requirements imposed on it. Moreover, this new tenure will promote a long-term relationship with the housing association or producer and a good business strategy for a long-term cooperation. At the same time, the housing producer will reduce its production costs by remodelling and recycling the components (Zairul and Geraedts, 2015).

As part of the solutions, the flexZhouse BM will offer more design options in the mass housing industry, offer financial solutions to the middle-income group through innovative leasing and further improve the quality of the products through an industrialization strategy. The flexZhouse BM also uses the principles of the circular economy as part of its strategy to provide innovative leasing to the customers. This chapter contributes to the scientific community by combining the idea of industrialized housing production with innovative leasing inspired by the circular economy principles. At present, few studies have paid specific attention to the integration of flexible housing with the circular economy. Therefore, this chapter fills a gap in the existing knowledge of industrialized housing and solves the identified problems through the development of a new BM. The new BM will benefit the government of Malaysia by finding a solution for green affordable housing schemes for the middle-income group (M40) and creating an alternative BM for the housing industry. In this chapter, the author will explain the conceptual framework of the BM based on theoretical framework on business model, flexible housing and circular economy (Fig. 7.1) formulation using circular economy principle and towards the end discusses the challenges and a way forward to introduce this new strategy in the Malaysian housing industry.

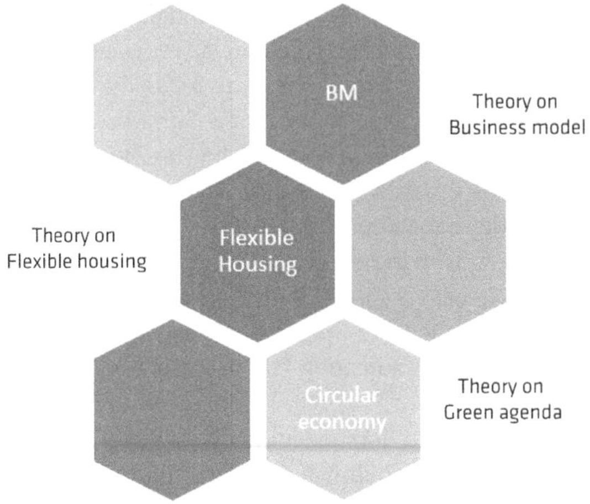

Figure 7.1. Theoretical framework for green affordable house (flexZhouse). Source: Zairul (2017).

Conceptual Framework of flexZhouse BM

The business model will be the next strategy to achieve competitiveness in the future. Adaptation to the new needs and open markets are pushing most companies to change its business model from time to time. In addition, the rise of the technology is shaping the future of a new business model for the housing industry in the future. Therefore, in this chapter, the author explains some of the components in the business model derived from several prominent scholars such as Chesbrough, 2010 and Osterwalder and Pigneur (2010). The following components will be discussed but not in particular order, which include; (1) value propositions; (2) target customers; (3) customer services/relationship; (4) revenue streams; (5) key resources; (6) cost structure; (7) partnership and (8) channels and (9) key activities.

i. Value Propositions

The value proposition will determine the target group for the product and services offered by the company. The flexZhouse BM may require new or different types of customers. A case study on Japanese prefabricated housing shows that housing producers have always targeted high-income groups. The Japanese market has developed clever production methods and introduced aesthetics in the housing prefabrication production, thus responding to the consumers' demand for quality, but affordability has of been less consideration (Noguchi, 2003). According to Barlow and Ozaki (2003), there is always a tendency for the customer to opt for certain basic standard design but to hope for individuation and customization. Borrowing the concept from other industries, the examples of a personal computer and a motor vehicle provide some ideas on how different needs of different target customers are processed and delivered. Depending on which groups are targeted, producing different value propositions could bring extra revenue for the company.

The flexZhouse strategy means the customer and the housing provider share risks and rights to the property. By combining an innovative leasing model with the circular economy principles, a company can increase its focus on the efficient management of the resources. This also generates opportunities for adjusting fees according to the services and products used by the customer. The flexZhouse BM therefore, should also incorporate a turnover formula, which defines the key resources needed to operate the business that later defines the pricing of the products and services provided by the company together with markups as well as gross and net profit margins. By combining an innovative leasing model with the principles of the circular economy, a company can increase its focus on the efficient management of the resources. This also generates opportunities for adjusting fees according to the services and products used by the customer.

Introducing mass-customization under the flexZhouse BM might pose significant challenges for the new business. However, anecdotal evidence suggests that there is a growing need for mass-customization and individuation in the housing market for middle-income groups in Malaysia (Daud and Hamzah, 2012). Therefore, the flexZhouse BM underpins the demands from new customers in the market and therefore, refines the present value propositions to its customers.

ii. Target Customer

The uprising trends in the market have called for a need to develop a new housing BM that focuses on flexible housing solutions. The flexZhouse BM will provide products for different customers and needs. It calls for better after-sales to maintain a healthy relationship with the customers. Hence, here, this chapter establishes the following conceptual contribution for the new BM:

- The customer will have multiple design options to choose from and have the freedom to change the exterior parts, interior elements and services that match their requirements and budget;
- The customer will be more active and participate at the beginning of the development and co-evolve the design;
- The customer will be able to change or modify (add or remove) certain components after a certain period of time; and
- The services that come with the products from the company will improve the customer relationship and prolong the business of the company.

iii. Customer Relationship

The flexZhouse provider should consider setting up several customer services as part of its marketing strategy to make maintenance work and after-sales activities more efficient and to improve service response time towards the customers. The flexZhouse after-sales services can be divided into three parts: (1) product or design-oriented; (2) focusing on service support system (e.g., reducing equipment repair time); and (3) minimizing risk (e.g., through extended warranties). For instance, the quality assurance is necessary to achieve customer's satisfaction, and this will bring returning customers especially in the housing industry where the market is customer driven.

iv. Revenue Streams

The innovative leasing is described as the customer choices on the products and services that suit their current financial situation. Based on the circular economy principle of Macarthur (2013) and Stahel (2008), the flexZhouse BM focuses on leasing the housing components and services that enable the manufacturer to retain ownership of the housing components and resources thus, contribute to its own future resource supply. In the case of flexZhouse, the housing module is expected to adopt a remanufacturing strategy and can be upgraded based on recent technology. The technology also allows the innovation that can lengthen the lifespan of the housing module. Further, the sustainable concerns are making a debut in the construction industry in general and towards housing industry specifically.

v. Key Resources

The affordability of the cost structure will be very much related to the arrangements of activities and resources, which means it will determine the price of the products. The new BM needs to completely change its way of working to match the circular economy principles. This is important because one of the factors driving the higher cost of conventional construction is the depletion of natural resources that leads to increase in prices for cement, steel and other construction materials (Begum et al. 2006). The concept of the circular economy includes reducing resource input by increasing stock and implementing the remanufacturing process. The key resources will be improved through investment on human resources and new equipment. However, the investment returns are still undetermined (Amit and Zott, 2012). A lesson learnt from the Japanese housing builders is that resources play the main role in the production line. The investments made by the Japanese housing builders through a big manufacturing plant, show houses and a learning centre for the buyers. In these markets, companies such as Sekisui House, Sekisui Heim, and Toyota build their industries on the provision of specialized technical skills of skilled workers (Gann, 1996). Therefore, the new flexZhouse BM's need a big investment in the manufacturing plant, off-site facilities and skilled workers to operate the centre.

vi. Cost Structure

The basis for offering the cost structure of the products is the resources needed to operate the business and to foster a partnership with other companies. By this simple concept, the cost structure for flexZhouse relies on the value propositions, partnership and resources used to operate the business. The adoption of an industrialized strategy in the housing industry will change the process of normal construction. Therefore, this change will impose a substantially different cost structure in comparison to conventional housing. Based on the revenue and innovative leasing discussed earlier, flexZhouse will introduce a closed-loop system for construction materials to reduce the operational cost by recycling and remanufacturing the housing components.

The idea of innovative leasing is crucial for the formulation of green affordable flexZhouse. The value configuration is closely related to other business elements. Changes in the value propositions will lead to certain changes in the lifecycle cost and affect the cost structure of the products (Mokhlesian and Holmén, 2012). flexZhouse may require completely different activities, partnerships and resources compared to the present practices in the housing industry. As a result, flexZhouse might cost the housing manufacturer more due to its infancy. Therefore, a new strategy that involves circular economy and closing the loop by optimizing the remanufacturing of the components could be the solution to the problem. The increasing popularity of the circular economy and sustainable construction in the housing industry might give an edge to flexZhouse to operate in the industry. However, flexZhouse will need support from partners and perhaps the government to realize it. The reconfiguration and remanufacturing of the components is an underpinning process for the flexZhouse BM.

Nevertheless, the uncertainty caused by the immaturity of the new BM is a concern, as it may lead to extra costs for either the manufacturers or the partners. However, there are ways to ensure the cost structure could provide customers with affordable prices. One way is to determine the economies of scale and the economies of scope. The cost of producing the components is expected to be reduced by the introduction of the circular economy and innovative leasing. In this new strategy (Fig. 7.2), the

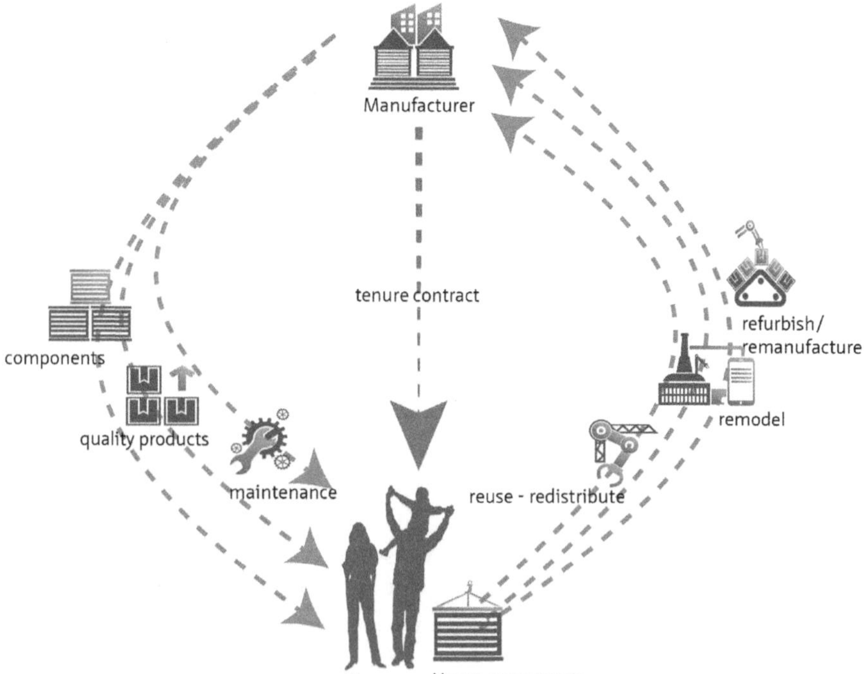

Figure 7.2. The circular economy strategy in the flexZhouse. Zairul (2017).

customer leases the components for a certain period and is only allowed to buy the units after a moratorium period. Because of this, the high costs of the production could be prolonged, and the manufacturer will benefit from the leasing fees and the customers will save on the cost of ownership.

In closing the loop of housing production (Fig. 7.3), manufacturers will have to change their mindset from being the sellers of the housing products and become the emancipators, provided by long-lasting upgradeable products of housing components. Their goal now will be to sell results rather than products, that is, performance and satisfaction rather than how many units of houses. Instead of purchasing the house, the middle-income group could lease it, paying a monthly fee based on how long they plan to use their housing component.

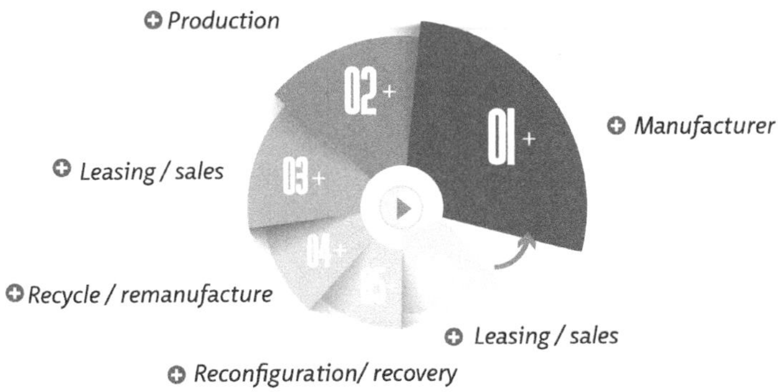

Figure 7.3. The circular loop in the flexZhouse BM. Zairul (2017).

vii. Partnership

Japanese industrialized housing manufacturers, as described by Barlow and Ozaki (2003) in their case study of Japanese industrialized houses, have shown that there were significant lessons to be learnt before the industry could move forward. It was reported that the factors that contribute to the success of Japanese housing industry are the size of the market itself, coupled with increasing demands from the population. The maintenance of the current rigid housing was seen to be costly as it required special craftsmen and this sometimes caused difficulty on site (Teck-Hong, 2012). Therefore, it is necessary to examine closely the potential of partnership involved in the process of the production and the key resources required to support the key activities. The complexity of the housing industry presents a challenge to monitor along the supply chain. A survey on the adoption of the industrialized systems showed that quality can be achieved through better supervision in the factory and thorough inspection of the product before it leaves the factory and is installed on site (Abas et al., 2013; Bari et al., 2012; Bildsten, 2014). However, the value of cross-industry learning does not stop at the techniques but also applies to knowledge transfer. Therefore, the idea tends to focus on the supply chain management strategy within the housing industry and reveal the opportunities to improve the time leads and the quality of the product.

viii. Channels

The supply chain includes a wide range of processes: extraction of raw materials; production stage in the factory; and the delivery and installation of the components through a systematic distribution process and marketing. Therefore, channels explain the continuous improvement of processes and relationship that are available to help the delivery of the product.

The advent of the technology and electronic commerce has further improved how the channels could reach the potential customer. At the early stage, the awareness of the product can be created by organizing and communicating information through social media, for example, Facebook and Twitter (Tucker et al., 2001). More traditional approaches to improving interactivity may also be used such as show houses and learning centres (Peterson et al., 1997). Channels include ways people can pay and perform the necessary transactions; this will further reduce paperwork and unnecessary complexity (Peterson et al., 1997). Channels also improve the flexibility of the communications for the customers through the means of internet and web browsing. In our case, flexZhouse will need to make use of the internet to increase awareness; promote the products and services; and spread the products' information.

ix. Key Activities

Literature supports industrialization attempts in the building sector with various objectives including reducing onsite activities (Vrijhoef and Koskela, 2005); flexibility in design (Habraken, 1972); concurrent engineering (Chimay et al., 2007); modular design (Gann, 1996); lean construction (Barlow and Childerhouse, 2003); and different understanding in a different context. Rinas and Girmscheid (2010) presented nine aspects to describe industrialized production, namely, (1) use of mechanical means and technologies; (2) use of high-tech systems and tools; (3) production in a constant process; (4) continued development of productivity; (5) standardization of products; (6) prefabrication; (7) rationalization; (8) modularization; and (9) mass production. Further, Yashiro (2009) divided the framework of the Industrialized Building (IB) system in Japan into several categories, including prefab houses of the 1940s, mass construction, component-based, mass customization, platform-oriented and service providing.

The author posits that different categories have a different supply chain maturity. This is supported by Ali et al. (2009). There are barriers to implementing the system mainly related to operating cost and financial barriers (Kamar et al., 2009). Several studies have argued that using an industrialized solution could increase the cost of the whole project and therefore discourage the utilization of the system (Abdul-Rahman et al., 2012). Despite the barriers, Japanese housing manufacturers are investing heavily to improve the flexibility of the housing design to meet their customer's new requirements and to achieve customer satisfaction. The old market of standardization has been changed to a more flexible one in terms of design. By the 1970s, the production of prefabricated housing units prefabricated had reached new levels of quality and resulted in satisfied customers. The customers were more prepared to accept factory-made housing and the provider increased its efforts to satisfy customer's quality expectations.

In Japan, the new concept of industrialized housing delivered by the Japanese housing providers offered many benefits in the planning and coordination of resource allocation, one of which was shortened time spent on site. In comparison to conventional housing, the Japanese manufactured housing industry managed to shorten the period of conventional construction from 120 days on site to only 40 days for the preparation of foundation works, building works, interior fit out and inspection (Barlow and Childerhouse, 2003). The new approach needs a new management decision in several fields. It is anticipated that some of (but not only) the following features will be part of that setup:

- The operating plant for manufacturing and remanufacturing will have to be located as close as possible to the development area. This will allow small repair and after-sales activity more efficient;
- The product will need to be designed for disassembly and technology improvement;
- Designers will need to think of a product that will utilize the raw materials efficiently and will be durable over a long period of time;
- A new job specification will arise, and this will boost the operation and management specialization;
- The users will learn how to take care of the product to enjoy the privilege of consuming the product, and any misuse will lead to penalties;
- The production will reduce the operational cost by recycling the stock; and
- The company will focus on the efficient management of the resources; creating money and wealth through the recurring money from the leasing activities; and adjusting fees according to the services and products used.

Previously, the author linked the innovative leasing to the problems of the linear economy and the chapter showed how depleting materials resources have led to an economic downturn and expensive housing stock. The production of housing product in a conventional way has caused significant loss of renewable resources and created more wastage. Here, innovative leasing has introduced a different BM that creates a sustainable loop of lifecycle chain that gives future house buyers a better option in the form of the new concept of housing 'ownership'. The new concept of housing ownership has changed the paradigm to the long-term (rental) income for the housing producer and the housing producer is now responsible for the product, which includes risk and cost of waste. In a nutshell, the housing producer will take responsibility for its own actions.

In the flexZhouse BM concept (Fig. 7.4), the company will emphasize on customer satisfaction, not only on making sales. The main revenue will come from the recurrent fees based on the leasing activities, coupled with maintenance activities. The next idea is to introduce usage fees based on the component that is included in the package prepared by the company. The flexZhouse BM shifts to a new skill that promotes a durable product for the housing component. The value creation is expected to be delivered through customer satisfaction rather than high sales. This can be done through the prudent use of the energy, resources and a high-quality housing component. In this case, the customer will have peace of mind, be satisfied with the product and contribute to a long-term relationship with the company. From the company's standpoint, this approach will benefit them as it reduces their resource

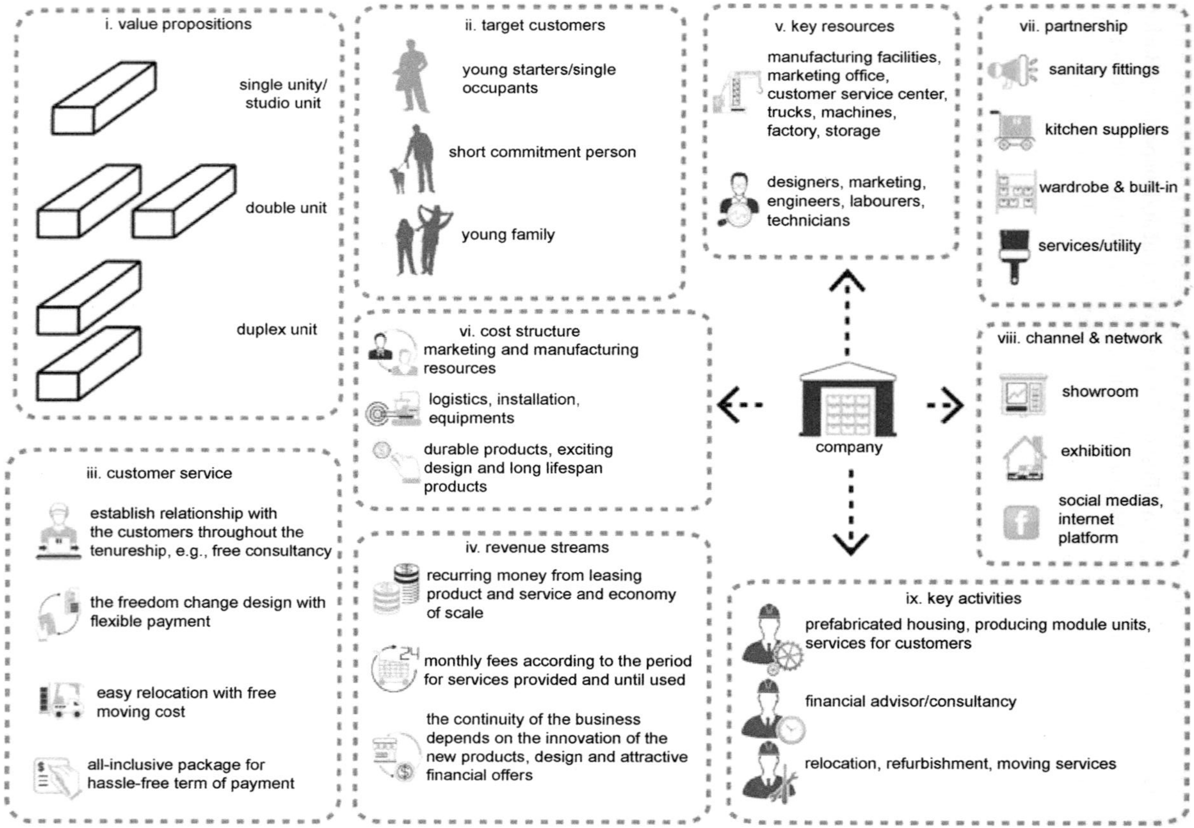

Figure 7.4. The flexZhouse BM concept (Adapted from Osterwalder and Pigneur (2010)).

consumption-investment but increase their revenues at the same time. In relation to this, unlike conventional housing system, the flexZhouse BM (Fig. 7.5) contributes to the green cause in the local housing scene and helps to reduce carbon footprints.

It is suggested that the government initially becomes the first party to initiate the project by acquiring the land and subsidizing the construction of the structure. The government under its housing agency will then construct the structure at the selected area or target area that has the potential for the scheme. Next, the government will invite tenders for the infill (housing units) and then the selected suppliers will produce and apply the concept on the site. The housing units or components will be leased to the potential customers and will have to be returned to the housing producer after the customer moves out. In the original contract, it is suggested that the minimum contract term for the units should be 12 months, and tenants should agree to give a minimum of two months' notice before relocating. In the event of changing the module unit, the tenant will be required to inform the suppliers three months in advance. All maintenance of the general services, common area and public utilities will be handled by the maintenance company appointed by the company who provides the infill.

In this framework, it is suggested that the monthly commitment of the users will be based on their financial capability and 30% income rule. The payment will consist of the rental payment for the structure, the infill and the building maintenance. The business life cycle (Fig. 7.6) shows the potential of the circular economy through the remodelling and refurbishing of the existing unit/module for the next customers.

In summary, the aim of this research is to offer an innovative BM and a way to provide a middle-income group in Malaysia with housing that is affordable and flexible in terms of design and offering. At the same time, this research aims to provide quality housing that improves the bureaucratic process.

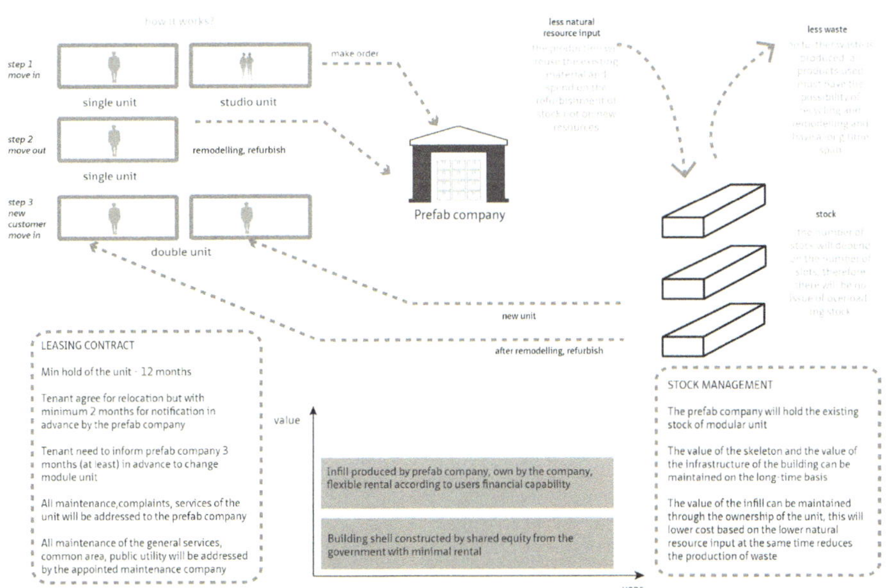

Figure 7.5. The flexZhouse BM structures. Zairul (2017).

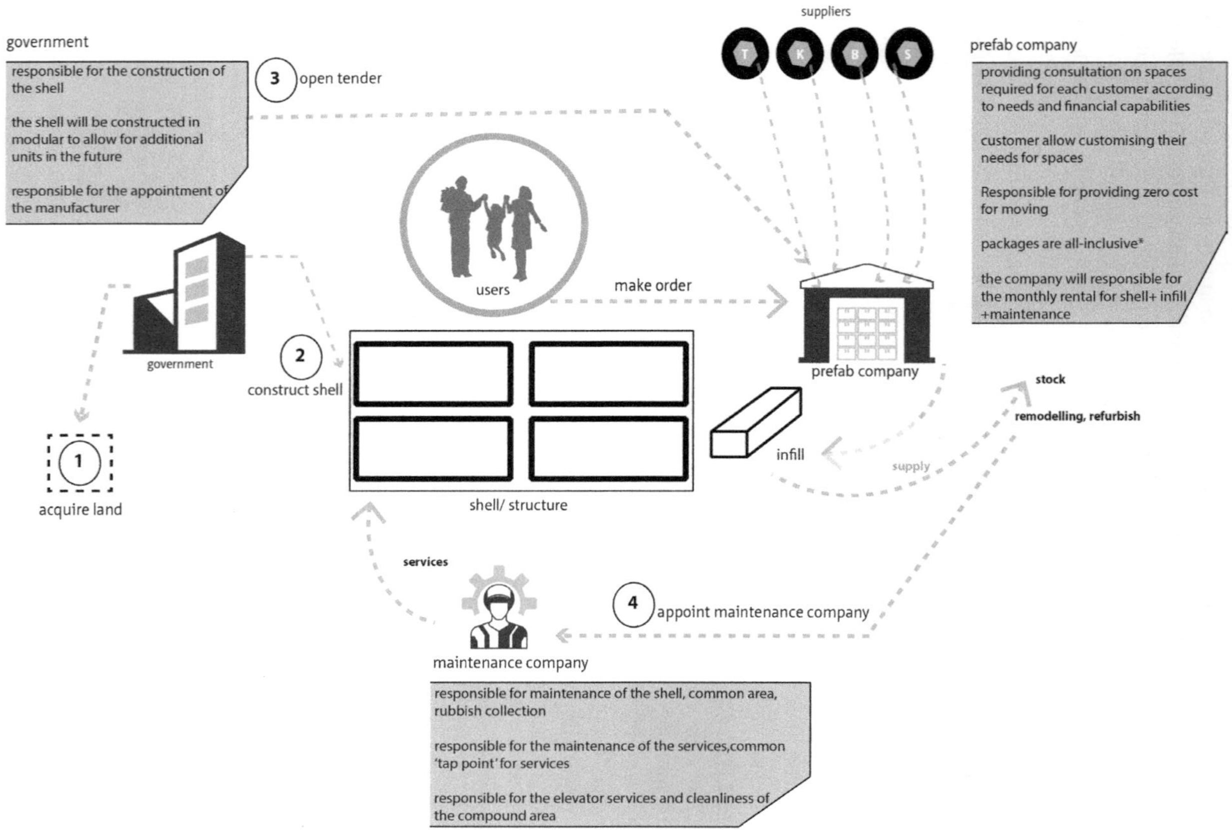

Figure 7.6. The flexZhouse BM life cycle chain. Zairul (2017).

A Way Forward

This chapter has presented a conceptual design for flexZhouse. A future study could look into details concerning the construction of the products and further clarify the technology that is available on the market, especially the mechanisms for moving the units in and out of the structure. Further study on the technical aspects of the products will help in terms of finalizing the production costs and the price of the unit to be sold on the market. flexZhouse requires strong technical supports. New skills require technology transfer from other countries. Awareness of the new technology and training should be given as part of the development of skilled and semi-skilled workers to operate the new BM. Given the current technology of the IBS system in Malaysia, flexZhouse will need a new paradigm to shorten the supply chain cycle. Lessons learnt from Japanese house builders on key resources are necessary to support flexZhouse.

This chapter has discussed the BM component to support the new BMs of flexZhouse. However, this chapter only provides a guideline about the new technology and how it can support flexZhouse. Hence, it is important for future studies to discuss such aspects as the type of machines, equipment, infrastructures and mechanisms that will be required to operate flexZhouse BM. Attention should be paid to novel techniques in the new system. The market must be aware of the new technology and how it can improve the existing problems to create a market that continues to support the new BM. The strategy must take into consideration the green aspects that are emphasized in the government policy.

In terms of economies of scale, it is necessary to distribute the network of the suppliers into a larger market. As discussed in the partnership components, the involvement of SMEs and subcontractors can reduce the problems associated with the presence of unskilled foreign workers in the country and can dispel the mistaken perception that flexZhouse is an expensive product. Nevertheless, there is scepticism among representatives of the industry, particularly concerning the skills that are required to operate such a new system and technology as flexZhouse will need more time, capital and resources at the beginning of the operation. Changing the way of working will create a serious resistance among current housing developers to flexZhouse. Sourcing skilled workers from outside will definitely cost more for their daily operations. However, some investment is necessary to achieve a successful business. As the financial implication creates the biggest obstacle, the initial investment is also necessary to help the business reaps profit in a feasible time.

During the early stage of the operation, the factory needs to be built at a suitable location since the logistics of the units is one of the main aspects of flexZhouse. Locations closer to the main highways are recommended to ease the transportation of the units to the site. The model of flexZhouse will resemble a car manufacturing plant. The housing units will be produced off-site and transported to the desired location once they are ready. Later, the units will be brought back to the factory for remanufacturing and reconditioning. Therefore, the biggest obstacle is the technology and skills required to operate flexZhouse. Finance is also a major obstacle as it may be difficult to convince the financial institutions to finance a project that has no local precedent.

Conclusion

The material presented in this chapter is part of the author's doctoral study titled "flexZhouse: a new business model for affordable housing in Malaysia" that focuses on the shortage of affordable housing for young starters in Malaysia. Although this chapter emphasized on the lifecycle chain of the mass housing development, one of its drawbacks is that it did not address the procedures for securing land or applying for planning permission (authority and land matters). This means that although flexZhouse provides proof of concept for the new BM, the notion is still in a preliminary stage of conceptual ideas. Introducing flexZhouse implies a radical change in the housing market. It is a new strategy, one that the housing industry in Malaysia needs because many previous proposals have proven unable to deliver affordable housing that offers flexibility and good quality housing to users. Furthermore, the proposed BM will support the government's mission to provide more affordable housing schemes for urban dwellers, especially the middle-income group along with more housing options for future customers.

Finally, flexZhouse will provide an alternative way for young Malaysians to own their own homes. The flexZhouse will add a new dimension to prefabricated housing and, as its name suggests, offer flexible options right from the beginning of the purchasing process. In contrast to the conventional housing system, flexZhouse will offer customers a choice of both interior and exterior housing design. The industrialized production will assure the quality of the housing and give customers peace of mind. The flexZhouse BM will be an alternative to current affordable housing and help to meet the needs of the population, especially young starters.

References

Abas, A., M. H. Hanafi and F. A. Ibrahim. 2013. Competencies factors of malaysian architectural firms towards the implementation of industrialized building system. Middle-East Journal of Scientific Research 18(8): 1048–1054. https://doi.org/10.5829/idosi.mejsr.2013.18.8.12471.

Abdul-Rahman, H., C. Wang, L. C. Wood and Y. M. Khoo. 2012. Defects in affordable housing projects in Klang Valley, Malaysia. Journal of Performance of Constructed Facilities, 28(April), 121025093213001. https://doi.org/10.1061/(ASCE)CF.1943-5509.0000413.

Abidin, N. Z., N. Yusof and A. a. E. Othman. 2013. Enablers and challenges of a sustainable housing industry in Malaysia. Construction Innovation: Information, Process, Management 13(1): 10–25. https://doi.org/10.1108/14714171311296039.

Afuah, A. 2013. The Theoretical Rationale for a Framework for Appraising the Profitability Potential of a Business Model Innovation. Ross School of Business Working Paper No. 1205. https://doi.org/10.2139/ssrn.2337057.

Akhtar, K. and M. Tabucanon. 1993. A framework for manufacturing strategy analysis in the wake of reducing product life-cycles: a case of a multinational in a newly industrializing count. Technovation 13(5): 265–281. Retrieved from http://www.sciencedirect.com/science/article/pii/016649729390001C.

Ali, A. S., S. N. Kamaruzzaman and H. Salleh. 2009. The characteristics of refurbishment projects in Malaysia. Facilities 27(1/2): 56–65. https://doi.org/10.1108/02632770910923090.

Amit, R. and C. Zott. 2009. Business model innovation: Creating value in times of change. Universia Business Review 3: 108–121. https://doi.org/10.2139/ssrn.1701660.

Bari, N. A. A., N. A. Abdullah, R. Yusuff, N. Ismail and A. Jaapar. 2012. Environmental Awareness and Benefits of Industrialized Building Systems (IBS). Procedia—Social and Behavioral Sciences 50(July): 392–404. https://doi.org/10.1016/j.sbspro.2012.08.044.

Barlow, J. and P. Childerhouse. 2003. Choice and delivery in housebuilding: lessons from Japan for UK housebuilders. Building Research & Information (December 2012): 37–41. Retrieved from http://www.tandfonline.com/doi/abs/10.1080/09613210302003.

Barlow, J. and R. Ozaki. 2003. Achieving "customer focus" in private housebuilding: Current practice and lessons from other industries. Housing Studies 18(1): 87–101. https://doi.org/10.1080/0267303032000076858.

Begum, R. A., C. Siwar, J. J. Pereira and A. H. Jaafar. 2006. A benefit–cost analysis on the economic feasibility of construction waste minimisation: The case of Malaysia. Resources, Conservation and Recycling 48(1): 86–98. https://doi.org/10.1016/j.resconrec.2006.01.004.

Bildsten, L. 2014. Buyer-supplier relationships in industrialized building. Construction Management and Economics 32(February 2015): 146–159. https://doi.org/10.1080/01446193.2013.812228.

Brege, S., L. Stehn and T. Nord. 2014. Business models in industrialized building of multi-storey houses. Construction Management and Economics 32(February 2015): 208–226. https://doi.org/10.1080/01446193.2013.840734.

Cagamas Berhad. 2013. Housing The Nation: Policies, Issues and Prospects. Kuala Lumpur: Cagamas Holdings Berhad.

Chesbrough, H. 2010. Business Model Innovation: Opportunities and Barriers. Long Range Planning 43(2-3): 354–363. https://doi.org/10.1016/j.lrp.2009.07.010.

Chimay, J. A., M. K. John and C.- D. Anne-Francoise. 2007. Concurrent Engineering in Construction Projects. Anne-Francoise, C.-D. (ed.). New York: Taylor & Francis.

CIDB. 2008. Malaysia Report. The 14th Asia Construct Conference (1): 1–4.

Daud, M. and H. Hamzah. 2012. Examining the potential for mass customisation of housing in Malaysia. Open House International 37(1): 16–27. Retrieved from http://umexpert.um.edu.my/file/publication/00007664_76462.pdf.

Economic Planning Unit. 2015. Rancangan Malaysia Kesebelas (Eleventh Malaysia Plan). Kuala Lumpur.

Gann, D. 1996. Construction as a manufacturing process? Similarities and differences between industrialized housing and car production in Japan. Construction Management & Economics (December 2012): 37–41. Retrieved from http://www.tandfonline.com/doi/abs/10.1080/014461996373304.

Habraken, N. 1972. Support: An Alternative to Mass Housing. London: Architectural Press (1972).

Kamar, K. A., M. Alshawi and Z. A. Hamid. 2009. Industrialised building system: The critical success factors. In I9th International Postgraduate Research Conference, pp. 29–30.

Kendall, S. H. 2012. YourSpaceKit: Off-Site Prefabrication of Integrated Residential Fit-Out.

Kley, F., C. Lerch and D. Dallinger. 2011. New business models for electric cars-A holistic approach. Energy Policy 39(6): 3392–3403. https://doi.org/10.1016/j.enpol.2011.03.036.

Macarthur, E. 2013. Towards the circular economy. Journal of Industrial Ecology 1: 4–8. https://doi.org/10.1162/108819806775545321.

Magretta, J. 2002. Why business models matter. Harvard Business Review 80(5): 86–92. https://doi.org/10.1002/1099-0690(200112)2001:23<4391::AID-EJOC4391>3.0.CO;2-D.

Mokhlesian, S. and M. Holmén. 2012. Construction Management and Economics Business model changes and green construction processes Business model changes and green construction processes, (September), 37–41.

Nawi, M. and A. Lee. 2013. Fragmentation Issue in Malaysia Industrialized Building System (IBS) Project. Journal of Engineering …, 9(1): 97–106. Retrieved from http://jestec.taylors.edu.my/Vol 9 Issue 1 February 14/Volume (9) Issue (1): 097–106.pdf.

Noguchi, M. 2003. The effect of the quality-oriented production approach on the delivery of prefabricated homes in Japan. Journal of Housing and the Built Environment 18: 353–364. https://doi.org/10.1023/B:JOHO.0000005759.07212.00.

Osterwalder, A. and Y. Pigneur. 2010. Business Model Generation: a handbook for visionaries, game changers, and challengers. John Wiley & Sons. https://doi.org/10.1523/JNEUROSCI.0307-10.2010.

Papargyropoulou, E. 2011. Sustainable Construction Waste Management in Malaysia: A contractor's perspective. Management and Innovation for a Sustainable Built Environment 3(June): 1–10. Retrieved from http://misbe2011.fyper.com/proceedings/documents/224.pdf.

Peterson, R. A., S. Balasubramanian and B. J. Bronnenberg. 1997. Exploring the implications of the internet for consumer marketing. Journal of the Academy of Marketing Science 25(4): 329–346. https://doi.org/10.1177/0092070397254006.

Raphael Amit and C. Zott. 2012. Creating value through business model innovation. MIT Sloan Management Review 53(53310): 41–49. Retrieved from http://sloanreview.mit.edu/article/creating-value-through-business-model-innovation/.

Ricart, J. E. 2011. How to Design A Winning. Harvard Business Review (February).

Rinas, T. and G. Girmscheid. 2010. Business model of the prefab concrete industry—a two-dimensional cooperation network. pp. 677–682. *In*: Challenges, Opportunities and Solutions in Structural Engineering and Construction.

Schaltegger, S., F. Lüdeke-Freund and E. G. Hansen. 2011. Business cases for sustainability and the role of business model innovation: Developing a conceptual framework. SSRN Electronic Journal 1(September 2015): 32. https://doi.org/10.2139/ssrn.2010506.

Stahel, W. 2008. The Performance Economy: Business Models For The Functional Service Economy. Handbook of Performability Engineering. Switzerland. Retrieved from http://www.springerlink.com/index/R2630P3T636Q1578.pdf.

Teck-Hong, T. 2012. Housing satisfaction in medium- and high-cost housing: The case of Greater Kuala Lumpur, Malaysia. Habitat International 36(1): 108–116. https://doi.org/10.1016/j.habitatint.2011.06.003.

Tucker, S. N., S. Mohamed, D. R. Johnston, S. L. McFallan and K. D. Hampson. 2001. Building and Contruction Industries Supply Chain Project (Domestic). Report for Department of Industry, Science and Resources (Vol. Doc 1). Retrieved from http://www.industry.gov.au/assets/documents/itrinternet/BC-SCMReport.pdf.

Vrijhoef, R. and L. Koskela. 2005. Revisiting the three peculiarities of production in construction. pp. 19–27 *In*: 13th International Group for Lean Construction Conference. Retrieved from http://usir.salford.ac.uk/9377/?utm_source=twitterfeed&utm_medium=twitter.

Yashiro, T. 2009. Overview of building stock management in Japan. pp. 15–32. *In*: Stock Management for Sustainable Urban Regeneration. Springer Japan.

Zairul, M. N. and R. Geraedts. 2015. New business model of flexible housing and circular economy. *In*: ETH (ed.). Zurich: ETH. https://doi.org/10.3929/ethz-a-010581209.

Zairul, M.N. 2017. FlexZhouse New business model for affordable housing in Malaysia.

Zott, C. and R. Amit. 2010. Business model design: An activity system perspective. Long Range Planning 43(2-3): 216–226. https://doi.org/10.1016/j.lrp.2009.07.004.

CHAPTER 8

A Composite Approach to Return on Investment (CROI)

Valuing Social and Affordable Housing

Judy A. Kraatz

Introduction

Access to safe and secure housing is a basic human right and need, and the associated social inclusion is now broadly recognized as an integral element to ensuring a sustainable built environment. Through having a stable home, people are better able to engage with family, education, employment, health services and with their local and broader community. Many governments provide social housing to ensure such access to those on very low and low incomes, through publicly funded and provided public housing, or in partnership with the Not-For-Profit (NFP) sector, in the form of community housing. Such partnerships are also used to ensure housing is available at affordable rents to those on low to moderate incomes.

The United Nations Economic Commission for Europe (UNECE) broadly states that social housing "is supplied at prices that are lower than the general housing market and it is distributed through administrative procedures" (Rosenfeld, 2015), typically requiring some form of state subsidy. The term social housing is used in Canada, Australia,[1] the Netherlands and the UK, other terms include public housing in the US and Israel; community housing in Denmark; and housing promotion in Germany. In Australia, affordable housing can be broadly defined as "housing that reduces or

Griffith University, Brisbane, Australia.

[1] In Australia, the Productivity Commission defines social housing as 'rental housing provided by not-for-profit, non-government or government organisations to assist people who are unable to access suitable accommodation in the private rental market.' (Productivity Commission, 2015) (p.G3). This sector in Australia includes public housing, community housing, as well as state-owned and managed Aboriginal and Torres Strait Islander (ATSI) housing (Romans, 2014).

eliminates housing stress for low-income and disadvantaged families and individuals in order to assist them with meeting other essential basic needs on a sustainable basis, while balancing the need for housing to be of a minimum appropriate standard and accessible to employment and services" (Affordable Housing Working Group, 2017). The effective and appropriate provision of public and community housing (Fig. 8.1) is increasingly difficult in light of current economic conditions, fiscal constraints and increasing housing needs and demands. This is made more problematic by policy commitments which "are now lagging behind housing needs and demands arising from associated processes of economic expansion, urbanization and increased inequalities" (Maclennan and Miao, 2017). This has been a growing global trend since the 1980s, especially since the Global Financial Crisis (GFC) of 2007–09, and particularly in countries where housing is predominantly provided and financed by the private market, with home ownership as an ultimate goal.

Achieving an economically, socially and environmentally sustainable framework for the provision of social and affordable housing is vital for three key reasons. Firstly, housing is a basic human right with far-reaching impacts on individual and national wellbeing and outcomes as detailed by the United National Rapporteur on Housing (Farha, 2017). Secondly, important links are emerging between housing and productivity in terms of processes, price outcomes and characteristics which warrant further investigation (Maclennan et al., 2015; Maclennan and Miao, 2017). And finally, access to safe and secure housing contributes to an equitable society, and to individual and national wellbeing, through identified links to non-housing outcomes such as social cohesion, education, employment, health and well-being, resource efficiency and urban amenity (e.g., neighbourhood security and transportation) (Rosenfeld, 2015). A lack of evidence exists however, which can be used by policy makers to demonstrate these broader benefits.

Contextualizing this broader evidence base is critical to realistically addressing current and future housing needs. Given current fiscal limitations on governments in many countries around the world it is unlikely that significant new funds will become available to address the considerable waiting lists for social housing. It is thus likely that the provision of social housing by governments will continue to be residualized and targeted to those with special or priority needs. This means that a significant cohort of people will remain in need of access to more affordable housing as a next step along the housing continuum, in order to engage more effectively with society through increased involvement in employment, education and the community. Understanding

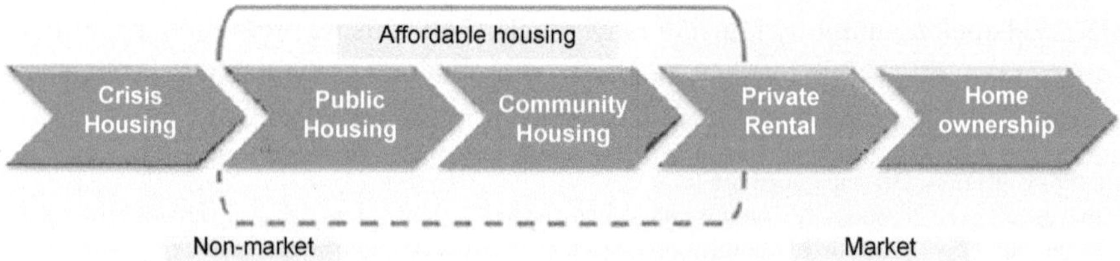

Figure 8.1. The housing continuum (Council on Federal Financial Relations, 2016).

and articulating the broader benefits is thus essential for policy makers to be able to make the case for continued investment, and to potentially attract investment from social and institutional investors.

Firstly a brief outline of the research methodology used to develop this composite approach to return on investment is provided. Then the need for evidence to support investment in this sector is discussed. Next the productivity-based conceptual model used to frame this investigation is described, followed by an examination of why current methods for determining the return on investment of social and affordable housing do not effectively capture these broader outcomes and benefits. The composite return on investment approach, designed to build an evidence base to support greater investment, from the public, private and not for profit sector is then presented, and finally a brief conclusion is provided.

Research Approach

This chapter presents a comprehensive approach to demonstrating that, by providing access to safe and secure housing, economic, social and environmental benefits are returned to the individual, the investor and society. This approach was designed as one element of the *strategic evaluation framework,* developed through research undertaken by the Sustainable Built Environment National Research Centre (SBEnrc) between 2014 and 2016. This included two collaborative research projects with university, government and industry partners. Early in the Rethinking Social Housing project[2] (2014–15), a productivity-based conceptual framework was developed which highlighted the potential for capturing broader benefits than those linked directly to the cost of providing housing (for example, improved engagement in education and employment, reduced demand on social services, and greater resource efficiency).

In the context of this conceptual framework, initial desktop research was undertaken, including limited reviews of Australian and international academic and industry literature on four key topics: (i) outcomes and indictors relevant to broad non-housing outcomes; (ii) links and associations which exist between housing and non-housing outcomes; (iii) return on investment methodologies with application to non-housing outcomes; and (iv) available and required data needed to provide an evidence base to support impact and outcome assessment. Early findings were tested through meetings and workshops with government and industry partners to test validity and practicability of the framework. As discussed later in this chapter, no single return on investment method was found which adequately reflected the broad returns to the individual, the community, society and the economy of having safe and secure housing. A follow-on research project, Valuing Social Housing[3] (2016–17), then further developed this aspect of the research, through a further more comprehensive review of the literature, and further testing with policy makers and those delivering social and affordable housing.

[2] http://sbenrc.com.au/research-programs/1-31-rethinking-social-housing-effective-efficient-equitable-e3/.

[3] http://sbenrc.com.au/research-programs/1-41-valuing-social-housing/.

The Need for Evidence

In Canada, 'the social and affordable housing sector comprises about 4–5% of the total' housing sector in that country (Carlson, 2014) while there were 140,000 families awaiting rent-subsidized housing in Canada in 2015 (Young, 2015). In the UK this sector represents 19% of the overall housing sector (Bourne, 2016) with a waiting list of 1.8 million in 2014. In the US, there are 1.2 million households living in public housing units (U.S. Department of Housing and Urban Development, 2016) with a shortfall of 5.3 million affordable housing units in 2013 (Rosenfeld, 2015). In the Netherlands, social housing organizations provide 2.4 million houses for 4 million people; out of a total national population of 16.8 million (Aedes, 2016). In 2010 in France 17.3% of the French population were tenants of social housing units (approximately 10 million people) (Wong and Goldblum, 2016) with 1.7 million applications for social housing in 2014 (Rosenfeld, 2015). In Australia, at June 2013, around 414,000 households, from a total population of approximately 23.1 million were living in social housing (Department of the Prime Minister and Cabinet, 2014), with 158,971 applicants on the public rental housing waiting list (Australian Institute of Health and Welfare, 2014).

Thus regardless of definitions, demand well surpasses current supply. Social housing as defined here is currently predominantly provided by the public sector, with limited current contribution from the private sector. In addition, much of the affordable rental housing in Australia is provided by Community Housing Associations (often with commonwealth government funding support), and through growing partnerships between government, not-for-profit organizations, and the private sector. To start to address current shortfalls, a need exists to justify intervention in the market through additional fiscal support, and social and private sector investment (which is currently under-represented in this aspect of urban infrastructure provision).

Better evidence is however required on the return on investment provided by this investment to leverage additional investment. To effectively do this, we need to go beyond the traditional specific housing outcomes to embrace externalities such as those identified in the nine domains identified in the Rethinking Social Housing research. The early review of literature which investigated the benefits of social housing revealed a plethora of outcomes and indicators. Through the course of that research these were distilled into nine domains, being community, economy, education, employment, environment, housing, health and wellbeing, social and urban amenity (Kraatz et al., 2015). Just as past research has confirmed the benefits and returns of greater resource efficiency in the design, operation and maintenance of housing, there is a need to gather evidence to better understand the layers of value which investment in social and affordable housing stock can potentially return across the housing supply chain.

Whilst providing better access to housing imposes a readily identifiable upfront cost to those providing the housing, the value of the return on this investment is less easily defined. Some of this return is readily quantifiable, such as the potential reduction in crisis accommodation. Some may be captured by shadow pricing or as externalities such as reduced health impacts of rough sleeping. Other aspects are difficult to quantify in any effective way, such as increased financial security for the tenant flowing on to greater engagement in education and employment and an improvement in intergenerational wellbeing. And these and other aspects may also have broader impact on national productivity and economic wellbeing (Kolstad et al., 2014; OECD and Ford Foundation, 2014).

The need thus exists to be able to provide evidence that a housing-based policy or program is achieving its desired direct impact, whilst capturing any other indirect impacts, which may often occur in other people or other agencies and organizations (i.e., reductions in: utilities bills for tenants, graffiti bills for local authorities; health services for allied departments). This may ultimately have impacts on budget distributions across and within government agencies and may potentially identify areas for trade-offs when decisions need to be made as to where funding is directed (Flatau et al., 2015). This is increasingly important in times of increased competition for funds, with agencies acknowledging 'a lack of quality data to measure and quantify many outcomes, and diverse views on how to measure outcomes, for example cash savings versus avoided costs' (The Office of Social Impact Investment, 2015). Importantly, such evidence can be used to provide a more comprehensive productivity context from which the broader return on investment can be assessed, to potentially leverage additional non-traditional private sector funding sources.

Productivity-Based Framework

Improving productivity is of key interest at all levels of government. It is proposed as a key lever for improving the supply of social housing through identifying and quantifying the productivity benefits which may arise from such investment, in terms of both housing and non-housing outcomes and impacts on individuals and society. The impact of productivity improvements in the design and construction of housing is widely addressed elsewhere (especially in terms of materials, energy and water efficiency), whereas those from a macroeconomic and fiscal perspective, in terms of the provision of social housing, are only recently being considered. For example how revenue increases if social housing could produce positive productivity impacts and the effect on other government expenditure (such as better health leading to less sick leave and more people working longer which adds to productivity and tax revenues) is still needed as an evidence base to support budget deliberations. There is now "a growing recognition by business and governments that housing outcomes may already be constraining national growth, or imposing undue expenditure costs on other budgets (e.g., homelessness effects on health)" (Maclennan et al., 2015).

In order to develop a more comprehensive understanding of the returns to individuals and society of investing in social and affordable housing, a productivity-based conceptual framework was developed (Fig. 8.2). In this productivity is considered as a "measure of the effectiveness of the use of resources in the production of defined outputs" (Maclennan, 2015). This can apply to economic, social and environmental efficiency and effectiveness.

This seeks to provide a more comprehensive rationale for policy making, delivery and evaluation of social housing programs which captures these broader outcomes and impacts, with productivity viewed through four lenses: (i) for the individual in terms of better engagement with employment, education, health resources and society; (ii)

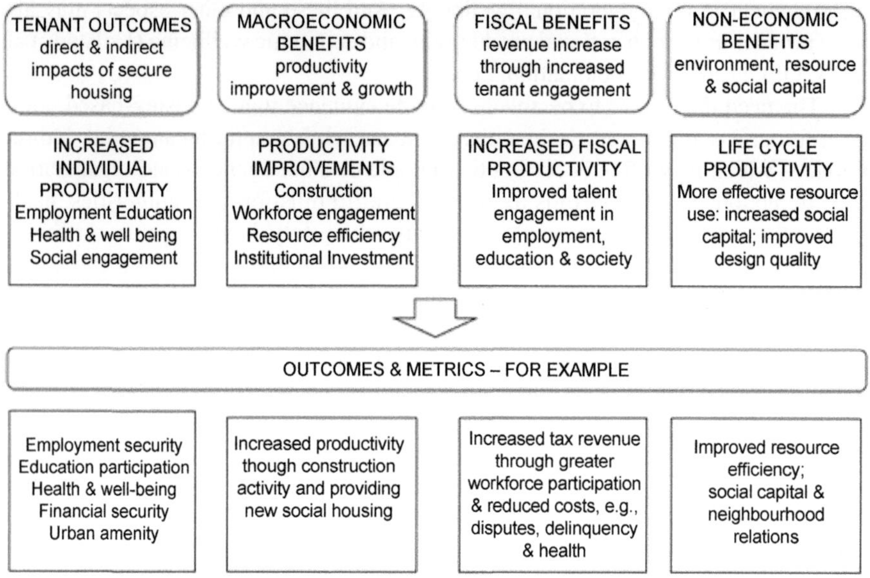

Figure 8.2. A productivity-based conceptual framework for social housing (Kraatz et al., 2015).

for the nation through macro-economic and/or (iii) fiscal benefits; and (iv) for non-economic productivity benefits associated with the creation of environmental or social capital (Kraatz et al., 2015). This conceptual framework thus provides four distinct avenues through which to further explore the potential return on investment of the non-housing outcomes associated with access to safe and secure housing.

Exploring Return on Investment

In the course of developing the *strategic evaluation framework* for social and affordable housing (Kraatz and Thomson, 2017),[4] a substantial array of tools and approaches were found to account for return on investment. These include: *cost benefit analysis* (CBA) looking ratio of housing costs to value of housing benefits; *cost consequence analysis* (CCA) looking at housing costs per tenant year; and *cost effectiveness evaluation* (CEE) looking at disaggregated housing costs and tenant outcome measures (Parkinson et al., 2014; Pawson et al., 2014). Several other approaches include: *financial feasibility analysis and post-occupancy evaluation* (Milligan et al., 2007); *results based accounting* and *social accounting* (*and auditing*), which is an approach to reporting that relates to the social, environmental and financial impact which an organization has had and considers the extent to which an organization is meeting its (usually pre-determined) social or ethical goals (New South Wales Office of Social Impact Investing, 2015); *social cost benefit analysis* (SCBA) which assesses the net value of a policy or project to society as a whole (HM Treasury, 2011); the *social*

[4] http://www.sbenrc.com.au/research-programs/1-54-procuring-social-and-affordable-housing-improving-access-and-delivery/.

impact value calculator, a simple excel tool to provide support to apply the values in the Social Value Bank to community investment activities (Campbell Collaboration, 2014); *social return on investment* (SROI) which is an approach that seeks to measure the impact of a project, program, social enterprise, non-profit organization or policy by analyzing the value created from the social outcomes and comparing these with the investment needed to generate these benefits' (Dunn, 2014b); and *wellbeing valuation analysis* (WVA) which builds on cost-benefit & SROI analyses, to provide values for the impact of a situation on the average person's well-being (Fujiwara, 2014).

It is argued that while these types of tools and methods may provide information on the economic costs and benefits of a program or activity, but they neglect an array of social and environmental outcomes that have value which is less easy to quantify. During the early research phase, two methods were considered as providing complementary understandings for establishing an approach to clarifying this broader return on investment: (ii) Social Return on Investment (SROI); and Wellbeing Valuation Analysis (WVA). The first of these, SROI, enables organizations to measure the broad benefit to society derived from their operations, in addition to the traditional monetary returns. This is becoming increasingly important with the rise of social benefits bonds and the like. The intent of this approach is to provide organizations with a ratio of inputs to impacts; and to calculate the dollar value of social impact compared to cost of the benefits provided (Ravi and Reinhardt, 2011; Kliger et al., 2011, Social Ventures Australia Consulting, 2012). This approach was initially developed in 2000, by the Roberts Enterprise Development Fund in the US, and first used in Australia in 2005 and then further endorsed in 2010 by the Productivity Commission aligning with their proposed Performance Measurement Framework.[5]

The second method, WVA, provides a measure for the benefit to an average individual of the services provided by an organization. The values established are consistent with the theory and principles underlying CBA & SROI. The Organisation for Economic Co-operation and Development (OECD) has been developing a well-being approach for several years (OECD, 2013). It is further used to measure the social value provided by housing associations in the UK, through assessing the impact on an average person's wellbeing of the broader non-housing benefits of access to safe and secure housing, and placing a dollar value on these (Fujiwara, 2014; Trotter and Vine, 2014; Trotter et al., 2014). This approach was developed in 2013–14 for measuring the social value of housing associations in UK, and emerged in response to the perceived lack of appropriate tools for quantifying social value on a sector-wide scale. It aims to estimates the impact of a good/service on subjective wellbeing, then uses this to calculate the exact amount of money that would produce an equivalent impact.

Whilst these two approaches address broader social return on investment for organizations, and include stakeholder involvement, what remains missing is an account of the potentially transformational nature of the broader impacts, to both an individual and to society. These two approaches do not capture impacts beyond that to the 'average' person, and do not reveal impacts to the broader society. For example they do not consider the impact of economic inequality in society, one manifestation of which is lack of access to safe and secure housing, is shown to lead to lower national

[5] http://www.pc.gov.au/research/supporting/national-agreements

productivity. Gurria notes that 'recent Organisation for Economic Co-operation and Development (OECD) research estimates that rising inequality has knocked as much as seven percentage points off cumulative Gross Domestic Product (GDP) growth in the US since 1990' (Gurria, 2015).The French Commission in the Measurement of Economic Performance and Social Progress note that traditional measures for non-market services such as social housing have been based on inputs rather than outputs: "an immediate consequence of this procedure is that productivity change for government-provided services is ignored, because outputs are taken to move at the same rhythm as inputs" (Stiglitz et al., 2009). Maclennan proposes that such productivity impacts can be linked to "poor housing and neighbourhood outcomes" and can "raise costs of non-housing programs aimed at raising human capacities" (Maclennan, 2008). He highlights housing and neighbourhood conditions all impact on "child socialization and learning as well as teenage job and university readiness" and has impacts on physical and mental health "through lowered levels of labour market participation, absenteeism, and reduced productivity" (Maclennan, 2008).

The Composite to Return on Investment Approach (CROI)

It is thus proposed that a single method to articulate return on investment does not capture the complex nature of the value returned to society and the individual. The proposed CROI approach aligns with this productivity-based conceptual framework, wherein productivity benefits are considered in terms of individual, macroeconomic, fiscal, and non-financial benefits. To address this, Table 8.1 illustrates how the composite approach proposed in this chapter can be used to provide a comprehensive overview of the potential productivity benefits through four lenses.

From this perspective, four elements are proposed to address this complexity. This can then be used to provide evidence to support: (i) greater investment by public agencies in social housing; and (ii) new investment by not for profits or public and private sector partnerships. These four elements can then be brought together to better understand and articulate the broad value of the provision of social housing.

Table 8.1. Value and productivity.

		Productivity Impacts			
	Nature of the value captured	**Individual**	**Macro-economic**	**Fiscal**	**Non-economic**[1]
SROI	Value to the organization and society		*	*	
WVA	Value to the average person's wellbeing	*	*		
Narratives	Value to the individual	*			*
Equity	Value of equity	*	*		*

Note: [1]For example, the creation of social & environmental capital.

Four elements of the composite return on investment approach: Social Return on Investment (SROI)—this method enables organizations to provide a ratio of impact to dollars input and/or an aggregated dollar return on investment for defined benefits to the organization and to society which may accrue from the provision of social and affordable housing. This is determined through: identifying key outcomes, indicators, and impacts; establishing financial proxies for these; and determining a dollar value for this benefit.

Wellbeing valuation—the OECD have been developing an approach to measuring physical and social wellbeing for several years. In the UK, a wellbeing valuation analysis method has been developed for community housing associations to measure the impact of their investment. This is done in terms of impacts on an average person's wellbeing of the broader non-housing benefits of access to safe and secure housing, and placing a dollar value on these.

Narratives—individual narratives can be used to understand the value to the individual of having a safe and secure home, and the relationships between this and other aspects of a person's life. The value a person places on a given amenity such as a home varies dependent on their life situation (Kolstad et al., 2014). Such narratives are currently captured in the Annual Reports of housing and service providers and also more increasingly in on-line digital stories. Developing narratives in a more rigorous way is proposed as a further avenue for capturing benefits as part of the CROI.

Equity—Comparing, understanding and aggregating the value different people place on social infrastructure, such as housing, can lead to understanding the broader value of equitable access to such resources. The French *Commission on the Measurement of Economic Performance and Social Progress*, the OECD and the *Intergovernmental Panel on Climate Change* all offer background theory which can contribute to this element (Stiglitz et al., 2009; Kolstad et al., 2014; OECD and Ford Foundation, 2015).

Element 1—Social Return on Investment Analysis (SROI)

Governments and organizations have increasingly sought to analyze and measure the non-economic value of their work in order to justify continued investment. SROI has been increasingly used to determine and quantify their social impact. This method allows social, environmental and other non-economic benefits and costs to be articulated in financial terms. These values can then be compared to the investment made, and the cost-effectiveness of a program or organization can be determined. A social return on investment can then be determined.

The United Kingdom has led the way in utilizing SROI within the government. This was the result of dissatisfaction that public contract allocations were too focused on cost efficiency and financial return (Harlock, 2012). In 2012 the *Public Contracts (Social Value) Act* was introduced, which required analysis of social value when determining contract allocation. It is essentially considered as a policy tool that enables the performance of NFPs and commercial operators to be more equally compared, by placing value on the less tangible, but valuable outcomes that NFPs provide.

While SROI provides a dollar-based value in financial terms, this should not be seen as a strictly financial return on investment, but rather as a financial representation of the value added. SROI evaluation "is best understood in the context of an endeavour to value wellbeing through measures other than classic economic indicators such as

GDP" (Ravi and Reinhardt, 2011). This method also provides a useful mechanism for tracking organizational change and can assist organizations to maximize their social value creation. Ravi and Reinhardt used this approach to measure the social value of community housing in Australia on a sector-wide scale. In another housing-based example this approach was adopted to discuss the premise "that investment in affordable housing for low-income women provides both micro and macro-economic benefits for cities and communities" (Kliger et al., 2011), finding that the "VHWA created $30 million Australian $s in value for its stakeholders and the Victorian community from an investment of $7.45 million Australian $s".

Several resources are available which provide further detail of this method (Nicholls et al., 2012, Social Ventures Australia Consulting 2012, The SROI Network (now Social Value UK) 2012) and http://1068899683.n263075.test.prositehosting.co.uk/about-us/the-sroi-project/.

Element 2—Wellbeing Valuation

In 2009, the French *Commission in the Measurement of Economic Performance and Social Progress* identified that "commonly used statistics may not be capturing some phenomena, which may have an increasing impact on the wellbeing of citizens" (Stiglitz et al., 2009). Their report indicates that data around subjective wellbeing remains limited, and that "national statistical systems need to build on these efforts (researchers and commercial data providers) and incorporate questions about various aspects of subjective wellbeing in their standard surveys". They proposed a shift in emphasis "from measuring economic production to measuring people's wellbeing", noting the gap which exists between the tradition approach of measuring Gross Domestic Product (GDP) data, to "what counts for common peoples" wellbeing.

The *Organisation for Economic Co-operation and Development* (OECD) have also recognized this shortfall in measurement approaches and have been actively developing methods and guidelines for several years around the measurement of wellbeing.[6] This relates to "how people experience and evaluate their life as a whole" (Organisation for Economic Co-operation and Development, 2013a). They have identified 11 dimensions related to material conditions and quality of life being (Fig. 8.3). These align with the nine domains identified in the SBEnrc Rethinking Social Housing research to address the broad impact of having access to safe and secure housing.

The *Intergovernmental Panel on Climate Change* (IPCC) also discuss several different approaches to wellbeing and its measurement defining a person's wellbeing as including "everything that is good or bad for the person—everything that contributes to making their life go well or badly" (Kolstad et al., 2014). Their work is particularly relevant in the following discussion on the value of equitable provision of social infrastructure such as housing.

In the UK, a wellbeing valuation methodology has been specifically developed for community housing providers to enable them "to measure the success of a social

[6] http://www.oecd.org/statistics/measuring-well-being-and-progress.htm.

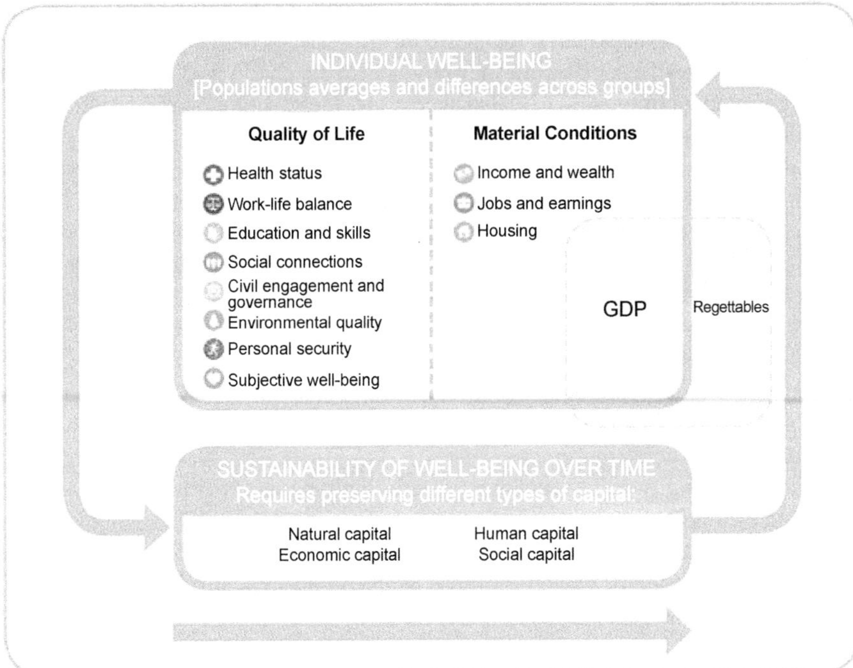

Figure 8.3. OECD Framework for measuring wellbeing and progress (Organisation for Economic Co-operation and Development, 2013).

intervention by how much it increases a person's wellbeing" (Trotter et al., 2015). The method was developed in response to the perceived lack of appropriate tools for quantifying social value on a large (i.e., sector-wide) scale. This approach draws upon both the SROI method and traditional cost benefit analysis (Fujiwara and Campbell, 2011; Fujiwara, 2014) and is now well developed (Fujiwara, 2013; Fujiwara, 2014; Trotter et al., 2014; Trotter et al., 2015). This analysis seeks to find 'from the data the equivalent amount of money needed to increase someone's wellbeing by the same amount'(Trotter et al., 2014). It provides headline wellbeing values for specific financial proxies for improvement in individual wellbeing for the average person based on their access to community housing. The approach estimates the impact of a good or service on people's subjective wellbeing, and then uses these estimates to calculate the exact amount of money that would produce the equivalent impact. Analysis draws on four UK datasets: (i) the *British Household Panel Survey*, a longitudinal survey of 10–15,000 people in the UK;[7] (ii) *Understanding Society,* which incorporated and replaced the previous, adding 60,000 new participants and a new set of variables;[8] (iii) *Crime Survey for England and Wales,* survey of all aspects of crime by the Office of National Statistics;[9] and (iv) *Taking Part*, which collects data in leisure, culture and sport.[10] Critically, this work provides a detailed investment decision-making framework

[7] https://www.iser.essex.ac.uk/bhps.
[8] https://www.understandingsociety.ac.uk/.

for housing associations in the UK. The *Social Impact Value Calculator* (Campbell Collaboration, 2014) has been developed to further assist community housing providers to estimate the broader social value they produce (Trotter and Vine, 2014). Community housing providers in the UK can also access the *Social Value Bank*[11] to undertake a valuation of their social impact.

Limitations of this approach include that it represents the average person, at a certain point in time, thus does not account for diversity within the social housing population. The former recognized by the IPCC in their discussion on temporal and lifetime wellbeing (Kolstad et al., 2014) and the latter by Fleurbaey (Fleurbaey, 2009). Stiglitz et al. also broaden this discussion suggesting that "surveys should be designed to assess the links between various quality-of-life domains for each person" (Stiglitz et al., 2009), that is how wellbeing in one aspect of a person's life (e.g., having access to safe and secure housing) affects other domains (e.g., access to education or employment). They refer to this as 'joint distribution' which furthers discussion around the need for attribution.

The following two elements thus seek to firstly better articulate the diversity of experience through the use of narratives (Element 3), and secondly highlight the need for better understanding the issue of understanding the benefits to the broader society of addressing housing provision for those unable to do so themselves (Element 4).

Element 3—Narratives

The potency of narrative is partially derived from at least three factors. First, narrative is intrinsically entertaining and engages the listener. Second, narrative humanizes and contextualizes an issue, moving it from the abstract policy discussion, for example, to a discussion of how the policy impacts or results from the experiences of real people. And third, each narrative contains embedded causal explanations that surreptitiously convey the speaker's/author's message or orientation (Salzer, 2000).

The intent of this element is to determine and account for the impact on a person's life of having access to safe and secure housing through narrative and story-telling. In this context, "narratives allow for implicit and explicit expression of the role of situational factors and context in explaining causal attributions" (Salzer, 2000). Such narratives can help account for the type, nature, scale and depth of impact which having access to safe and secure housing can have on a person's life experiences in other domains (McCreless and Trelstad, 2012). The depth of impact being "the amount or intensity of change experienced", or the "change in subjectively experienced wellbeing" (McCreless and Trelstad, 2012).

There is now a significant amount of literature which explores the theoretical underpinnings and development of this approach (Abbott, 1992; Salzer, 2000). Salzer provides a useful definition of narrative as "a symbolized account of actions of human beings that has a temporal dimension. The story has a beginning, a middle, and an

[9] http://www.crimesurvey.co.uk/.

[10] https://www.gov.uk/government/collections/taking-part.

[11] http://www.hact.org.uk/social-value-bank.

ending . . . The story is held together by recognizable patterns of events called plots. Central to the plot structure are human predicaments and attempted resolutions" (Sarbin, 1986, in (Salzer, 2000). Whereas, Abbott discusses the different approaches to this method and highlighting distinctions between techniques which address "actual interactions between identifiable actors" and those which "are most effective with successions of things happening to a single actor" (Abbott, 1992).

In the context of the CROI approach it is proposed that value be determined from qualitative narratives structured around a rigorous framework, and gathered through surveys, interviews and case studies. This may be facilitated by the use of mobile technologies for data gathering, to produce accessible rich narratives which can be readily communicated to decision-makers and investors.

Element 4—Value of Equity

This final element adds a further layer in order to firstly articulate the value to society of ensuring a minimum quality of life for all, and secondly recognize that indicators must not be based on the 'average' or 'representative' person. With housing outcomes increasingly seen as "a major reinforcer of wealth and income inequalities in some advanced economies" (Maclennan and Miao, 2017), this element draws together key findings from several different fields to indicate the need for further investigation in this area to establish a comprehensive understanding of the return on investment of safe and secure housing.

Fleurbaey provides a valuable overview of several approaches to providing better measurement around such social welfare functions (whether grounded in welfarism, liberalism or perfectionism approaches) (Fleurbaey, 2009). He highlights the need for continuing research in this area, and further theoretical research is much needed to explore this in the context of social housing. Consideration here is through three lenses: the OECD approach to inclusive growth (OECD and Ford Foundation, 2015); the French *Commission on the Measurement of Economic Performance and Social Progress* approach to understanding the diversity of experiences in terms of individual wellbeing (Stiglitz et al., 2009); and the *Intergovernmental Panel on Climate Change* (IPCC) discussion of distributive justice and differential value (Kolstad et al., 2014).

"Inequalities and the problems to which they give rise have a spatial dimension. Better transport and housing infrastructure can spur growth and improve inclusiveness in cities, providing vital access assets for economically deprived areas to high-quality jobs and education" (OECD and Ford Foundation, 2015).

OECD and Inclusive Growth. The OECD inclusive growth approach is defined as "economic growth that creates opportunity for all segments of the population and distributes the dividends of increased prosperity, both in monetary and non-monetary terms fairly across society" (OECD and Ford Foundation, 2015). That report maintains that inequality in non-income outcomes (such as education, employment opportunities, access to infrastructure and health conditions) can undermine long term growth. Included in this approach is multidimensional wellbeing, that is, current wellbeing (material living conditions and quality of life) and wellbeing over time, including for future generations, across economic, natural, human and social capital. It also considers the need to include the non-monetary dimensions of wellbeing and assess the impact of policies on different social groups in terms of employment, health and educational issues and outcomes. For example those most disadvantaged often live shorter lives and experience difficulty breaking away for problematic educational and employment outcomes (see also (Ianchovichina and Lundstrom, 2009, Organisation for Economic Co-operation and Development (OECD), 2014)).

The Commission identifies that 'the first cross-cutting challenge for quality-of-life indicators is to detail the inequalities in individual conditions in the various dimensions of life, rather than just the average conditions in each country'(Stiglitz et al., 2009). Their approach also identifies the need to acknowledge diversity of experience, the linkages between the various domains (or dimensions) of a person's life (joint distribution), and measuring change over time. All of this presents challenges from a statistical and data gathering perspective. Fleurbaey suggests that whilst such comparison across individuals and cultures is problematic, "looking for patterns in the data may be able to provide a viable avenue for such investigation" (Fleurbaey, 2009). Stiglitz et al. also suggest that "average measures of income, consumption and wealth should be accompanied by indicators that reflect their distribution". For example, as illustrated in Fig. 8.4, those "in the bottom quintile of the distribution of equivalent income report worse health and a higher incidence of unemployment compared to people identified as 'worse-off' based on either their consumption expenditure or their subjective life-evaluations" (Stiglitz et al., 2009).

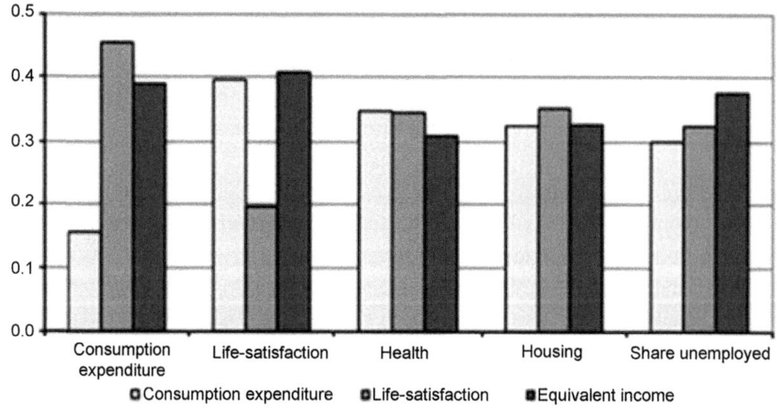

Note. Data refer to people considered as "worse-off" (i.e., in the bottom quintile of the distribution) according to three different measures of their quality of life: (i) household consumption expenditure (adjusted for the number of people in each household); (ii) life-satisfaction (based on the question, "*To what extent are you satisfied with your life in general at the present time?*" with answers on a five-point scale); and (iii) a measure of equivalent income, based on four "functionings", i.e., self-reported health, employment status, quality of housing, and having incurred wage arrears. For each of these three measures of quality of life, the figure plots the average levels of various factors shaping quality of life among the "worse-off" based on one measure relative to those based on all others.

Source: Fleurbaey, M., E. Schokkaert and K. Decancq. 2009. "What good is happiness?", CORE Discussion Paper, 2009/17, Université catholique de Louvain, Belgium. Computations based on data from the *Russia Longitudinal Monitoring Survey*.

Figure 8.4. Characteristics of the most deprived people according to different measures of quality of life, Russia in 2000 (Stiglitz et al., 2009).

International Panel on Climate Change (IPCC). The IPCC discussion provides a further lens through which to examine this final element, and captures knowledge and data relevant to the impact on individual outcomes, for specific circumstances (e.g. abilities, point in time, etc.), and in given locations. This also provides an avenue to compare one person's wellbeing with another. Kolstad et al. discuss an approach which aggregates a person's wellbeing at a point in time to create lifetime wellbeing for individuals which can then be aggregated across people to determine an overall value to society. These authors also present the idea of distributive justice which "suggests that equality of wellbeing does have value ... The resulting ethical theory is called prioritarianism ... according to prioritarianism, improving a person's wellbeing contributes more to social welfare if the person is badly off than if they are well off" (Kolstad et al., 2014) (Fig. 8.5).

The issues associated with valuing equity as part of the CROI approach are many and complex. There is however, significant and viable theory and evidence to support further research in this area including building an understanding on how housing and other policies relevant to the eight other non-housing domains impacts across the

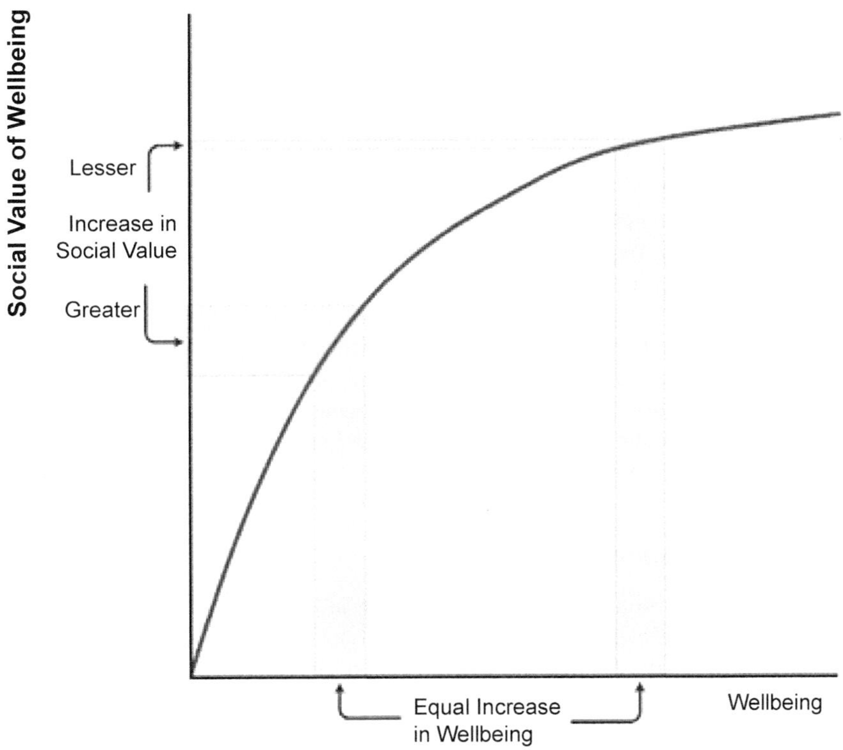

Figure 8.5. The prioritarian view of social welfare. The figure compares the social values of increases in wellbeing for a better-off and worse-off person (Kolstad et al., 2014).

population. Also important is building a more complete picture of economic growth and its impacts and benefits across a country's population.

Conclusions

This chapter examines how to better capture the broader impacts of having safe and secure housing so that government agencies, NFP providers and the private sector can better justify investment. When considering the value of improving access to social and affordable housing through the lens of productivity, return on investment can be examined from several perspectives, from that: of the individual; of the broader economy through macro-economic returns; of fiscal returns and savings to government; and non-financial returns by way of social and environmental capital.

This way of approaching this issue has underpinned the development of a strategic evaluation framework which includes the Composite Return On Investment (CROI) presented for discussion in this paper as a key element. The intent being to develop a more complete evidence base for demonstrating the benefits of investment in social housing. Whilst SROI and WV are good tools for capturing aspects of the broader return on investment of providing safe and secure housing, their limitations do not enable a comprehensive understanding across time and society of the benefits. Narratives have provided a contextualized understanding of life circumstance throughout recorded history, and understanding equity impacts across the whole of society can assist with gaining social acceptance of the need for the provision of social and affordable housing.

Using a composite approach can thus potentially enhance understanding and contextualize the temporal, diversity and equity dimensions needed to better represent the benefits of and return on investment in such housing. This chapter thus presents a first tentative step towards establishing a broader and more comprehensive CROI approach. The issues associated with realizing it are many and complex. There is however, significant and viable theory and evidence to support further investigation. This would require bringing together the theoretical strands presented here, and then identifying ways in which statistical data and analysis can be used to inform this element. Emerging data capture techniques may facilitate this in a way not possible until now.

Acknowledgement

This research has been developed with funding and support provided by Australia's Sustainable Built Environment National Research Centre (SBEnrc) and its partners. Core Members of SBEnrc include Aurecon, BGC Residential, Queensland Government, Government of Western Australia, New South Wales Roads and Maritime Services, New South Wales Land and Housing Corporation, Curtin University, Griffith University and Swinburne University. Core Project Partners were Queensland Government Department of Housing and Public Works, Western Australian Housing Authority, New South Wales Land and Housing Corporation, Curtin University, and Griffith University. The National Affordable Housing Consortium was a Project Partner and Professor Judy Yates chaired our Project Steering Group.

References

Abbott, A. 1992. From causes to events notes on narrative positivism. Sociological Methods & Research 20(4).

Aedes. 2016. Dutch Housing in a Nutshell: Examples of Social Innovation for People and Communities The Hague, Brussels.

Affordable Housing Working Group. 2017. Supporting the implementation of an affordable housing bond aggregator. C. o. F. F. Relations, Commonwealth of Australia.

Australian Institute of Health and Welfare. 2014. Housing assistance in Australia.

Bourne, R. 2016. The UK Doesn't Need More Social Housing—But We Do Need to Build More Homes. The Telegraph.

Campbell Collaboration. 2014. The production of a systematic review. Retrieved 20 October 2014, from http://www.campbellcollaboration.org/resources/research/the_production.php.

Carlson, M. 2014. Alternative Sources of Capital for the Social/Affordable Housing Sector in Canada. Canada.

Council on Federal Financial Relations. 2016. Innovative Financing Models to Improve the Supply of Affordable Housing. Canberra, Australia.

Department of the Prime Minister and Cabinet. 2014. Reform of the Federation White Paper: roles and repsonsibilities in housing and homelessness. Canberra, Commonwealth of Australia.

Dunn, J. 2014b. Social Impact Investing: The returns can be simple to see yet hard to define." AFR Special report Retrieved 20 April 2014, 2015, from http://business.nab.com.au/afr-special-report-social-impact-investing-8140/.

Farha, L. 2017. Report of the Special Rapporteur on adequate housing as a component of the right to an adequate standard of living, and on the right to non-discrimination in this context. H. R. Council, UN General Assembly.

Flatau, P., K. Zaretzky, S. Adams, A. Horton and J. Smith. 2015. Measuring Outcomes for Impact: in the community sector in Western Australia. Social Impact Series. B. Foundation. Perth, Australia.

Fleurbaey, M. 2009. Beyond GDP: The quest for a measure of social welfare. Journal of Economic Literature 47(4): 1029–1075.

Fujiwara, D. and R. Campbell. 2011. Valuation Techniques for Social Cost-Benefit Analysis: Stated Preference, Revealed Preference and Subjective Well-Being Approaches: A Discussion of the Current Issues. H. Treasury. London, UK, HM Treasury.

Fujiwara, D. 2013. The Social Impact of Housing Providers. UK, Housing Associations Charitable Trust (HACT).

Fujiwara, D. 2014. Measuring the Social Impact of Community Investment: The Methodology Paper. London, Housing Associations Charitable Trust (HACT).

Gurria, A. 2015. Addressing growing inequality through inlcusive growth: Insights for the US and beyond. Retrieved 6/1/2017, from http://www.oecd.org/economy/addressing-growing-inequality-through-inclusive-growth-insights-for-the-us-and-beyond.htm.

Harlock, J. 2012. Impact measurement practice in the UK third sector: A review of emerging evidence. Birmington, Third Sector Research Centre.

HM Treasury. 2011. The Green Book: Appraisal and Evaluation in Central Government. H. Treasury. London, UK.

Ianchovichina, E. and S. Lundstrom. 2009. Inclusive Growth Analytics: Framework and Application, The World Bank Economic Policy and Debt Department.

Kliger, B., J. Large, A. Martin and J. Standish. 2011. How an innovative housing investment scheme can increase social and economic outcomes for the disadvantaged. State of Australian Cities. Sydney, Australia, UNSW.

Kolstad, C., K. Urama, J. Broome, A. Bruvoll, M. C. Olvera, D. Fullerton, C. Gollier, W. M. Hanemann, R. Hassan, F. Jotzo, M. R. Khan, L. Meyer and L. Mundaca. 2014. Social, Economic and Ethical Concepts and Methods. Climate Change 2014: Mitigation of Climate Change. Contribution of Working Group III to the Fifth Assessment Report of the Intergovernmental Panel on Climate Change. O. Edenhofer, R. Pichs-Madruga, Y. Sokona et al. Cambridge, United Kingdom and New York, NY, USA, Cambridge University Press.

Kraatz, J. A., J. Mitchell, A. Matan and P. Newman. 2015. Rethinking Social Housing: Effective, Efficient and Equitable—Final Industry Report. Brisbane, Australia.

Kraatz, J. A. and G. Thomson. 2017. Valuing Social Housing: Final Research Report. Brisbane, Australia, Sustainable Built Environment National Research Centre.
Maclennan, D. 2008. Housing for the Toronto Economy. Toronto, Canada.
Maclennan, D. 2015. How to Understand the Links Between the Housing System and Productivity Growth. AHURI. Sydney, Australia, AHURI.
Maclennan, D., R. Ong and G. Wood. 2015. Making connections: housing, productivity and economic development. AHURI, Australian Housing and Urban Research Institute at RMIT University, at Curtin University.
Maclennan, D. and J. Miao. 2017. Housing and capital in the 21st century, Housing. Housing, Theory and Society 34(2): 127–145.
McCreless, M. and B. Trelstad. 2012. A GPS for Social Impact: Root Capital and Acumen Fund propose a system for program evaluation that is akin to GPS. Stanford Social Innovation Review(Fall).
Milligan, V., P. Phibbs, N. Gurran and K. Fagan. 2007. Approaches to evaluation of affordable housing initiatives in Australia. National research venture 3: housing affordability for lower income Australians. Melbourne, Australian Housing and Urban Research Institute (AHURI). 7.
New South Wales Office of Social Impact Investing. 2015. Social Impact Investment Policy: Leading the way in delivering better outcomes for the people of NSW. Retrieved 6 June 2016, from http://www.osii.nsw.gov.au/.
Nicholls, J., E. Lawlor, E. Neitzert and T. Goodspeed. 2012. The Guide to Social Return on Investment. T. S. Network. London, UK, UK Government.
OECD and the Ford Foundation. 2014. All on Board: Making Inclusive Growth Happen, OECD.
Organisation for Economic Co-operation and Development. 2013. Measuring Well-Being and Progress. Paris, France, OECD.
Organisation for Economic Co-operation and Development. 2013a. OECD Guidelines on Measuring Subjective Well-being, OECD Publishing.
Organisation for Economic Co-operation and Development (OECD). 2014. Framework for Inclusive Growth Paris, France.
Parkinson, S., R. Ong, M. Cigdem and E. Taylor. 2014. Wellbeing Outcomes of Lower Income Renters: A Multilevel Analysis of Area Effects: Final Report. Melbourne, Australian Housing and Urban Research Institute (AHURI).
Pawson, H., V. Milligan, P. Phibbs and S. Rowley. 2014. Assessing Management Costs and Tenant Outcomes in Social Housing: Developing a Framework. Melbourne, Australian Housing and Urban Research Institute (AHURI).
Ravi, A. and C. Reinhardt. 2011. The Social Value of Community Housing in Australia, Community Housing Federation of Australia (CHFA), PowerHousing Australia and Bankmecu.
Rosenfeld, O. 2015. Social Housing in the UNECE Region: Models, trends and challenges. Geneva Switzerland, United Nations Economic Commission for Europe.
Salzer, M. S. 2000. Toward a Narrative Conceptualization of Stereotypes: Contextualizing Perceptions of Public Housing. Journal of Community & Applied Social Psychology 10: 123–137.
Social Ventures Australia Consulting. 2012. Social Return on Investment Lessons learned in Australia. Perth, Australia, Investing in Impact Partnership.
Stiglitz, J. E., A. Sen and J. P. Fitoussi. 2009. Report by the Commission on the Measurement of Economic Performance and Social Progress. Paris, France: 292.
Stiglitz, J. E., A. Sen and J.-P. Fitoussi. 2009. Report by the Commission on the Measurement of Economic Performance and Social Progress. Paris, France, Commission on the Measurement of Economic Performance and Social Progress.
The Office of Social Impact Investment. 2015. Social Impact Investment Policy: Leading the way in delivering better outcomes for the people of NSW. Sydney, Australia, NSW Government.
The SROI Network (now Social Value UK). 2012. A guide to Social Return on Investment. UK, The SROI Network: 110.
Trotter, L. and J. Vine. 2014. Value calculator. London, UK, HACT.
Trotter, L., J. Vine, M. Leach and D. Fujiwara. 2014. Measuring the Social Impact of Community Investment: A Guide to using the Wellbeing Valuation Approach. London, UK, HACT Housing.
Trotter, L., J. Vine and D. Fujiwara. 2015. The Health Impacts of Housing Associations' Community Investment Activities: Measuring the Indirect Impact of Improved Health on Wellbeing an Analysis of Seven Outcomes in the Social Value Bank. Simetrica and HACT. UK: 12.

U.S. Department of Housing and Urban Development. 2016. HUD's Public Housing Program." Retrieved 1 July 2016, from http://portal.hud.gov/hudportal/HUD?src=/topics/rental_assistance/phprog.

Wong, T. C. and C. Goldblum. 2016. Social housing in France: A permanent and multifaceted challenge for public policies. Land Use Policy 54: 95–102.

Young, L. 2015. 140,000 Canadian families are waiting for housing. Here's what the parties plan to do." Retrieved 23 May 2016, from http://globalnews.ca/news/2268505/140000-canadian-families-are-waiting-for-housing-heres-what-the-parties-plan-to-do/.

CHAPTER 9

Contribution of Affordable Housing Projects to Green Network in Compact Cities

A Hong Kong Case

Shulin Shi

Introduction

After the industrial revolution during the 18th to 19th century, the world has experienced a great shift in urbanization. By 2015, 54.9% of total population in the world have lived in urban areas (Central Intelligence Agency, 2017). And the world's urban population is expected to nearly double by 2050 (United Nations, 2017a). Besides benefits and flourish brought by this process to human society, sprawl of urban area, especially with a compact development mode, has also led to numerous interdependent environmental problems such as natural habitat depletion and fragmentation, biodiversity loss, pollutions and water-related problems (United Nations, 2012; 2015).

On the other hand, scholars and practitioners have been making great effort to avoid or diminish such problems. For instance, at early stage of urbanization, pioneers in landscape architecture like Fredrick Law Olmsted proposed the concept and plan of greenways by integrating linear system of green and open spaces in urban areas for Boston and Brooklyn (Jongman and Pungetti, 2004; Makhzoumi and Pungetti, 2005). Even today, the urban park network in these cities are still considered very successful planning examples in providing aesthetic, recreational and environmental benefits to urbanites (Fabos, 2004; Jongman, 2004; Jongman and Pungetti, 2004). Since then, there have been different concepts and practices aiming at balancing urban development and environmental needs all around the world, especially in developed countries, such

Technological and Higher Education Institute of Hong Kong, HKSAR.
Email: sprucysky@hotmail.com; shishulin@vtc.edu.hk

as green belts, ecological networks and green networks (Ersoy, 2016). Among these concepts, green network has integrated the needs of ecosystem and humans together (Barker, 1997; Jongman and Pungetti, 2004; Külvik et al., 2008; Forest Research, 2011). The green network approach acknowledges green space in urban areas as functionally connected systems of all kinds of green and open spaces, provide great aesthetic and recreational capacities, and help improve urban environmental quality for all (Barker, 1997; Tzoulas and James, 2010).

In other words, green network aims to help eliminate conflicts between urban ecosystem and human society and fulfill their different needs at the same time. This is well consistent with the general concern of sustainable development regarding urban settlements (United Nations, 2012; 2017b). However, to maintain a certain area and a well configured green network would not be easy for a compact city, especially a highly developed one. Land, is a precious resource for almost any kind of development in such cities. When market demand is far beyond property supply, most private developers would tend to fully develop the land to maximize profit. Greening or landscape would probably be the last thing to be provided in those projects. Under such circumstance, the government plays a critical role to maintain the balance between development and conservation, and to ensure a proper configuration of green network within urban area. In Hong Kong, Affordable Housing (AH) projects developed by the Hong Kong Housing Authority (HA) contribute a lot to green network within densely built urban areas. This chapter will discuss their environmental and social benefits at city, district, and class levels, together with their contributions to sustainability of the city.

Situation in Hong Kong

Hong Kong is a city located at the south-eastern tip of Pearl River Delta, China, with a land territory of 1,106.34 km^2 (HKSAR Government, 2017). Due to the subtropical climate, summers here are humid and hot, while winters are cool and dry. Transitional seasons of spring and autumn are mild and short. The average annual temperature is around 22–23°C, with around 34°C absolute daily maximum and 7°C absolute daily minimum. It rains a lot, with an annual rainfall of 2,318 mm (HKSAR Observatory, 2017). Theoretically, it is suitable for a wide spectrum of plants and animals.

Since the 1840s, urban development in Hong Kong has adopted a high-density mode, and this has moved towards an exceptionally high-rise direction since 1960s (Tian et al., 2014). It is a typical compact city, most of its 7.39 million population live in urban areas on only 25% of its territory (Central Intelligence Agency, 2017; HKSAR Census and Statistics Department, 2017). Although Hong Kong has also preserved 40% of its land as different kinds of conservation areas such as country parks and designated areas (HKSAR Government, 2017), natural resources within developed urban areas have been damaged significantly and typically replaced by infrastructures and high rise buildings. By 2016, open space within developed area was only 3.39 m^2 per capita (HKSAR Planning Department, 2017), which is fairly low compared to other compact cities in the world. These earn the city a name of "concrete jungle".

Such compact development in Hong Kong is not only a consequence of property market, but also a result of lacking requirement on private sectors, especially private developers, to provide open space or greenery. The most common practice among private developers was to reach the cap of allowed plot ratio so as to maximize profit. This has led to the typical block pattern that buildings are close to each other, at least at podium level, leading to narrow sidewalks and lanes. In most old districts, there is little space for street trees (Fig. 9.1).

Fortunately, besides private developments, a large proportion of housing projects are AH projects developed by the HA. The first such project was established in 1954. Although the original purpose was to accommodate poor people who lost their homes in a big fire in Shek Kip Mei, the government later on provided homes for poor people who cannot afford private housing as an important aspect of welfare, and founded Housing Authority in 1973 to implement majority of AH projects. By the end of March 2016, there were almost 2.14 million Hong Kong people (about 30% of the total population) lived in Public Rental Housing/Tenants Purchase Scheme (PRH/TPS) flats. Besides, 298,600 flats of Subsidized Home Ownership (SHO), mainly Home Ownership Scheme (HOS) had been provided to low income population (HKSAR Information Services Department, 2016).

Different from private developments, most of these AH projects have been incorporated with a certain proportion of open spaces with green and landscape elements. Since issuing of Hong Kong Planning Standards and Guidelines in 1982, AH projects have followed it to provide at least 1 m^2 open space per person based on planned population in each project (Yeung, 2003). After new practice notes to foster a quality and sustainable built environment with specific requirements on provision of greenery for new developments was launched in 2011 (HKSAR Building Department, 2011), corresponding standard for AH developments has been changed to assign minimum 30% of total site area for a site larger than 20,000 m^2, or minimum 20% of

Figure 9.1. Typical street view in Hong Kong old districts (taken by author).

total site area for a site with area between 1,000 m^2 and 20,000 m^2 as green coverage (HKSAR Council for Sustainable Development, 2010). With such high open space provision standards, together with large scope of development, AH projects have greatly contributed to green network within urban areas in Hong Kong.

Based on the latest projection of housing demand, about 460,000 units of housing supply has been adopted for the period of 2016 to 2026. Among these projected new developments, around 60% would be AH, respectively with 200,000 PRH units and 80,000 SHO flats (HKSAR Government, 2017). Considering the potential of future development, AH projects would probably continuously provide more urban open space. Nonetheless, there is no specific plan or projection on developing new public parks. In such a situation, it is foreseeable that besides public parks, AH projects would become another major contributor of urban greeneries in Hong Kong. These would become part of the green network and play important roles in sustainable development in Hong Kong. They will also influence urban ecosystem and residents directly and indirectly.

Green Network's Ecological Values

In the context of landscape ecology, structure of urban ecology is commonly discussed based on three fundamental concepts, namely, matrix, patch and corridor. Flows of energy, materials and organisms between them are taken as reference (Forman and Godron, 1986). Among many indicators of good ecological network, connectivity between matrix, patches and corridors are broadly considered critical. It has been the focus of many studies in compact urban areas, as disperse of organisms greatly rely on these natural resources among artificial urban fabrics, which would be very sensitive to sprawl of urban area and urban constructions. According to a study on habitat loss and fragmentation of cross-border watershed in Hong Kong during 30 years (Xie, 2013), the extent and intensity of urbanization have been identified as significant causes of habitat isolation. Loss of habitat and its connectivity consequentially leads to degradation of ecosystem services value. In fact, such cross-border area is much less developed compared to most urban areas in Hong Kong. It can be imagined how serious disperse of organisms has been intervened with by those densely built-up urban areas. Therefore, besides carefully controlling urban sprawl, it is also critical to closely monitor green network at city and district scales in developed areas, so as to ensure sound configuration of landscape patches and corridors for a well-connected ecosystem.

Urban Green Network at the City Scale

Hong Kong consists of Hong Kong Island, Kowloon, New Territories, and numerous islands, with 18 districts in total. In this city, typical developed areas occupy the lower flat coastal or reclaimed lands, while conserved areas are mainly slopes with high ecological value or relatively steep gradient thus not suitable for dense development (HKSAR Town Planning Board, n.d.; HKSAR Planning Department, 2016). Figure 9.2 shows spatial distribution of developed urban areas and natural matrixes in Hong Kong in 1973 and 2016. It can be seen that within a half century, a large area of natural matrixes have been eaten away by urban sprawl and cut into islands.

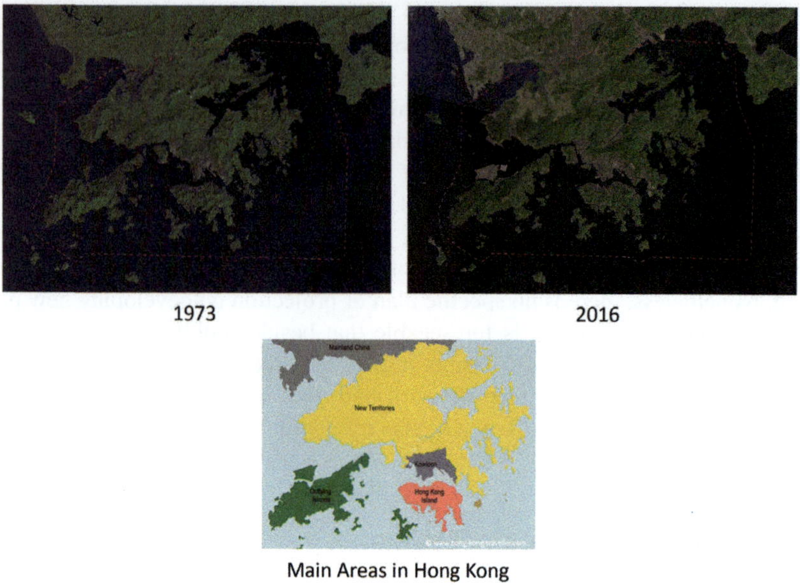

Figure 9.2. Spatial distribution of developed urban areas and natural matrixes in Hong Kong (1973 and 2016) (Sources: USGS Landsat; http://www.hong-kong-traveller.com/geography-of-hong-kong.html#.Wddf5fmOwQ8).

Among these, Kowloon looks quite striking. As one of the earliest human settlements in Hong Kong, its concentrated urban fabrics through time separate matrixes in the north and south. When zooming in to examine Kowloon area (Fig. 9.3), it can be seen that its west, south, and east edges are defined by the Victoria

Figure 9.3. Distribution of major parks and green spaces in Kowloon based on Google map and map from 123rf.com (Sources: https://www.google.com.hk/maps?hl=en&tab=wl; https://de.123rf.com/photo_23935768_hong-kong-verwaltungs-karte.html).

Harbour, mostly with clear cut and tall retaining walls. Along the northern edge of Kowloon, a vehicular trunk road cuts through slopes along the foot of Lion Stone Hill and extends towards the east and west New Territories. These constructed features greatly bound this area out from surrounding natural matrixes, and seriously hinder ecological interchanges. Within the Kowloon area, the situation appears even worse: according to a study assessing green space fragmentation in the whole of Hong Kong (Tian et al., 2011), districts in Kowloon have the highest population density and lowest green coverage in the countryside and urban fringe, compared to other districts. These areas are considered as extremely compact living environments. Under such circumstance, greeneries within urban area become very critical in this area, as they are almost the only resources to enhance landscape and ecological connectivity and disperses. In other words, sufficient and well-configured greeneries are very much in need in Kowloon area.

Greenery within urban areas of Hong Kong mainly include open spaces like public parks, gardens, plazas, sitting-out areas, and roadside plantations (HKSAR Planning Department, 2016). There are 13 major parks in Kowloon (locations shown in Fig. 9.3). They are unevenly distributed in Kowloon area, skewing towards the northern part of it. These major parks, such as Lion Rock Park and Jordan Valley Park, would contribute to bridge the ecological matrix and compact urban area, therefore have relatively higher ecological values. On the other hand, those located in western and southern Kowloon, like Lai Chi Kok Park, Tung Chau Street Park, and Kowloon Park, appear much more isolated from each other and matrixes. Although they also contribute to the ecosystem in the compact urban area as major patches, they cannot form a well-connected green network on their own. In such a situation, urban greeneries on a smaller scale would also play an important role in making the entire system work as corridors or stepping stones to connect these major parks and matrixes.

Urban Green Network at District Scale

When further zooming in to district scale, local green network with smaller patches becomes the focus. Kowloon consists of five districts: Sham Shui Po District, Wong Tai Sin District, Yau Tsim Mong District, Kowloon City District, and Kwun Tong District. Among these five districts, Sham Shui Po and Yau Tsim Mong Districts are very mature dense urban areas in Hong Kong. And both of them have only a few major parks and scattered open spaces compared to other districts in Kowloon. It is interesting to know whether and how urban green network could be constructed or expanded in these two districts.

Since this chapter applies a specific focus on AH developments and their contribution to urban green network, AH developments in these two districts need to be considered. It turns out that the scope and number of AH developments are quite different between these two districts. There are 16 PRH/TPS Estates and 6 HOS/PSPS Courts in Sham Shui Po District, while only 1 PRH/TPS Estates and 2 HOS/PSPS Courts in Yau Tsim Mong District. Considering that all these projects were developed by the Housing Authority in Hong Kong, and follow similar planning and design standards, they provide greeneries in similar ways. Under such circumstances, it is clear that AH developments in Sham Shui Po District would have contributed to the district's green network in a much more significant way than in Yau Tsim Mong District. Hence it is more meaningful to study the green network within AH developments in Sham Shui Po District.

Geo-information of all these AH projects and HOSs were incorporated into ArcMap. As shown in Fig. 9.4, distribution of AH projects is quite uneven within Sham Shui Po district. The majority of these projects are concentrated in the middle and eastern parts of the district. Most AH projects are close to large open spaces such as public open spaces and non-public greening. Under such circumstances, these open spaces together with greeneries within AH projects have the potential of creating a well-connected green network, which may connect greenbelts and country parks in the north and the coast in the south, thus enhance movement of energy and organisms cross the district and between matrixes.

Actually, in this district, most buildings that are surrounded by greeneries other than public open spaces belong to AH projects (Fig. 9.5). Such projects include 16 PRH/TPS Estates, covering about 70 ha with 133 blocks and 6 HOS/PSPS court sites covering about 5.9 ha with 24 blocks in total. These AH projects are all developed through 1970s to date (HKSAR Housing Authority, 2017a). Based on information from Statutory map (HKSAR Town Planning Board, n.d.) and Google map satellite

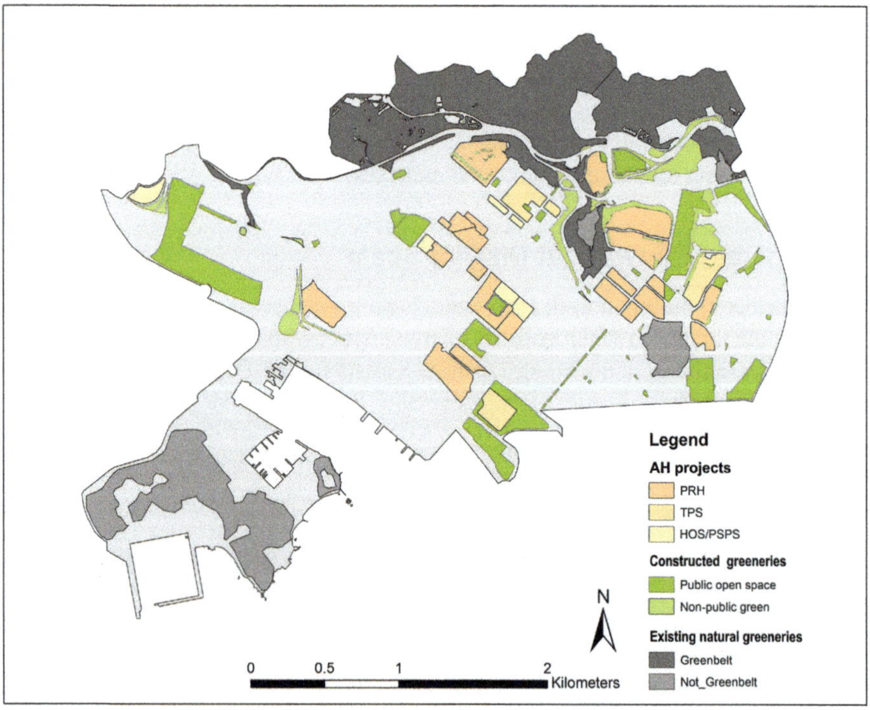

Figure 9.4. Distribution of AH projects and constructed greeneries outside AH projects in Sham Shui Po District (based on iB5000 digital topographic map produced by Survey and Mapping Office/Lands Department, Government of HKSAR).

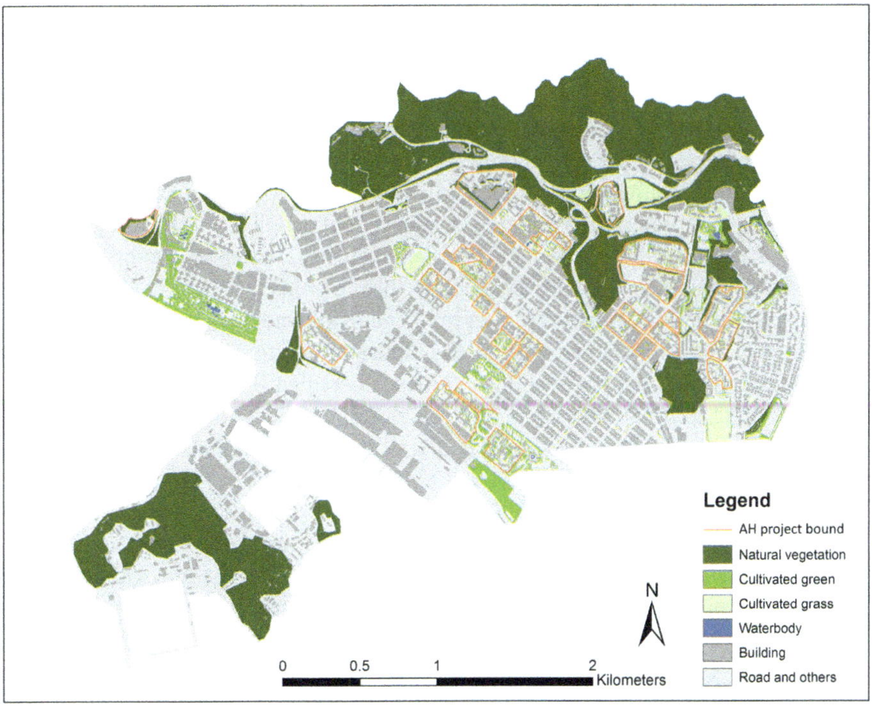

Figure 9.5. Provision of greeneries within AH projects in Sham Shui Po District (based on iB5000 digital topographic map produced by Survey and Mapping Office/Lands Department, Government of HKSAR).

images ("Google map," 2017), spatial distribution of these projects and ground level green space within them was incorporated into Geodatabase in ArcMap. It turns out that the total area of AH developments is 81.34 ha in this district, while total area of greenery in these developments is 9.30 ha (11.43% of 81.34 ha). Although this proportion is below the lower bound of 20% as specified in *Measures to foster a quality and sustainable built environment* (HKSAR Development Bureau, 2010), it is much higher than that provided by private development. In fact, most private development before 2011, especially industrial ones, do not provide any greening at all. The figure is also higher than the overall proportion of greenery in this district. According to the same Geodatabase with information of Public Open Space (POS), there are 66.09 ha POSs out of 733.56 ha constructed area in this district. Within these POSs, only 46.06 ha is greenery. In other words, only 7.5% ((46.06 + 9.30)/733.56*100%) of the constructed area is covered by greeneries in this district. Among these, around 1/6 is greeneries within AH projects. It is clear that compared to POSs, AH projects provide quite a large proportion of greenery in Sham Shui Po District. This greatly contributes to green network, supports the local ecosystem, and benefits their residents much more than other types of developments.

Green Network's Social Values

According to Biophilia hypothesis, human-beings have innate attachment to nature and other forms of life (Kellert and Wilson, 1995; Wilson, 1984). People have long sought natural scenic beauty and outdoor experience, which become stronger under the process of urbanization (McAndrew, 1993). Concerning nature or natural elements' contribution to urbanites, numerous studies have shown that natural elements and environments are preferred to urban ones, and benefit people in various ways (Berg et al., 2007; Kaplan, 1983; Kaplan et al., 1972). Such preferences tend to be stronger in highly dense cities and especially beneficial to health problems caused by high density, such as elevated blood pressure, heart rate, and skin conductance, as well as raised arousal level and stress (Lawrence, 2002). Spacious, green, and quiet environments are found desired by people in compact cities (Wiersinga, 1997).

According to urban landscape related studies, contact with nature is needed by people to escape the urban bustle, relax and possibly contemplate or enjoy aesthetic and amenity benefits in nature (Fuller et al., 2007; Gonzalez et al., 2010; Harper et al., 2005; Matsuoka and Kaplan, 2008). It was found by Ulrich (1979) that compared to American urban scenes lacking natural elements, nature scenes dominated by green vegetation helped stressed subjects feel significantly better. Although such differences would be less significant for unstressed people in normal arousal states, the effects of being exposed to nature could be much more positive than to urban ones (Ulrich, 1981). According to another study which involved 250 782 subjects registered with 104 general practices, higher percentage of green space in people's living environments improves their perceived general health significantly for all age groups (Maas et al., 2006).

Natural elements, especially well designed and maintained ones, can also improve actual and perceived safety within urban areas, especially nearby residences. It is found that if a building is located in a greener area, fewer property and violent crimes would be reported by residents. Meanwhile, residents living in such environments report lower level of fear, fewer incivilities, and less aggressive and violent behavior (Donovan and Prestemon, 2012; Kuo and Sullivan, 2001a; 2001b). People are even willing to pay more for urban green spaces or conserve nearby forests (Kaplan and Kaplan, 2003). Besides safety concerns, greeneries nearby homes also benefit children's cognition development. A study conducted by Wells (2000) found that children's level of cognitive functioning tended to improve after their homes were relocated to areas with better greeneries.

Contact with nature can also help people restore attention. For instance, in a workplace with windows, sunlight penetration showed significant direct effect on job satisfaction, intention to quit, and general well-being. A view of natural elements such as trees, vegetation, plants and foliage is also found to be helpful on these aspects (Leather et al., 1998). Besides, attention fatigued people would prefer natural environment much more than urban ones (Staats and Hartig, 2004).

As revealed by World Happiness Reports since 2012, the rank of Hong Kong has wavered around 70 among 150 studied countries and slightly dropped down from 2005–07 to 2014–16 (Helliwell et al., 2017). According to local surveys on happiness conducted by Chu Hai College of Higher Education and Lingnan University (Center for Public Policy Studies, 2016), overall happiness index for the people of Hong Kong was 67.6 (out of 100) in 2016, dropped significantly compared to 70.0 (out of 100) in 2015. When discussing relationship between working hours and happiness index, happiness declined significantly when working hours increased. Specifically, happiness index for those who worked 39 hours or less every week was 7.06, while it was 6.26 for those who worked 60 hours or more per week. This indicates a significant impact of pressure and stress from work on happiness. If there were more greeneries, they would be released somehow from such stress and might have had higher happiness index.

Strikingly, in contrast to the general declining trend, those with a monthly household income between HK$10,000 and HK$29,999 showed a slight rise on self-reported happiness than in 2015. Such low income families would probably fulfill the monthly income criteria to apply for AHs, which is capped by HK$11,250 for one person and HK$27,050 for four persons family, etc. (HKSAR Housing Authority, 2017b). Some of them may actually live in AHs. Considering that these residents are generally lower classes economically, they may not be able to enjoy good materialistic life quality. In such a situation, mental health and social support become more critical to their general well-being. Considering the favorable provision of open spaces with landscape designs in AH estates and well-being benefits of natural elements as discussed above, the improved happiness among low income subjects could partially be the result of such supportive living environment. Residents in AH estates would have more opportunities to contact natural elements and other people, and have more interpersonal interactions, when they move around in the estates. Although there are few systematic studies on well-being benefits of greeneries in AH estates to their residents in Hong Kong yet, such benefits could be supported by the huge amount of findings on similar topics in other places (Bonaiuto et al., 1999). Therefore, it could be generally safe to say that greeneries provided in AH projects helps promote the residents' perceived happiness.

Urban Green Network Under Two Scenarios at Class Level

From the above discussions, it can be confirmed that greeneries in AH projects are commonly considered as effective proxy for habitat, beneficial to ecosystem in urban areas, as well as residents' well-being (Park, 2011). According to mounting studies on this topic, class level analysis has received abundant attention as the variation of natural elements would directly and significantly influence structure of ecosystem (Groot et al., 2002; Lamarque et al., 2011; Wallace, 2007). To further discuss contribution of such green network in Sham Shui Po District, natural coverage on ground is studied at the class level. Specifically, based on status quo in the district, urban greeneries are categorized as natural vegetation, cultivated vegetation, cultivated grass, and waterbody (detailed descriptions in Table 9.1). Following this, each patch of greenery within this district, including those within AH projects, falls into one category.

In addition, in order to clarify contribution of greenery within AH projects to the district's ecosystem, an extreme scenario (scenario-1) was virtualized in ArcMap for comparison. In this scenario, all AH projects are replaced by buildings following the conventional private development model in rest areas of this district (Fig. 9.6).

Table 9.1. Natural coverage in Sham Shui Po District.

Category	Description
Natural vegetation	Original vegetation, mainly trees and shrubs on slopes
Cultivated vegetation	Trees or shrubs planted and maintained for landscape purposes
Cultivated grass	Grassland created and maintained for landscape or sport purposes, with few or no vegetation on top
Waterbody	Water features in greenery. Swimming pools excluded

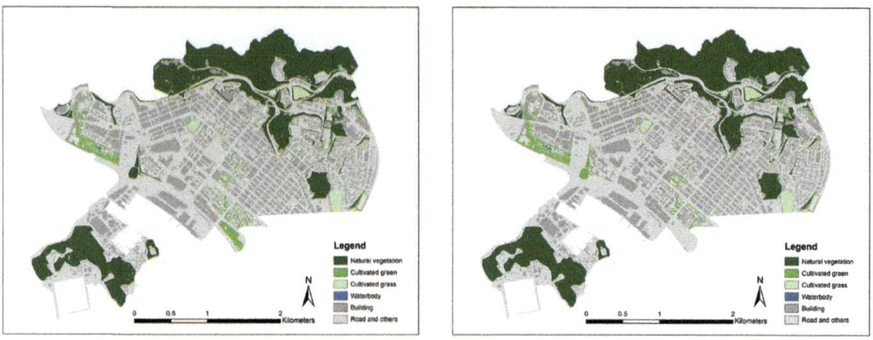

Status quo | Scenario-1

Figure 9.6. Greeneries in Sham Shui Po District: status quo and scenario-1 (based on iB5000 digital topographic map produced by Survey and Mapping Office/Lands Department, Government of HKSAR).

It represents the situation that no AH project was developed in this district, while all constructions were developed by private developers who provided minimum greenery in the projects as mentioned above. Specifically, most existing unpreserved natural vegetation and cultivated greeneries on AH sites are removed and the land is occupied by buildings and paved circulations. The effect of such a change on greeneries in this district was compared between the status quo and scenario-1.

Analysis was conducted with the software of FRAGSTATS (McGarigal and Marks, 1995). Specifically, the vector data (ArcMap shape files) were converted to raster data with pixel size of 3 m x 3 m. Land coverage of each category in Table 9.1 together with other land coverage that hinder ecological disperse were included, as shown in the image (Fig. 9.6). In FRAGSTATS, four metrics were employed to analyze the greening system in Sham Shui Po District: class area (the total area of a given category), patch density (the number of patches under each category per square kilometer), edge density (the total length of all edge segments per hectare for the concerned category) and mean patch size (the average area of the corresponding patch category). Results of FRAGSTATS analyses for both actual land cover and that in scenario-1 are shown in Table 9.2.

In Table 9.2, it can be seen that in Sham Shui Po District, natural vegetation takes up the largest proportion of greenery. This mainly consists of the greenbelt in the north and vegetation on small hills in the southwest, which are largely conserved from urban development. Besides, a small portion of natural vegetation is vegetated slopes. Such slopes are distributed mainly along the mountain foot in the northern

Table 9.2. Results of FRAGSTATS analyses.

Metrics	Actual land cover	Scenario-1 land cover
Class area (hectare)	Natural vegetation: 226.8774 Cultivated vegetation: 35.2296 Cultivated grass: 8.1333 Waterbody: 0.7236	Natural vegetation: 224.5761 Cultivated vegetation: 28.2339 Cultivated grass: 8.1288 Waterbody: 0.6498
Patch density (number of patches per km^2)	Natural vegetation: 10.0867 Cultivated vegetation: 134.5603 Cultivated grass: 1.395 Waterbody: 1.7169	Natural vegetation: 8.9946 Cultivated vegetation: 61.5705 Cultivated grass: 1.4991 Waterbody: 1.392
Edge density (meters per hectare)	Natural vegetation: 77.575 Cultivated vegetation: 133.263 Cultivated grass: 5.2118 Waterbody: 2.2019	Natural vegetation: 73.5697 Cultivated vegetation: 95.703 Cultivated grass: 5.2876 Waterbody: 1.9146
Mean patch size (hectare)	Natural vegetation: 2.4136 Cultivated vegetation: 0.0281 Cultivated grass: 0.6256 Waterbody: 0.0452	Natural vegetation: 2.6735 Cultivated vegetation: 0.0491 Cultivated grass: 0.5806 Waterbody: 0.05

part of the district. Vegetation on such slopes is not preserved by Hong Kong Planning Standards and Guidelines (HKSAR Planning Department, 2016). Therefore, they are removed from AH sites and covered by buildings in scenario-1. This results in a 2.3 ha reduction of natural vegetation in this district.

The second largest portion of greeneries in this district is cultivated greening, mainly referring to greeneries in POSs and developments, especially in AH projects, as well as some roadside greening. Replacing AH projects with conventional private ones removes existing greenery within AH sites, and reduces 1/5 (19.86%) cultivated vegetation in this district. On the other hand, scenario-1 does not affect cultivated grass and waterbody much (Table 9.2). In other words, the "lost" greeneries mainly contain trees and shrubs. Such elements play much more important ecological roles in urban area than grass, as they provide another layer and space in canopies which would be much less disturbed by human beings, especially motor vehicles. Also, canopies of trees could connect with each other above ground level and provide connected corridors or even habitats for animals like birds. These can effectively promote disperse and movements of organisms, as well as ecological diversity within the compact urban environment (Zipperer et al., 1997). If they are removed, such movements and diversity would be affected greatly.

Besides, there are significant decreases on patch density and edge density of cultivated vegetation, from status quo to scenario-1. By removing 20% of cultivated vegetation in this district, cultivated vegetation patch density is reduced by 54.2%, while the patch edge is reduced by 28.2%, although the mean patch size is increased greatly by 74.7%. These indicate that cultivated vegetation in AH projects are mostly small patches, compared to those outside AH sites in this district. Although these patches are small in size, they play an important role as stepping stones and corridors in the densely developed urban area, providing functional connections between matrixes and larger habitats scattered around the developed area. This is very critical for urban species, especially terraneous ones that have much more difficulties in crossing densely constructed structures and buildings if not relying on urban greeneries, as most spaces outside greeneries are intensively constructed due to high population density. For some species which are quite adaptive and tolerant to urban environment and disturbances, or those requiring smaller habitats, they would even obtain enough living resources within such small patches. Therefore, it is critical to have these patches within AH sites in the non-matrix area to support genic flow and biodiversity.

On the other hand, these green elements within AH projects also play important roles in urbanites' daily life. First of all, green or natural elements are commonly preferred to urban or constructed elements by people with different background as discussed above. Well-being benefits of natural elements have been proven with mounting evidence from different aspects. Beautiful and different leaves and flowers bring people various aesthetic and sensory satisfaction throughout the year. They also provide different kinds of recreational spaces inside the city area. These vibrate and soften the city from all kinds of hard constructions, and also help release people from stress and restore attention in daily lives. In addition, trees, especially those with spread canopies are very critical for a subtropical city like Hong Kong, as they provide abundant shades to protect people from severe sunshine and help cool the ground down during most of the year, and reduce urban heat island effect. According to previous studies, temperature in vegetated urban area could be reduced by 1–4.7°C and even as much as more than 7°C in summer (Schmidtc, 2006; Onishi et al. 2010). Furthermore, they help improve air quality by reducing and offsetting air pollution (e.g., NO_2, PM_{10}, O_3, CO, SO_2, etc.), and greenhouse gas emissions (Pugh et al., 2012; Zhao et al., 2010). Such contribution of green network could be more effective if implemented at broader spatial scales (Baro et al., 2014).

Summary

Through the above discussion, especially virtual changes on green network from the status quo to scenario-1 in Sham Shui Po District, it is clear that greeneries provided by AH projects play an important role within compact urban area in Hong Kong. They are not only critical parts of the urban ecosystem, but also very supportive to citizens' daily life. In other words, they contribute in many ways to maintain balance between ecosystem and human society, as well as within each of the two systems. In addition, the strategy of reserving a certain proportion of land for ground level greeneries in AH projects is proven to be effective and contributive, compared to maximum constructions in private developments. Since such AH projects are usually mass development of large pieces of lands, they have the opportunity to plan ahead distribution of such green network within their sites and corresponding to greeneries outside the sites. Meanwhile, as conservation and development of other major greeneries within the city are also taken care of by government departments, it would be much more convenient for the Hong Kong Housing Authority to collaborate with them in order to plan a well-structured green network at city scale. Considering all these advantages and contribution of green network construction by AH developments in Hong Kong, such development mode could be a good reference for other compact cities.

Although this chapter only discusses green network at city, district and class levels, some components of green network also exist at the building level, especially with an increasing trend of establishing green walls and green roofs in Hong Kong. These green elements, although usually in small scales, could also contribute to green network in cities. Discussions on them would also be inspiring for future development of compact cities.

Acknowledgements

Base digital topographic map (iB5000) for this chapter was funded by Research on Sustainable Living: Phase Two of the Capacity Building Plan (Ref. no.: UGC/IDS25/16). The author would also thank Miss Cheung Wai Sum Suzanna and Miss Tang Lai Yi for map preparation.

References

Barker, G. A. 1997. A Framework for the Future: Green Networks with Multiple Uses in and Around Towns and cities. Retrieved from London: http://publications.naturalengland.org.uk/publication/77041.

Baro, F., L. Chaparro, E. Gomez-Baggethun, J. Langemeyer, D. J. Nowak and J.Terradas. 2014. Contribution of ecosystem services to air quality and climate change mitigation policies: The case of urban forests in Barcelona, Spain. AMBIO 43: 466–479. doi:10.1007/s13280-014-0507-x.

Berg, A. E. V. D., T. Hartig and H. Staats. 2007. Preference for nature in urbanized societies: Stress, restoration, and the pursuit of sustainability. Journal of Social Issues 63(1): 79–96.

Bonaiuto, M., A. Aiello, M. Perugini, M. Bonnes and A. P. Ercolani. 1999. Multidimensional perception of residential environment quality and neighborhood attachment in the urban environment. Journal of Environmental Psychology 19: 331–352.

Center for Public Policy Studies. 2016. Happiness study in Hong Kong: Happiness index 2016. Retrieved 22 September 2017, from Lingnan University https://www.ln.edu.hk/cpps/08_highlight/08-happiness.html

Central Intelligence Agency. 2017. The world factbook. from Central Intelligence Agency https://www.cia.gov/library/publications/the-world-factbook/geos/xx.html.

Donovan, G. H. and J. P. Prestemon. 2012. The effect of trees on crime in Portland, Oregon. Environment and Behavior 44(1): 3–30. doi:10.1177/0013916510383238.

Ersoy, E. 2016. Landscape ecology practices in planning: Landscape connectivity and urban networks. In M. Ergen (ed.). Sustainable Urbanization: InTech. Retrieved from http://www.intechopen.com/books/sustainable-urbanization.

Fabos, G. J. 2004. Greenway planning in the United States: its origins and recent case studies. Landscape and Urban Planning 68(2): 321–342.

Forest Research. 2011. Green Networks and People: A Review of Research and Practice in the Analysis and Planning of Multi-Functional Green Networks. Retrieved from http://www.snh.org.uk/pdfs/publications/commissioned_reports/490.pdf.

Forman, R. T. T. and M. Godron. 1986. Landscape Ecology. New York Wiley.

Fuller, R. A., K. N. Irvine, R. Devine-Wright and K. J. Gaston. 2007. Psychological benefits of green space increase with biodiversity. Biology Letters 3: 390–394.

Gonzalez, M., S. Ladet, M. Deconchat, A. Cabanettes, D. Alard and G. Balent. 2010. Relative contribution of edge and interior zones to patch size effect on species richness: An example for woody plant. Forest Ecology and Management 259: 266–274.

Google map. 2017. Retrieved 6 October, 2017, from Google https://www.google.com.hk/maps?hl=en&tab=wl.

Groot, R. S. D., M. A. Wilson and R. M. J. Boumans. 2002. A typology for the classification, description and valuation of ecosystem functions, goods and services. Ecological Economics 41: 393–408.

Harper, K. A., S. E. Macdonald, P. J. Burton, J. Q. Chen, K. D. Brosofske and S. C. Saunders. 2005. Edge influence on forest structure and composition in fragmented landscapes. Conservation Biology 19: 768–782.

Helliwell, J., R. Layard and J. Sachs. 2017. World happiness report 2017. Retrieved from New York.

HKSAR Building Department. 2011. New practice notes to foster a quality and sustainable built environment. Retrieved from http://www.bd.gov.hk/english/documents/news/20110131ae.htm.

HKSAR Census and Statistics Department. 2017. Population. Retrieved 30 September, 2017, from Government of Hong Kong SAR https://www.censtatd.gov.hk/hkstat/sub/so20.jsp.

HKSAR Council for Sustainable Development. 2010. Report on the public engagement process on building design to foster a quality and sustainable built environment. Hong Kong SAR: Hong Kong SAR Government.

HKSAR Development Bureau. 2010. Measures to foster a quality and sustainable built environment. Hong Kong: HKSAR Government Retrieved from http://www.devb.gov.hk/filemanager/article/en/upload/6209/LegCo%20Brief%20GFA%20eng.pdf.

HKSAR Government. 2017. Policy Address outlines comprehensive plans to boost housing and land supply. Retrieved from http://www.info.gov.hk/gia/general/201701/18/P2017011800512.htm.

HKSAR Government. 2017. Hong Kong—the facts. Retrieved from https://www.gov.hk/en/about/abouthk/facts.htm.

HKSAR Housing Authority. 2017a. Estate locator. Retrieved 22 September, 2017, from Hong Kong Housing Authority http://www.housingauthority.gov.hk/en/global-elements/estate-locator/index.html.

HKSAR Housing Authority. 2017b. Income and asset limits. Retrieved from https://www.housingauthority.gov.hk/en/flat-application/income-and-asset-limits/index.html.

HKSAR Information Services. 2016. Hong Kong: The Facts (Housing). Retrieved from http://www.housingauthority.gov.hk/en/about-us/publications-and-statistics/hong-kong-the-facts-housing/index.html.

HKSAR Leisure and Cultural Services Department. 2017. Facilities & Venues Search. from Leisure and Cultural Services Department, the Government of HKSAR http://www.lcsd.gov.hk/en/facilities/facilitiessearch/phoneaddress.php?cat=MPARK&dist=all&keyword.

HKSAR Observatory. 2017. Extract of annual data (1884-1939 & 1947-1960). Retrieved 30 September, 2017, from Government of Hong Kong SAR http://www.hko.gov.hk/cis/yearlyExtract_e.htm.

HKSAR Planning Department. 2016. Hong Kong planning standards and guidelines. Hong Kong SAR: Planning Department, Government of HKSAR.

HKSAR Planning Department. 2017. Land utilization in Hong Kong 2016. Retrieved from http://www.pland.gov.hk/pland_en/info_serv/statistic/landu.html.

HKSAR Town Planning Board. (n.d.). Statutory Planning Portal 2. from Town Planning Board, HKSAR.

Jongman, R. H. G. 2004. The Context and Concept of Ecological Networks. pp. 7–33. *In*: Jongman, R. H. G. and G. Pungetti (eds.). Ecological Networks and Greenways Concept, Design and implementation. Cambridge, UK: Cambridge University Press.

Jongman, R. H. G. and G. Pungetti (eds.). 2004. Ecological Networks and Greenways Concept, Design and Implementation. Cambridge, UK: Cambridge University Press.

Kaplan, R. 1983. The role of nature in the urban context. pp. 127–162. *In*: Altman, I. and J. F. Wohlwill (eds.). Behavior and the Natural Environment. New York Plenum Press.

Kaplan, S., R. Kaplan and J. S. Wendt. 1972. Rated preference and complexity for natural and urban visual material. Perception & Psychophysics 12(4): 354–356. doi:https://doi.org/10.3758/BF03207221.

Kellert, S. R. and E. O. Wilson (eds.). 1995. The Biophilia Hypothesis (New edition ed.). Washington: Island Press.

Kaplan, S. and R. Kaplan. 2003. Health, supportive environments, and the reasonable person model. American Journal of Public Health 93(9): 1484–1489.

Külvik, M., M. Suškevičs and K. Kreisman. 2008. Current Status of the Practical Implementation of Ecological Networks in Estonia. Retrieved from http://www.ecologicalnetworks.eu/documents/publications/ken/EstoniaKENWP2.pdf.

Kuo, F. E. and W. C. Sullivan. 2001a. Aggression and Violence in the Inner City: Effects of Environment via Mental Fatigue Environment and Behavior 33(July): 543–571.

Kuo, F. E. and W. C. Sullivan. 2001b. Environment and crime in the inner city: does vegetation reduce crime? Environment and Behavior 33(May): 343–367.

Lamarque, P., F. Quetier and S. Lavorel. 2011. The diversity of the ecosystem services concept and its implications for their assessment and management. Comptes Rendus Biologies 334: 441–449.

Lawrence, R. J. 2002. Healthy residential environments. pp. 394–412. *In*: Bechtel, R. B. and A. Churchman (eds.). Handbook of Environmental Psychology. New York: John Wiley & Sons, Inc.

Leather, P., M., D. Pyrgas, Beale and C. Lawrence. 1998. Windows in the workplace: sunlight, view, and occupational stress. Environment and Behavior 30(November): 739–762.

Maas, J., R. A. Verheij, P. P. Groenewegen, S. D. Vries and P. Spreeuwenberg. 2006. Green space, urbanity, and health: how strong is the relation? J. Epidemiol Community Health 60: 587–592. doi:10.1136/jech.2005.043125.

Makhzoumi, J. and G. Pungetti. 2005. Ecological Landscape Design and Planning: The Mediterranean Context.

Matsuoka, R. H. and R. Kaplan. 2008. People needs in the urban landscape: Analysis of landscape and urban planning contributions. Landscape and Urban Planning 84(1): 7–19. doi:10.1016/j.landurbplan.2007.09.009.

McAndrew, F. T. 1993. Environmental psychology. Pacific Grove, Calif.: Brooks/Cole Publishing Company.

McGarigal, K. and B. J. Marks. 1995. FRAGSTATS: Spatial pattern analysis program for quantifying landscape structure. Retrieved from Portland.

Onishi, A., X. Cao, T. Ito, F. Shi and H. Imura. 2010. Evaluating the potential for urban heat-island mitigation by greening parking lots. Urban Forestry & Urban Greening 9(4): 323–332.

Park, S. H. 2011. Ecological connectivity assessment and urban dimensions: A case of Phoenix metropolitan landscape (Ph.D.). Arizona State University.

Pugh, T. A. M., A. R. MacKenzie, J. D. Whyatt and C. N. Hewitt. 2012. Effectiveness of green infrastructure for improvement of air quality in urban street canyons. Environ. Sci. Technol. 46: 7692–7699. doi:dx.doi.org/10.1021/es300826w.

Schmidtc, M. 2006. The contribution of rainwater harvesting against global warming. London, UK: Technische Universität Berlin, IWA Publishing.

Staats, H. and T. Hartig. 2004. Alone or with a friend: A social context for psychological restoration and environmental preferences. Journal of Environmental Psychology 24(2): 199–211.

Tian, Y., C. Y. Jim, Y. Tao, and T. Shi. 2011. Landscape ecological assessment of green space fragmentation in Hong Kong. Urban Forestry & Urban Greening 10: 79–86.

Tian, Y., C. Y. Jim and H. Wang. 2014. Assessing the landscape and ecological quality of urban green spaces in compact city. Landscape and Urban Planning 121: 97–108.

Tzoulas, K. and P. James. 2010. Peoples' use of, and concerns about, green space networks: A case study of Birchwood, Warrington New Town, UK. Urban Forestry and Urban Greening 9(2): 121–128.

Ulrich, R. S. 1979. Visual landscapes and psychological well-being. Landscape Research 4(1): 17–23.

Ulrich, R. S. 1981. Natural versus urban scenes-some psychophysiological effects. Environment and Behavior 13(5): 523–556.

United Nations. 2012. The future we want. Retrieved from.

United Nations. 2015. Transforming our world: The 2030 agenda for sustainable development. Retrieved from https://sustainabledevelopment.un.org/post2015/transformingourworld/publication.

United Nations. 2017a. New urban agenda. Retrieved from http://habitat3.org/the-new-urban-agenda/.

United Nations. 2017b. UN Habitat III Conference. Retrieved from http://www.un.org/sustainabledevelopment/habitat3/.

Wallace, K. J. 2007. Classification of ecosystem services: Problems and solutions. Biological Conservation, 139(3): 235–246. doi:https://doi.org/10.1016/j.biocon.2007.07.015.

Wells, N. M. 2000. At home with nature: Effects of "greenness" on children's cognitive functioning Environment and Behavior 32(November): 775–795.

Wiersinga, W. 1997. Compensation as a strategy for improving environmental quality in compact cities. Amsterdam: Bureau SME.

Wilson, E. O. 1984. Biophilia. Cambridge: Harvard University Press.

Xie, Y. 2013. Habitat loss and fragmentation under urbanization: the spatiotemporal dynamics of causes, processes andconsequences at landscape level. (Ph.D.). University of Hong Kong, Hong Kong SAR. Retrieved from http://hdl.handle.net/10722/191205.

Yeung, Y. M. (ed.). 2003. Fifty Years of Public Housing in Hong Kong: A Golden Jubilee Review and Appraisal: Chinese University Press for the Hong Kong Housing Authority, Hong Kong Institute of Asia-Pacific Studies.

Zhao, M., Z. Kong, F. J. Escobedo and J. Gao. 2010. Impacts of urban forests on offsetting carbon emissions from industrial energy use in Hangzhou, China. Journal of Environmental Management 91: 807–813.

Zipperer, W. C., S. M. Sisinni, R. V. Pouyat and T. W. Foresman. 1997. Urban tree cover: An ecological perspective. Urban Ecosystems 1: 229–246.

CHAPTER 10

FlexZhouse

Innovation in Flexible Housing for Modern Affordable Housing Solution in Malaysia

Mohd Zairul

Introduction

The flexible house was first coined in the Netherlands in the 1980s due to the changing of social household's structure and lifestyles. A study on existing mass housing development by (Omar et al., 2012) found that house buyers prefer housing customization and want an alternative to current mass housing developments in terms of design and space flexibility. In most emerging countries, individuation and customization in mass housing have become popular due to the increased changes in lifestyle (Hentschke et al., 2014). In most cases, certain spaces in a house may become obsolete over time due to the changing needs of its occupants.

Back in 2008, Kendall mentioned that in this modern era, we need buildings that are changeable and adaptable because we cannot afford to destroy and rebuild the entire building to accommodate the new requirements. Today, this new demand constitutes a new synergy in the housing architecture. Kendall later mentioned that customers nowadays expect long-lasting investment that results in a balanced change and stability of the building stocks (Kendall, 2008). Therefore, a substantial part of the flexZhouse components is designed purposely to cater the changing users and needs as well as to minimize the construction wastes and dependency on raw construction materials.

At the moment, house buyers are no longer interested in standard designs produced by the housing developers (Noguchi, 2003; Yashiro, 2014). Personalization has always been associated with extra cost and can potentially increase the housing price (Barlow and Childerhouse, 2003; Gann, 1996). House buyers often find it difficult to physically adapt a shelter to their requirements and often resort to physically modify their existing

Universiti Putra Malaysia.

dwellings, which in turn, causes so much waste and environmental burdens (Wong and Li, 2010). Understandably, individual spaces in the house may become obsolete at times due to the changes in users' needs.

The flexZhouse concept adapts the idea of infill and support by Habraken (1972) and further supported by Schneider and Till (2006). The concept has become popular recently through the idea of open architecture by that identifies the multiple elements that are essential in supporting the idea of sustainable or green building. Both 'support' and 'infill' are defining the boundary it disseminates. The novelty of the flexZhouse for the housing industry is the creation of the longest possible lifespan with the highest potential use of value and at the same time, reduce the dependence on material resources and energy as much as possible (Zairul, 2015; Zairul, 2017).

The flexZhouse model extends the meaning of 'support' and 'infill' by Habraken (1972). 'Support' means a base building that acts as the structure in which the services of the building will be installed. In terms of the structure, flexZhouse recommends the use of Reinforced Concrete (RC) or steel frame for a lightweight construction. The 'infill' comprises of flexible units where each unit means a 'box' or a 'modular unit' (also called as 'a box frame' in Industrial Building System terminology). Each unit is produced using a steel frame structure and is fabricated offsite. The shape of the unit is approximately that of a container but not necessarily the size of an ISO container. However, the sizing of its base is derived from the ISO container to ease the logistics. The infill will be available in different modular sizes. The process will start with customers selecting additional features or additional requirements from a menu to add to the basic unit or part. At this stage, the customer will be able to customize their design as allowed by the manufacturers. The process will then go into production, installations of units, site work and finally hand-over. The advantage of the module is that it can be 'added' or 'removed' and even relocated to other places.

This chapter discusses the conceptual framework of flexZhouse based on a thesis entitled flexZhouse: A New Business Model for Affordable Housing in Malaysia. The conceptual framework was developed based on the results of a focus group method where young starters in Malaysia were asked to express their needs and aspiration in future housing. There were four focus groups with six to seven participants in each group. In order to gauge the potential house buyers' needs and aspiration for flexZhouse, the following question was asked: What type of products/services can be offered to you (young starters)? The feedback to this question contributed to the value propositions that include the choices and options for the design; the different needs of different groups of customers; and the new strategy for better customer services. This chapter aims to present this feedback that became the basis for the formulation of the flexZhouse conceptual idea as discussed next.

The Characteristic of Prefab Housing

Further, the multiple stages and different characteristics in the construction industry have triggered a lengthy discussion and debate in the industry whether the sequential and fragmented activities in the current housing industry or in construction, in general, might be replicated in other industries, or vice versa (Love et al., 2004). It is necessary to integrate the physical differences, the actors involved and its operation. Barlow and Ozaki (2003) argued that the housing industry differs from the manufacturing industry in several respects, such as the size of the production, the higher degree of production complexity and higher operational cost. However, the important lesson that the housing industry can learn from the manufacturing industry is its effective and efficient forms of integration and the repetitive actions and management of processes leading to the production with respect to time, cost and quality (Ali et al., 2009). Nevertheless, one of the main reasons for the housing industry to adopt the strategy used in manufacturing is to reduce the carbon footprint of the conventional construction and promote sustainability. The new approach needs a new management decision in several fields.

Conceptual Framework of flexZhouse

The proposed flexZhouse arose from the idea of 'open building' with 'flexible infill' connotes the flexibility of the unit or infill to fill the spaces between two load bearing columns within the unit. In this module, the notion of empowerment is reflected in the choices of the infill and the spaces inside the unit. The unit will allow users to take control of their own design during the design stage. This is a way to support the idea of involving the user at the beginning of the design development chain and promote interaction, participate and co-evolve with the layout (Till et al., 2006). The flexibility of the infill is the capability of being produced systematically in a factory and designed for ease of assembly and disassembly.

The conceptual idea started with a design brief developed in a design workshop and a focus group discussion attended by a group of architects in Malaysia. The results of the focus group will be a conceptual idea that leads to the formulation of the conceptual framework. The design brief called for an exploration of structure and infill concepts. The structure or the support could be made from Reinforced Concrete (RC) materials or steel structure of columns and beams. The building's services will run through the structure and here, infill is defined as a modular box structure of a standard size, which can be added on and dismantled. Logistically speaking, although the size was derived from the size of a container, the design brief allowed the designers to use any sizes that are habitable and movable on the road (see Fig. 10.1).

The flexZhouse model requires the design brief to include a unit that can be easily relocated. The box or the components will be equipped with bedrooms; kitchen equipment complete with piping; bathroom with all necessary utilities; and choices of bedroom designs. The first idea to support the customizable options is to design a module/unit that is adaptable to the current and future needs. The size of the module is similar in size to that of a normal container house with a cubic grid of 3.5 metres by 8.0 metres (approximately 28 m^2). This module is further arranged into several types (Fig. 10.2), which include a studio unit (28 m^2); a single unit with one bedroom (28 m^2); and two extended units, namely a double unit (56 m^2) and a duplex unit (56 m^2) with subdivided floor with a staircase as part of the circulation area. The interior of the unit is further divided into several basic parts, which includes a bedroom, a bathroom, a living area as well as a choice of a kitchen type and a balcony.

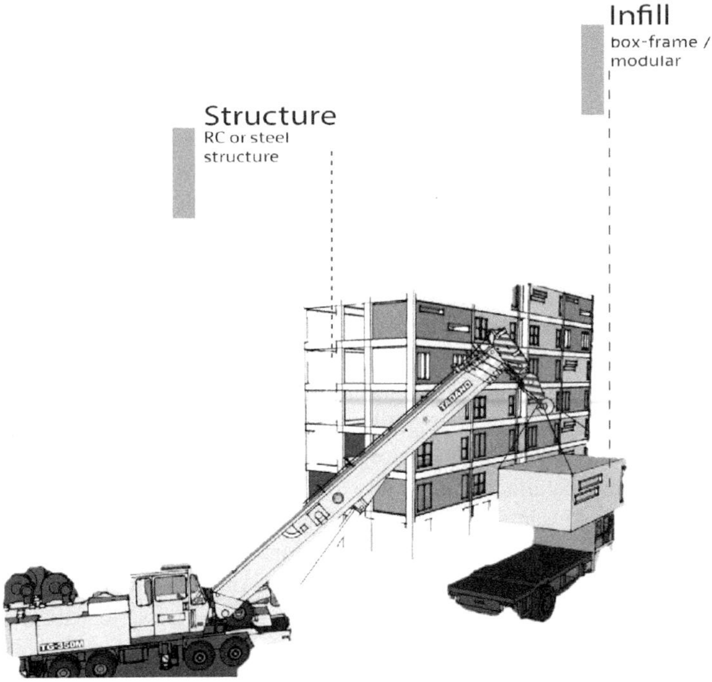

Figure 10.1. The idea of flexZhouse using 'structure' and 'infill' component. Source: Zairul (2017).

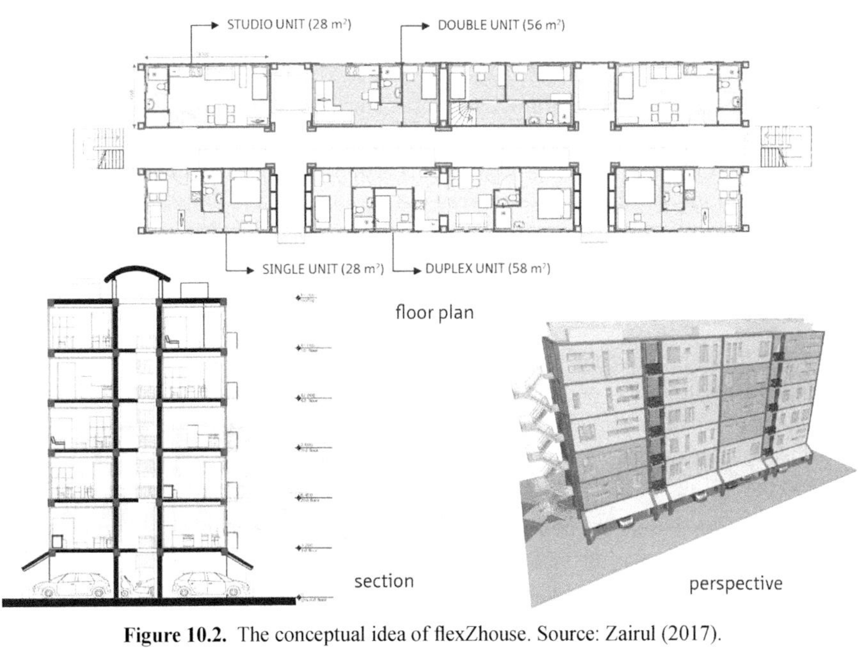

Figure 10.2. The conceptual idea of flexZhouse. Source: Zairul (2017).

Although the proposed model benefits from container technology, the modular boxes may not necessarily use shipping containers. Each module will be developed and manufactured according to the client's requirements in terms of the number of rooms, internal fit-out and external façade (based on standardized dimensions and structure). In this system, the user will be able to change the design after an agreed period of time. In terms of the sequence, the customer will place the order at the factory after choosing from the available designs and customizing the internal and external features. The concept of flexZhouse suggests for leasing to be introduced at the beginning of the contract. The leasing will then be determined after both parties have agreed to the terms and conditions. At an agreed time, the unit will be transported to the site and the user moves in. After a certain time (as per conditions agreed in the contract), the user will be able to change both the inside and the outside of the unit. The new unit will be transported to the site, replacing the old one, which will be sent back to the manufacturer for upgrades or downgrades.

For this concept to work, the factory should be located to support the transport of the module from manufacturing facility to the allocated site. The initial idea shows the notion of a very basic RC structure with the support of RC slab with metal rail to assist the installation of the units into the 'slots'. Traditionally, people move to a new place when their personal needs change over time, but with this concept, the idea of growing and shrinking with the same spaces is possible with the support of the manufacturing techniques. The idea of this flexible dwelling is that it can be transported using trailers, and assembled within one day. The flexibility of the unit allows the user to change the interior and exterior façade of the house according to their own preferences. It is argued that it is unlikely that the flexibility of the house would depress the housing market and limit the continuing sales of housing. This is because leasing the units and allowing changes to be made to them will prolong the lifecycle of the units and provide continuous investment for the housing developers. In this new concept, the housing developers are expected to enjoy increased revenue through reducing their investment in raw materials. Through this new concept, the principle of circular economy is supposed to repress the wastages as the units are recyclable and can be remanufactured for future use. The idea of leasing at the beginning is to help the young starters to save for their housing deposits at the beginning of their career.

The seven steps of the conceptual flexZhouse process are described below:

Step 1: The company constructs the basic structure of the flexZhouse. The structure is made from either RC or steel and the relevant services (mechanical and electrical conduits, water supply and sewerage piping, and risers for the services and amenities) are installed. The structure is in compliance with the modular sizes for the infill components (Fig. 10.3);

Step 2: The customers are helped to choose the type and size of the unit they want as well as the lease that is affordable for them (Fig. 10.4);

Step 3 & 4: The customers select a unit from the available designs; add the components, furniture and electrical fittings; change the layout of the bathroom, kitchen style, façade, type of windows and doors; and add a balcony as an accessory. A Virtual Reality (VR) simulation of the unit is provided for easy visualization of the unit (Figs. 10.5 and 10.6);

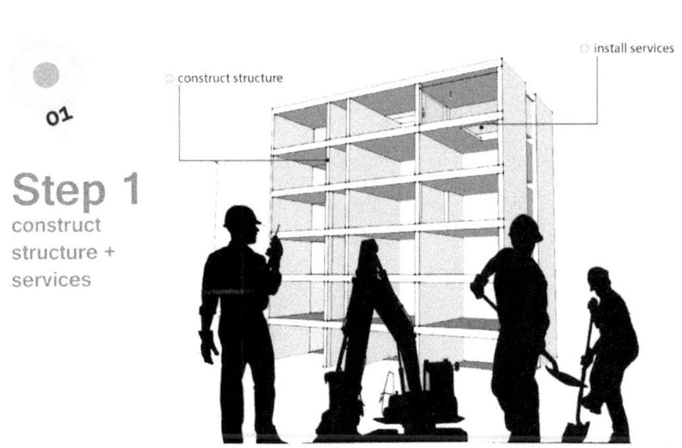

Figure 10.3. Step 1—The construction of the structure and the installment of services. Source: Zairul (2017).

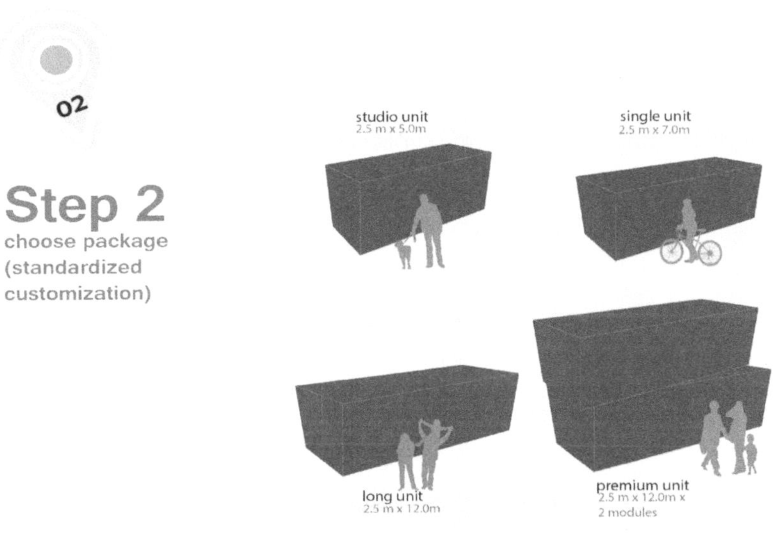

Figure 10.4. Step 2—Customer to choose package available. Source: Zairul (2017).

Step 5: The customers are given slots on a first come, first served basis. The level and position of the unit are subjected to the type and sizes of the units as well as the availability of the space (Fig. 10.7);

Step 6: The customers finalize the design with the consultant; agree on the final cost of the products (leasing); sign the agreement; and decide on the moving and delivery day (Fig. 10.8);

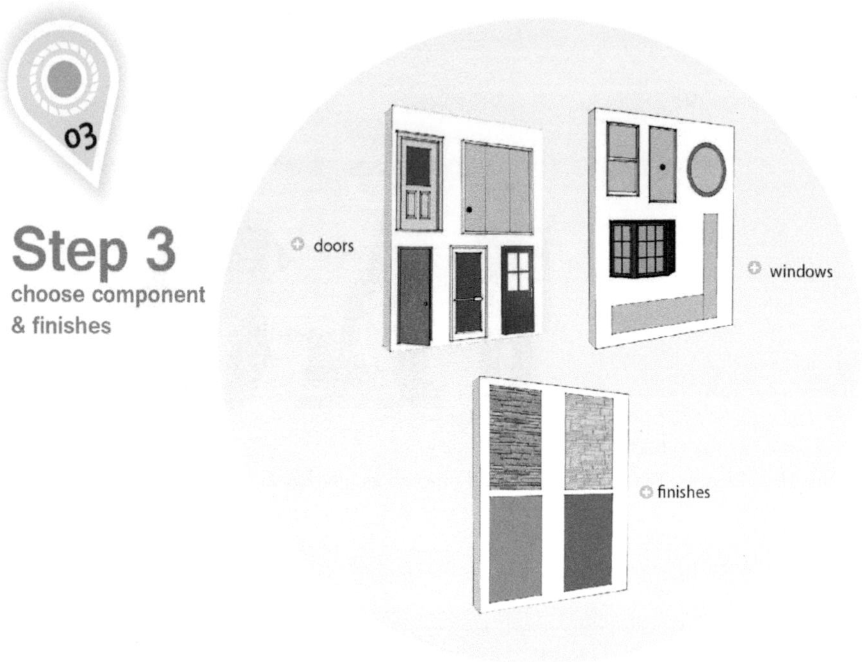

Figure 10.5. Step 3—Customer to select component and finishes. Source: Zairul (2017).

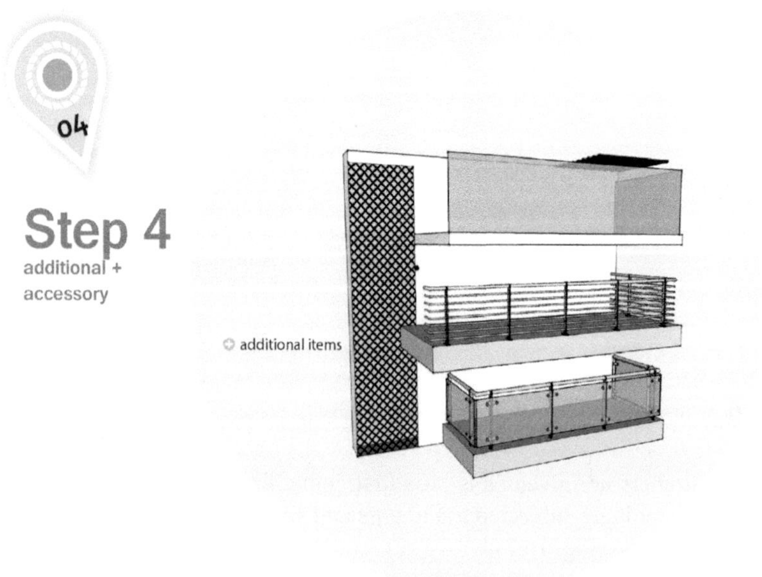

Figure 10.6. Step 4—Customer to select additional accessory. Source: Zairul (2017).

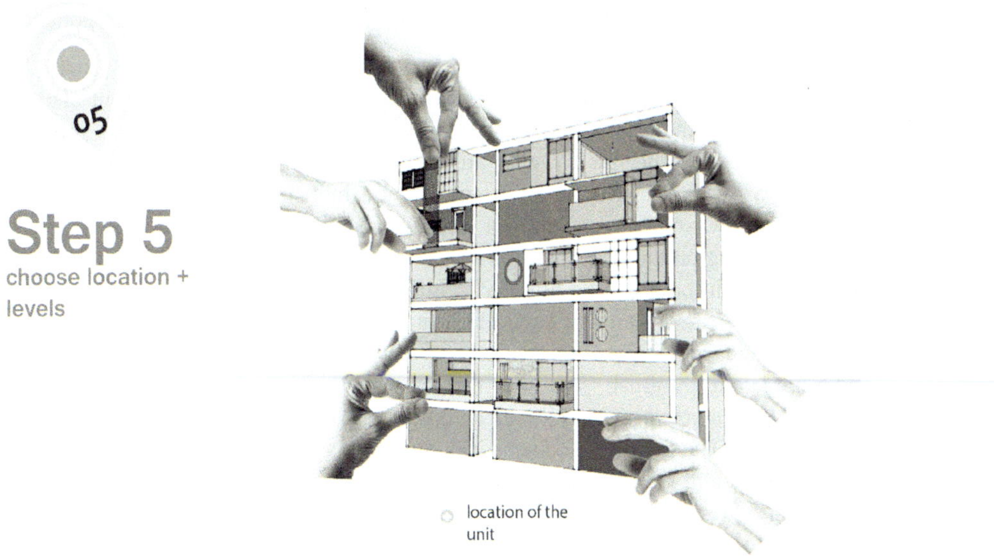

Figure 10.7. Step 5—Customer to choose a location. Source: Zairul (2017).

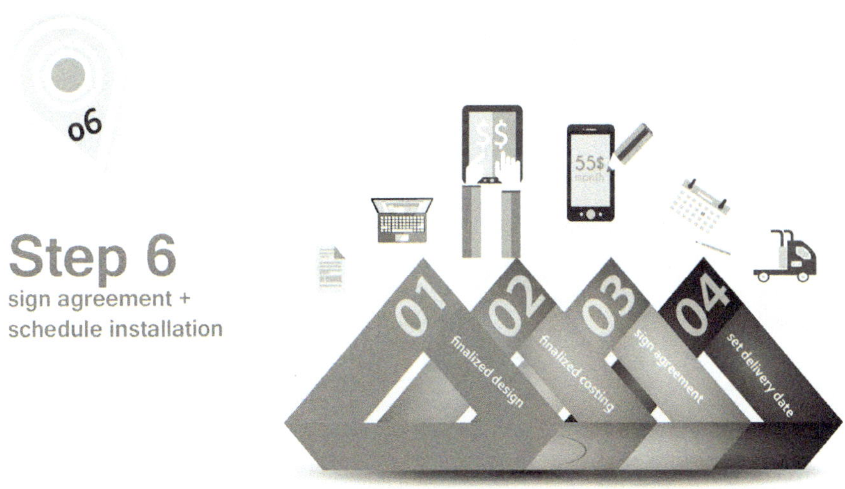

Figure 10.8. Step 6—Customer to sign the agreement and scheduled an installation time. Source: Zairul (2017).

Step 7: After a minimum stay of 18 months, the customers can change the components (add/reduce them) and agree to the relocation process to allow another component in the case of unit reshuffling (Fig. 10.9).

The unit's support system will be designed not only for comfort but also for efficiency. Each module will feature a fully appointed living room with modular wraparound couch and flat TV with an ample built-in cabinet. The kitchen will be equipped with high-tech appliances, including a built-in refrigerator, an induction stove, a microwave oven as well as a washing machine and a dryer for an upgraded unit. Depending on the size, the dining area will be equipped with a restaurant-style booth. The bedrooms will be fitted with fully furnished beds with an ample storage underneath. The bedroom will have a built-in wardrobe and be upgradable, according to the package. The bathroom for each unit will have a large shower, a full vanity unit and a porcelain low-flush toilet. For the upgraded version, the user might have the option to have a private sundeck with an outdoor shower (if the location permits such arrangement). The facilities such as air-conditioner and water boiler will be optional for customers and could be added as part of the chosen package. The lighting will be built into the walls and the ceiling to reduce maintenance needs. It will be able to last for up to 100,000 hours of continuous use. The interior and exterior of the house will be customizable in terms of materials, colour combinations and specifications, according to the financial capability of the customers. Figure 10.10 illustrates the flexZhouse concept along its life cycle.

For example, Ronald, 24 years old, single, profession as an engineer, chooses package 1. During the design process, he chooses his own furniture from the options offered to him by the manufacturer. He changes the exterior façade of the unit and the

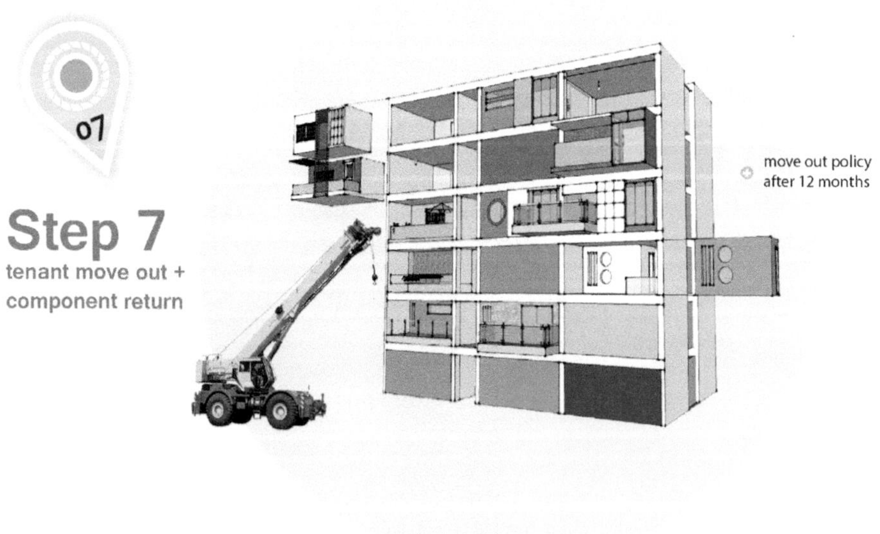

Figure 10.9. Step 7—Removing or relocation component process. Source: Zairul (2017).

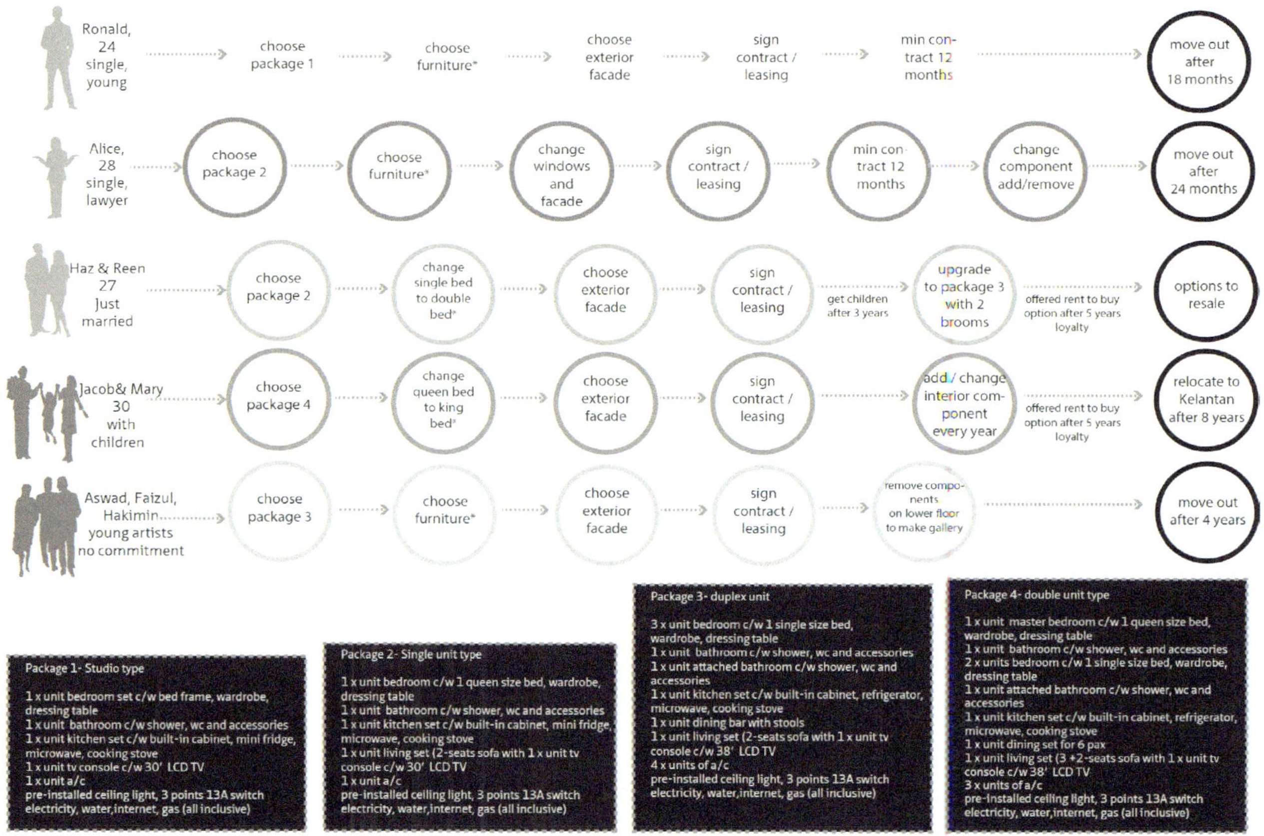

Figure 10.10. Illustration on flexZhouse concept.

type of windows. After the cost of the customization has been calculated, he agrees with the leasing price of the unit with a minimum tenure of 12 months. Later, he gets a new job in a new place, and he decides to move out after 18 months. He returns the components to the manufacturer for refurbishment.

Alice, 28 years old, single, lawyer, chooses package 2. She changes the furniture and the façade of the unit according to her needs. She signs the contract with the company and, after 18 months, she decides to upgrade the components of her house. However, due to certain changes in her life, she decides to move out after 24 months. All the units are returned to the factory for remodelling and refurbishment for the next customer.

In the third case, Haz and Reen, both 27, newly married couple, are initially interested in package number 2, but change the furniture in the bedroom to double bed instead of single. They amend the exterior façade and sign the contract. Three years later, they have children and decide to upgrade the module to package 3, which has additional bedrooms. After 5 years they are entitled to a loyalty program, which gives them the option to buy the units from the manufacturer. The cost of the unit depends on the market price and additional discount from the company. The fourth case illustrates the flexible house and the ability to be relocated to different places. After 5 years, the couple decides to buy the unit and relocate to a different location that has the same structure run by the same company. In the final scenario, the users decide to change the unit into an art gallery and to use only the upper floor for accommodation. The flexibility of the building allows the users to modify the usage of the unit according to their future plans.

Based on the focus group responses, all respondents unanimously agreed that the idea of the unit was to use the ISO container as a basis (although not necessarily be built from shipping containers) and must be easily transportable. Although the system will benefit from container technology, the majority of the sizes and modules could be wider, longer and taller than a standard ISO container. For this case, the unit will arrive at the site with the mechanical, electrical and sanitary systems already incorporated. It was suggested that thermal insulation provided by a ceramic material could be covered with a spray similar to that used by National Aeronautics and Space Administration (NASA) for space shuttles. Both the roof and the walls will be made from prefabricated metal panels clad with sheets of acrylic and fabric to insulate against the tropical climate of Malaysia. The different designs proposed by the respondents during the session provided design alternatives of the flexZhouse. The inputs were taken into consideration along with their pros and cons for the final revision of the flexZhouse. Some of the illustrations produced earlier show only the typical layout of the unit and module; but the positions of the openings will depend on the unit's location (end or intermediate unit).

A Way Forward for flexZhouse

This chapter presents the main findings of a design workshop and a focus group discussion among Malaysian architects on the innovations in housing architecture through the introduction of flexZhouse. The flexZhouse aims to solve the housing quality issue through off-site production and a fabrication process that involve high precision, and a product installation that has close tolerances. The quality is assured by the factory and, in contrary to the conventional construction, the workmanship of the flexZhouse products is at its best.

Nevertheless, the flexZhouse requires strong technological support. The supply chain involves skilled and semi-skilled workers to operate the machines as well as personnel with automotive background. Learning from the case studies of the Japanese prefab companies, flexZhouse uses a minimal labour force, adopts high technology and integrates an automation system in the housing prefabrication. New skills require technology transfer from other countries. Although not impossible, awareness should be increased and training should be given as part of the development of skilled and semi-skilled workers to operate the flexZhouse. Given the current technology in the IBS system in Malaysia, the flexZhouse needs a new paradigm to shorten the supply chain cycle. Lessons learnt from Japanese house builders on key resources will support the flexZhouse.

The flexZhouse challenges the Malaysian housing industry to adopt a new perspective. At present, the industry is associated with immobile and inflexible physical units. The flexZhouse demonstrates that housing can be mobile and flexible and the market needs to adjust to the norms of prefab housing concept. Another concern is that the market regards prefab housing as temporary structures, thereby leading to insecurity and uncertainty about its potential. Implementing the flexZhouse will require a detailed consideration of the community interaction so as to promote a culturally standard model.

The new market of young starters is said to easily adapt to changes. However, before the new concept can go further, a new market survey should be done to assess whether the concept could be implemented in the industry. Further marketing analysis and strategy could help the introduction of the new business model to the market. The configuration of the housing should take into consideration the cultural, religious and community desires to ensure that the social needs are served and delivered. Some sceptics among our respondents include the phases that the flexibility is needed in their lives. Customization might not be needed at the very beginning; it can come a bit later once the commitment and needs change in the future. This is the main reason for the basic unit of 127.83 square metre (called as the studio unit) to be introduced. This unit could be a starter unit for young starters and also act as an additional module for an extension package.

In summary, the flexZhouse opens up the possibility of customer involvement during the early design stage. This will definitely add a new dimension to the mass housing industry in Malaysia. Customization does not necessarily mean expensive. As revealed by the case study of Sekisui Heim, the adoption of 'standardized customization' allows more input from the customer at an early stage. Customers from all walks of life have different needs. Therefore, although the concept is still premature, the idea of flexZhouse provides flexibility in terms of design preferences and user's affordability.

Conclusion

Earlier in Chapter 7, the author explains on how a company does business with clients, partners and vendors through a new business model of flexZhouse, and in this chapter, the author explains on the 'proof of concept' on how the flexZhouse could be introduced to the housing market. Ultimately, it provides innovative solutions for modern housing in Malaysia.

References

Ali, A. S., S. N. Kamaruzzaman and H. Salleh. 2009. The characteristics of refurbishment projects in Malaysia. Facilities 27(1/2): 56–65. https://doi.org/10.1108/02632770910923090.

Barlow, J. and P. Childerhouse. 2003. Choice and delivery in housebuilding: lessons from Japan for UK housebuilders. Building Research & Information (December 2012): 37–41. Retrieved from http://www.tandfonline.com/doi/abs/10.1080/09613210302003.

Barlow, J. and R. Ozaki. 2003. Achieving "Customer Focus" in Private Housebuilding: Current Practice and Lessons from Other Industries. Housing Studies 18(1): 87–101. https://doi.org/10.1080/0267303032000076858.

Gann, D. 1996. Construction as a manufacturing process? Similarities and differences between industrialized housing and car production in Japan. Construction Management & Economics (December 2012), 37–41. Retrieved from http://www.tandfonline.com/doi/abs/10.1080/014461996373304.

Habraken, N. 1972. Support: An alternative to Mass Housing. London: Architectural Press (1972).

Hentschke, C., C. Formoso, C. Rocha and M. Echeveste. 2014. A Method for Proposing Valued-Adding Attributes in Customized Housing. Sustainability 6(12): 9244–9267. https://doi.org/10.3390/su6129244.

Kendall, S. 2008. Why Open Architecture, and Why Design Exercises? pp. 11–14. *In*: CIB W104 - Open Building Implementation and CIB W110—Informal Settlements and Affordable Housing: Education for an Open Architecture.

Kendall, S. H. 2012. YourSpaceKit: Off-Site Prefabrication of Integrated Residential Fit-Out.

Love, P. E. D., Z. Irani and D. J.Edwards. 2004. A seamless supply chain management model for construction. Supply Chain Management 9(1): 43–56. https://doi.org/10.1108/13598540410517575.

Noguchi, M. 2003. The effect of the quality-oriented production approach on the delivery of prefabricated homes in Japan. Journal of Housing and the Built Environment 18: 353–364. https://doi.org/10.1023/B:JOHO.0000005759.07212.00.

Omar, E. O., E. Endut and M. Saruwono. 2012. Before and After: Comparative Analysis of Modified Terrace House. Procedia—Social and Behavioral Sciences 36(June 2011): 158–165. https://doi.org/10.1016/j.sbspro.2012.03.018.

Schneider, T. and J. Till. 2006. Flexible housing: opportunities and limits. Architectural Research Quarterly 9(2): 157. https://doi.org/10.1017/S1359135505000199.

Till, J., T. Schneider, T. Jeremy and S. Tatjana. 2006. Flexible housing: the means to the end. Arq: Architectural Research Quarterly 9(3-4): 287. https://doi.org/10.1017/S1359135505000345.

Wong, J. K. W. and H. Li. 2010. Construction, application and validation of selection evaluation model (SEM) for intelligent HVAC control system. Automation in Construction 19: 261–269. https://doi.org/10.1016/j.autcon.2009.10.002.

Yashiro, T. 2014. Conceptual framework of the evolution and transformation of the idea of the industrialization of building in Japan. Construction Management and Economics 32(1-2): 16–39. https://doi.org/10.1080/01446193.2013.864779.

Zairul, M.N. 2017. flexZhouse New business model for affordable housing in Malaysia.

Zairul, M. N. 2015. New industrialised housing model for young starters in Malaysia: Identifying problems for the formulation of a new business model for the housing industry. pp. 291–298. *In*: Asia-Pacific Network Housing Research Conference. Asia-Pacific Network Housing Research.

CHAPTER 11

Energy Efficient Design Strategies for Affordable Housing

Pingying Lin[1,*] and *Hao Qin*[2]

Introduction

Energy consumption by the building sector accounts for up to 40% of the total energy consumption in developed countries, which reflects the great potential of the building sector in achieving the target of energy saving and sustainability. In response to that, substantial attention has been paid to the strategies of improving building energy efficiency and reducing building energy consumption. Green building movement was initiated to address the negative environmental and resource impacts of the buildings. Promotion of green building design and construction by mandatory policies, energy standards and codes, and related incentives has gained momentum in many countries. Actions initiated by the government play a prominent role in popularizing green building technologies. Affordable housing, which is normally supplied by the government, could be a desirable sample for showcasing how the energy target can be achieved through the application of various green design strategies.

Energy conservation has been regarded as the most critical issue in green building design. It is not only because of the adverse environmental impacts on the planet but also the increased crisis on energy deficiency. In order to facilitate the delivery of green buildings, many green building rating systems have been developed worldwide, including the UK's BREEAM (Building Research Establishment Assessment Method), the United States' LEED (Leadership in Energy and Environmental Design), Australia's Green Star, China's GBL (Green Building Label), etc. Energy is often the most important category in green building rating systems, accounting for a great proportion of the credits. As shown in Fig. 11.1, Energy and Atmosphere in LEED version 4 for BD+C: New Construction and Major Renovation accounts for 33% (U.S.

[1] College of Architecture and Urban Planning, Tongji University
[2] Beijing Mobike Technology Co., Ltd. Email: hhaoqin@outlook.com.
* Corresponding author: pingyinglin@outlook.com

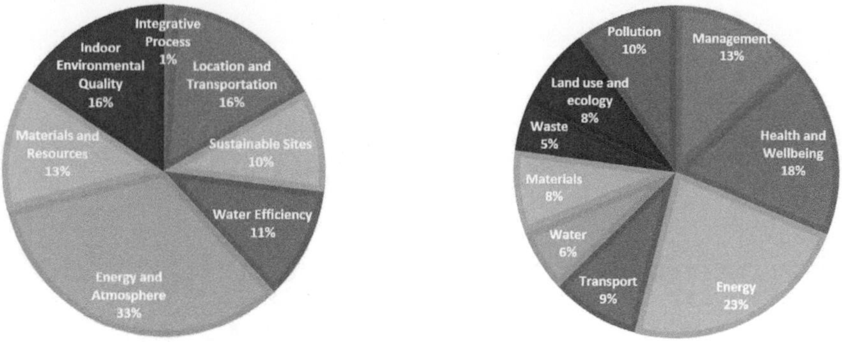

(a) LEED v4 for BD+C:New Construction and Major Renovation

(b) BREEAM New Construction 2016

Figure 11.1. Credits distribution of LEED and BREEAM.

Green Building Council, 2013), while Energy in BREEAM New Construction 2016 accounts for 23% (BRE Global, 2016). The energy performance standards adopted by these rating systems vary with local contexts, although they are all more stringent than ordinary building energy codes. Buildings certified by LEED could achieve an energy saving of 18–39% over non-certified buildings, while BREEM certified projects consume 6–30% less energy than non-certified projects (Doan et al., 2017). Therefore, in order to deliver green affordable housing, energy conservation should be the main concern for the design professionals who are responsible for decision making at the building design phase.

Affordable housing has been an important part of residential housing stock in many countries and cities, which are also under the pressure of offering more units. One of the world's biggest public housing program has been implemented in Hong Kong, wherein almost 30% of the citizens live in over 760,000 public housing units and another 15% live in subsidized sale flats (HKHA, 2017). In order to accommodate the increasing needs, the Hong Kong government has set a goal to provide 200,000 public rental housing units and 80,000 subsidized sale flats in the next decade. In the UK, around 32,630–66,700 affordable homes have been offered each year in the recent decade (DCLG, 2017). With such a sizable development of affordable housing worldwide, there will be great opportunities to ameliorate the energy deficiency and mitigate the environmental impacts by delivering green affordable housing. It is evidently challengeable to achieve the building energy performance target with a common problem as funding limit in developing affordable housing.

This chapter is dedicated to illustrating the energy efficient design strategies for green affordable housing. The energy use pattern in residential buildings is introduced first. Next the specific energy efficient design strategies, which are classified into three groups, namely passive design, energy efficient technologies, and renewable energy solutions are introduced. With respect to the financial limitation for affordable housing, passive design strategies are considered to be the most cost-effective approach to green building, which are also the main focus of this chapter.

Energy use in Residential Buildings

In terms of buildings' life cycle, building energy consumption involves two kinds of energy, operational energy and embodied energy. Operational energy refers to the energy used during the operation period to sustain a comfortable indoor environment through heating, cooling, lighting, and ventilation, etc. The embodied energy represents the energy consumed during the manufacture, demolition and disposal processes of building materials. The assessment of embodied energy depends on various factors, such as system boundaries, transport distance, primary and delivered energy, data sources, completeness of the data and technology of the production processes, etc. (Dixit et al., 2010). This chapter will focus on operational energy, which is more likely to be controlled by proper building design. Energy reduction on building operation contributes significantly to the conservation of the total energy use over the whole life of a building.

The energy end-use patterns of residential and non-residential buildings vary substantially. A commercial building is one of the most energy consuming non-residential building types. Its energy use depends on the internal loads, such as intensive occupation and consequent activities, lights and equipment. The energy consumption of residential buildings is determined by envelope loads (interaction with outdoor climate through building envelope), which leads to a much larger part of the energy used for heating and cooling to meet thermal comfort requirement. Normally, space heating and cooling, domestic hot water, and lighting, account for a majority of residential building energy use, which can be verified by the residential energy end-use data shown in Fig. 11.2. It can be seen that space conditioning is the largest energy consumer in both US and Hong Kong, accounting for 54% in the US and 25% in Hong Kong. It was followed by water heating (18%) and lighting (6%) in the US. As for Hong Kong, space conditioning, hot water, and lighting together constitute 49% of the total energy

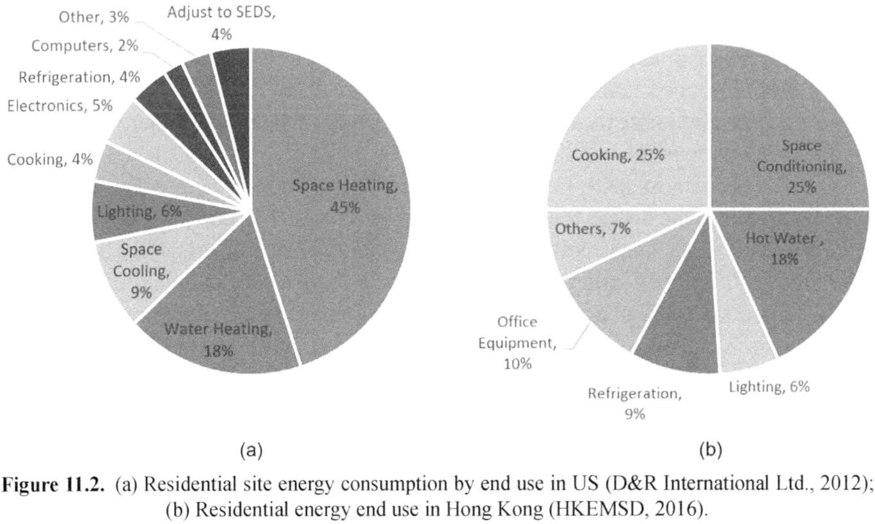

Figure 11.2. (a) Residential site energy consumption by end use in US (D&R International Ltd., 2012); (b) Residential energy end use in Hong Kong (HKEMSD, 2016).

use, which is much lower than the US (78%). The variance could be largely due to climatic difference and the consequent comfort requirement difference between these two areas. Except for the above three main consumers, cooking, office equipment, and refrigeration together account for an important part of the total energy (44%) in Hong Kong, which mainly depends on the appliance and would not be affected by building design and major service system. The identification of energy end-use pattern provides critical implication for the potential of reducing energy consumption through integrated building design and optimization of building service system.

Energy Efficient Design Strategies

Achieving a substantial reduction of energy consumption is a critical challenge for the high-performance green building. The green building movement worldwide has initiated a radical change in how buildings are designed and managed. A series of innovative strategies have been developed to help create a low/zero energy profile for different types of buildings. These strategies can be categorized into several groups: optimize the passive solar design and the thermal performance of the building envelope to minimize the building energy loads; utilize energy efficient building service systems (HVAC, lighting, lift and escalators, etc.), to minimize energy consumption; adopt alternative renewable energy.

The design of energy-efficient green building requires synthetical optimization of various innovative strategies. Trade-offs shall be made with respect to external factors, such as financial condition and climate context. For affordable housing, energy efficient design should not only significantly reduce operational energy costs but also minimize the increase in capital costs.

Passive Design Strategies

Passive design has gained more recognition as an important strategy to address the increasing environmental concern in affordable housing development. It requires much less upfront financial investment than other advanced energy efficient technologies, which help to ameliorate the contradiction between the need to solve energy deficiency and the tight financial condition of affordable housing. Passive design has been considered as the most cost-effective approach to achieve a high-performance building. Passive design is the design that takes advantage of natural resources, such as sunlight and wind flow to sustain a comfortable building indoor environment, where the heating, cooling, lighting, and ventilation systems of the building can be operated and managed with the minimum use of active energy sources. Passive design is the first critical step that architects can take towards energy efficient green building. It occurs at the early design stage when the most important decisions concerning building performance and sustainability can be made. The promotion of passive design helps to deliver climate responsive and energy efficient building projects.

Passive design requires a full understanding of the complexity of building energy performance and the influential factors. It involves complex compliance between the optimal design solutions from many aspects, such as orientation, window glazing area, building shape, thermal massing and ventilation paths, etc., in response to the climate conditions. Single passive design measure may contribute to energy saving for heating demand, while results in extra energy demand for other comfort requirements. Consequently, energy performance of the building should be evaluated systematically.

There is no general or normative passive design strategy that is suitable to be implemented in all kinds of climate backgrounds. Passive design strategies shall be carefully identified and applied based on thorough examination of the advantages and disadvantages of prevailing climate and local topography.

Building Form

The decision on building shape is usually made at the very beginning stage of conceptual design. Building shape has a significant impact on the energy consumed for the heating and cooling since it regulates the total exterior surface area that interacts with the outdoor environment (Pacheco et al., 2012). Buildings with simple and compact plan shapes (e.g., rectangle) will consume less energy on heating than those with more complex layouts such as an L-shaped or T-shaped plan. It is found that the optimization of building shape in conjunction with orientation can lead to heat energy savings by 8% (Aksoy and Inalli, 2006).

Until now, there have not been many building regulations and green assessment systems that give credit directly for building designs with energy efficient shapes. The Passivhaus standard developed in Germany, as a pioneer, designates efficient building shape as one of the five key design strategies aiming for high energy efficiency (BRE, 2011). The NHBC Foundation in the UK published a research on the energy efficiency of building shape and form (NHBC Foundation, 2016). A variable, form factor, was proposed to parameterize the compactness of building shape. It is defined as the ratio of the total heat loss area of walls, roofs, floors, and openings to the habitable floor area of all stories. The basic principle of building shape selection in terms of energy efficient design is to minimize building surface. Lower form factor means less building surface exposed, which is proved to be a design solution with higher energy efficiency. Figure 11.3 shows the example of how different designs with the same floor area can have different heat loss area. As shown in Table 11.1, this research compares several typical types of homes in the UK, all with the same habitable area, which illustrates that a bungalow, with the highest form factor, shows the least energy efficiency.

There have been some similar factors developed to describe the building form. The surface-to-volume ratio is more frequently used by researchers. It is defined as the ratio of the exposed building envelope to building volume. Compared to form factor, it also considers the vertical dimension. The Shape Factor (SF), defined as the ratio of building length to building depth, was developed to describe the building plane layout. Aksoy and Inalli (2006) compared the buildings with different SF values (1/1, 2/1 and 1/2) and found that building with a square shape (SF = 1/1) had more advantages in heating energy conservation.

However, reducing the building envelope has two conflicting effects on energy conservation. Although it can cut down the heat loss, the level of daylight and natural ventilation would be sacrificed as well (Ratti et al., 2005). The decision should be made according to the prevailing local climate. For those cities at high latitudes, where

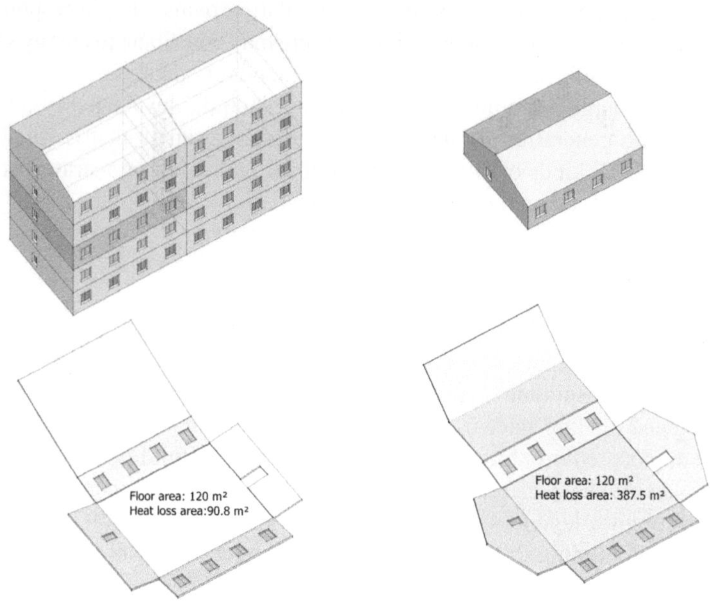

Figure 11.3. Design scenarios with the same floor area, but different heat loss area (adapted from NHBC Foundation, 2016).

Table 11.1. The types of homes and their Form Factors in the UK (adapted from NHBC Foundation, 2016).

	Type	Form factor	Efficiency
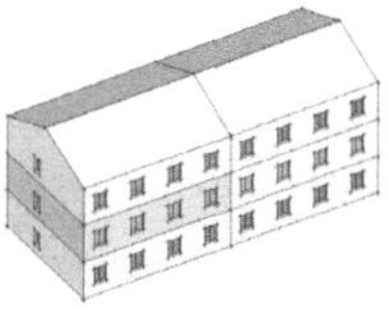	End mid-floor apartment	0.8	Most efficient
	Mid-terrace house	1.7	
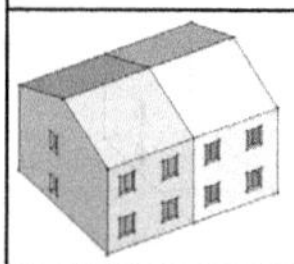	Semi-detached house	2.1	
	Detached house	2.5	
	Bungalow	3.0	Least efficient

heating load plays a major role in the energy demand, the heat lost via the building envelope cannot be compensated by the solar heat gain when the building surface is increased. In these cases, heat conservation through minimizing exposed building surface should be favored over the consideration of daylight and ventilation. On the contrary, for those cities at low latitudes where excessive solar heat gain is the main problem, design scenarios with more building surface that encourage natural ventilation may be adopted. For energy efficient passive design, heating, cooling, lighting, and ventilation should be considered simultaneously in energy consumption assessment.

Orientation

With respect to passive solar building design, orientation, as a fundamental aspect, is usually the first issue that would be considered during the design process, since it can be easily addressed and implemented. Optimizing building orientation not only helps to cut back energy demand but also reduces the need for more advanced passive systems (Pacheco et al., 2012). The building orientation directly affects the level of solar heat received on the building façade and the daylight transmitted through the windows into the inner space. It has been investigated that orienting a building to the south leads to the best performance both in gaining solar heat in the winter and controlling solar radiation in the summer (Tang, 2002). In most climates, passive solar buildings normally align the long side to the east-west axis and assign windows to the south-facing side (for the northern hemisphere), aiming to increase the solar heat gain in the winter and thus reduce the heating energy load. In summer when the sunlight angle is much higher, the southern facing windows with shading partitions are able to block the unwanted summer heat while still capturing a proper level of daylight, which could result in a considerable energy reduction in cooling and lighting. In building design, building orientation is often considered in conjunction with building shape. Aksoy and Inalli (2006) found that a building with a square plane tends to have the highest heating demand when it was positioned at an intermediate angle of 45°. The minimum heating energy can be achieved when one of the building facades faces south.

Envelope

To achieve the target of substantial energy reduction, it is vital to optimize the thermal performance of building envelope. Building envelope refers to the building skin, where various energy flows are transmitted between the indoor and outdoor environment. It determines the indoor thermal environment and consequently the extra energy demand from mechanical heating and cooling systems, ventilation, and artificial lighting. Building envelope consists of walls, window glazing area, roofs, and skylights. The walls manage the amount of solar heat gain through heat conduction, reflection, and infiltration, while the glazing area controls daylighting through incident sunlight and cooling as well as air purifying through air ventilation. Therefore, the envelope is a decisive component that determines the amount of energy a building will use during its operating period. It has been estimated that energy losses through building envelope account for 10–25% of the total energy for most buildings (Kubba, 2017).

Thermal Mass and Insulation

Thermal mass is an important characteristic of building mass to store heat and sustain the indoor environment against the outdoor temperature fluctuations. In a hot-arid climate zone with considerable diurnal temperature variation, building wall devised in the passive sense should have high thermal mass with adequate thickness to store the solar heat and prevent the heat transmission in the daytime, and thereafter release the heat when the air temperature drops dramatically in the night.

U-value, referred to as heat transfer coefficient, is used to reflect the ability of the building component as insulators, which means the ability to impede the heat transmitting between the indoor and outdoor environment. Lower U-value of a building element represents more effectively that the heat is prevented from transmitting through the material. Building energy codes and green building evaluation system usually set the requirements of maximum U-values for different building components as a prescriptive measure to reduce building energy load. The maximum U-values of 0.2 W/(m^2.k) for building roofs and 0.3 W/(m^2.k) for building walls are stipulated in the 2013 UK building regulations.

Generally, higher standards for wall thermal resistance are proposed for cold climate area. Adding a thermal insulation layer in the building wall is a widely adopted strategy. Research revealed that thermal insulation could also contribute to a reduction of cooling energy in hot climates. Bojić and Yik (2005) investigated the cooling energy load of a typical high-rise public housing estate in Hong Kong by simulation and found that increasing the insulation level of the envelope and the partitions would help to save the annual cooling energy by up to 38%. The position of thermal insulation layer should be decided through accurate examination in response to local climates. Bojic et al. (2002) compared the influence of positioning the insulation layer near the interior surface, in the middle and near the exterior surface. The simulation results indicated that a thermal insulation layer with a 50 mm thickness positioned near the interior wall surface could lead to a reduction in the annual cooling energy by 9.1% and in the peak cooling load by 10.5%. However, it was found that after the thickness threshold of 50 mm for insulation layer is met, the rising rate of energy saving would drastically drop when the layer thickness increases. It means an insulation thickness over 50 mm would result in a much higher initial cost for insulation material and a slightly higher energy reduction, while an overly thin insulation layer with a low initial cost would cause much higher energy use. Consequently, an integrated analysis should be conducted to identify the optimal insulation layer thickness that meets both objectives of cost-effectiveness and energy efficiency in good balance.

Windows/Glazing

Windows influence the building energy load in several ways: introducing sunlight into the room for illuminance; controlling direct solar heat gain; allowing wind flow to pass through the rooms for ventilation and fresh air. The performance of windows is usually evaluated by various factors, the Solar Heat Gain Coefficient (SHGC), the Visible Transmittance (VT), the U-value, and the air infiltration or leakage of the window assembly (Kibert, 2013). Figure 11.4 shows a typical energy performance

Figure 11.4. Energy performance label of windows by National Fenestration Rating Council in the United States (Source: National Fenestration Rating Council).

label of a window product certified by National Fenestration Rating Council in the United States, illustrating the values of main factors.

Solar heat gain via the windows is determined by the position and size of the windows, as well as the glazing type selected. SHGC is expressed as a value between 0 and 1. It is the fraction of incident solar radiation, covering directly transmitted radiation, as well as absorbed radiation, which is later released into the room. Lower SHGC value of a window represents less solar energy it permits. Normally, the SHGC value of the east and west facing windows should be relatively lower since they are exposed to high-intensity solar energy during the morning and the afternoon (Kibert, 2013). The size of the windows is often regulated in building regulations by the window to wall ratio, which is defined as the ratio of the building's glazing area to its total wall area.

The most effective way to cut down the energy consumption of lighting is to maximize the use of daylight. As a critical sustainable strategy for high-performance building, the introduction of daylight not only contributes to substantial energy reduction but also improves the occupants' health in the physical and psychological sense. The Visible Transmittance (VT), with a value between 0 and 1, is an important variable that measures the percentage of the visible spectrum (390 to 700 nanometers) that pass through the glazing. The higher VT value, the more visible sunlight can be transmitted into the building. However, increasing the sunlight incidence would also mean more solar heat gain, which could cause more energy used for cooling the building. Trade-offs must be made between admitting daylight and controlling solar heat gain. In response to that, glass with low emissivity coating (low-E coating) has been developed to reflect the majority of ultraviolet and infrared light without compromising the level of visible light.

Natural Ventilation

Natural ventilation, also known as passive ventilation, refers to the natural air movement driven by the pressure differences, which passes through the building to cool and ventilate the indoor space without relying on the fans or other mechanical systems. Natural ventilation not only provides a major opportunity for a massive energy reduction on cooling but also supplies sufficient cool and fresh air to sustain indoor air quality, which improves the occupants' comfort both thermally and psychologically. Natural air movement provides effective cooling for occupants because air flow enhances the evaporation of perspiration and thus makes people feel cooler, even though the actual air temperature remains the same. For hot and humid climates, natural ventilation is regarded as one of the most critical passive design strategies, as it can dramatically reduce the building cooling load which is a major part of the total building energy use, at the same time maintain a comfortable or tolerable level of air temperature and humidity in the building.

Natural ventilation is most often adopted in residential buildings. But how to maximize the natural ventilation to keep a comfortable indoor environment with none or very little energy used for the mechanical system, especially in summer, needs thorough investigation. In general, wind effect ventilation and stack ventilation (buoyancy ventilation), as two major kinds of natural ventilation, are often exploited for building cooling. Wind effect ventilation is caused by air pressure differences. The wind flow hits on a building façade that is directly facing the wind (known as the windward side) and creates higher pressure, while the leeward side of building façade experiences pressure. This pressure difference will induce the air to flow through the building horizontally and cool the indoor space (Aflaki et al., 2015). Stack ventilation, or buoyancy ventilation, can be driven by the difference in temperature and humidity. In a building, original cool air would be warmed up due to human activity, then rises up vertically. This effect is utilized to stimulate stack ventilation by arranging openings at a higher level, inducing the warm air to be vented. A low-pressure area in the room would be formed to suck in new fresh air from the openings arranged at the lower level. Although the wind speed caused by stack ventilation is lower than wind effect ventilation, it does not rely on the background wind, which makes it workable with relatively stagnant air condition.

ASHRAE standard 55 (ANSI/ASHRAE, 2013) suggests that a wind velocity of 0.2 m/s would be appropriate for indoor environments, with the maximum being 1.5 m/s for human comfort. The design of natural ventilation should consider the outdoor micro-climate (prevailing wind direction, wind speed, etc.), and geographic condition, which are the major issues that the design should respond to. Building design can affect the effectiveness of natural wind effect ventilation in the following ways (EMSD, 2017):

- The orientation of envelope openings (orient the envelope openings to the perpendicular angle of the summer prevailing wind)
- Building form and dimensions (the room depth should be as small as possible to maximize the proportion of the space that can be reached by fresh air)
- Types and sizes of openings
- External elements (elements that help to increase the wind pressure difference surrounding the openings)

The design considerations that affect stack ventilation are summarized as follows:

- The size of inlet and outlet openings
- The height difference between inlet and outlet openings
- Vertical space connecting multiple floors of the building (atria, shafts, and chimneys, etc.)

Except for the conventional type of air ventilation through the building envelope, there is also another type of air flow through adventitious openings on the envelope (Etheridge, 2015). Adventitious openings refer to all the other openings that are not created on purpose, for example, the gap occurring between different building components due to the constraints of construction material and technologies. The air flow through adventitious openings is normally called air leakage or infiltration, which is undesirable due to energy loss in the process. Air leakage is assessed as the rate of leakage per m^2 of external envelope per hour at an artificial pressure differential through the envelope of 50 Pa. Air leakage usually occurs at the junctions between different components of building envelope, such as junctions between walls, floors, doors and window frames, etc. To address this, the concept of airtightness is introduced to evaluate how effectively the building envelope is sealed. It is defined as the ability of the building to resist air leakage. Great efforts have been taken to promote airtight houses in the UK. In an experiment conduct in the UK, they found that an energy saving of up to 15% could be achieved by improving the airtightness from 11.5 to 5 $m^3/(m^2 \cdot hr)$ @50 Pa (Coxon, 2013).

Shading Devices

Optimizing shading devices is an important energy-efficient design strategy, especially for a hot climatic region. Shading devices can help to prevent overheating by controlling the incidence of sunlight and improve occupants' visual comfort by controlling glare. Deliberately designed shading devices would lead to a remarkable reduction in the building peak cooling load. Nonetheless, shading devices may cause more energy use at other times inversely as they probably block the daylight penetration and cause inadequate illuminance. Excessive shading should be avoided with an aim to control the initial cost and balance the energy demand for cooling and lighting.

There have been many different types of shading devices, which can be classified as fixed shading devices and movable shading devices. Fixed shading devices include overhangs, side fins, and egg-crate, etc. (Fig. 11.5). Movable shading devices include blinds, shades, and deciduous plants, etc. The basic principle of selecting a shading device is to prevent the excessive sunlight from passing through the window in the hot season, while not blocking the desirable solar heat in the cold season. It is also important that the shading device can help to improve the visual comfort while not compromising the view. The selection of shading device type is dependent on various aspects, such as building type, building form, façade orientation, seasonal sun path, as well as aesthetic and economic concerns.

With respect to the seasonal and diurnal change of sun path, it is commonly agreed that single horizontal shading device (overhang) works best on the façade facing the south, protecting the window from high altitude sunlight during the middle of the day. Vertical side fins perform better on east and west facades. As east and west facades are exposed to the rising and setting sun at a much lower angle, which cannot be effectively blocked by overhangs. For east and west facades, side fins in combination with overhangs, or egg-crate, can provide optimum shading throughout the day. The optimal length of an overhang depends on the size of the glazing area, while the width of an overhang depends on the prevailing climate and the consequent energy demand. To deliver a proper design of shading device, it is important to analyze the position of the sun during both the cooling and heating season. Shading Coefficient (SC) is defined as the ratio of solar heat gain (direct sunlight) passing through a certain fenestration system (window, skylight, etc.), to the solar heat gain passing through 3 mm Clear Float Glass. It is an indicator of shading effect of the fenestration system compared to the baseline. Its value ranges from 0 to 1. Lower SC value represents less solar heat is transmitted through the fenestration system. The American Society of Heating, Refrigerating and Air-conditioning Engineering (ASHRAE) standard counts SC as one of the factors that should be considered in the assessment of building heating and cooling load.

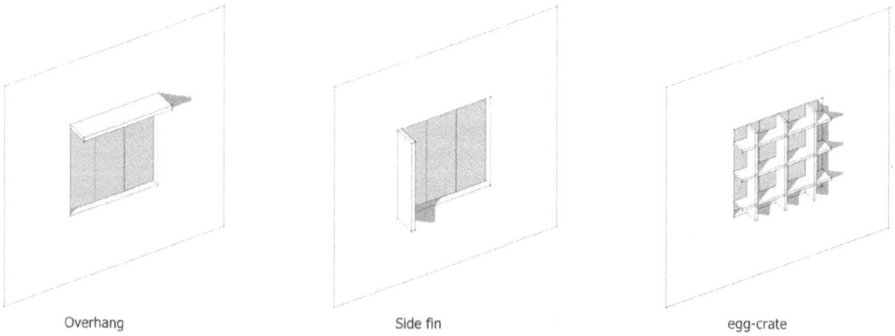

Figure 11.5. Typical fixed shading devices.

Reflective Roof

Reflective roof, also known as cool roof, refers to the roof with high solar reflectance coatings. It is a highly cost-effective strategy that can be easily implemented for both new constructions and sustainable refurbishment of existing buildings. A reflective roof can reflect a major part of solar heat back into the sky. The part of solar radiation absorbed by the roof will later emit in the form of longwave radiation attributed to the high thermal emittance of cool roofs. High solar reflectance (low solar absorption) and high thermal emittance make it possible to lower surface temperature and reduce the heat transmission through the roof into the interior space. A dramatic reduction in building cooling load and outdoor urban heat island intensity can be achieved by simply using the cool material, especially for hot climates. However, a cool roof may also increase the heating load during the cold season. Before implementing a cool roof, the proportion of heating and cooling energy for a given building should be compared and analyzed. It is particularly beneficial to use a cool roof at the regions that cooling energy load accounts for the majority of the total energy use. For those areas that similar amounts of energy are used for heating and cooling, a switchable cool roof is proposed to be applicable since higher reflectance coating can help to decrease the cooling load in hot seasons and lower reflectance coatings in conjunction with thermal insulation layer can minimize the heating energy loss in cold seasons (Testa and Krarti, 2017).

Many energy standards worldwide have set specific requirements for roofs with cool coatings. ASHRAE 90.1-2013 requires that cool roofs must have a minimum 3-year aged reflectance of 0.55 and a minimum thermal emittance of 0.75 (ASHRAE, 2013). A simple approach for adopting a cool roof strategy is to paint the roof with a light color. The solar reflectance for a black surface is 0, while a white surface is 1. Lighter surfaces have higher solar reflectance than dark-colored surfaces.

Green Roof/Green Wall

Greenery has long been considered as a valuable sustainable strategy both at the building and urban scale. Greening systems on the buildings not only improve the building performance but also enhance the urban environmental quality with aesthetical, environmental and ecological benefits. Green roofs and green walls are the two most common ways to integrate greenery with sustainable building envelope design. A green roof is a roofing assembly with multiple layers. From the top to down, the layers include vegetation, growing medium (substrate), filter, drainage layer, root barrier and waterproof material layer. According to the difference in vegetation types, construction, maintenance, cost, and use, green roofs can be categorized into three types: extensive, semi-intensive and intensive roofs (Table 11.2). The performance of green roofs can be impacted by a number of factors: vegetation species (leaf area index, foliage height, form, biological attributes), greenery coverage, substrates (thickness, water content), and the local climate conditions (Raji et al., 2015).

The existence of a green roof can contribute to building energy saving in multiple ways (Okeil, 2010). It increases the thermal insulation and thus considerably reduces the building heating and cooling energy loss. In summer, a lower surface temperature can be maintained by the evapotranspiration effect and shading effect provided by the vegetation of a green roof, consequently reducing building peak power demand

Table 11.2. Classification of green roofs (adapted from IGRA, 2017).

	 Extensive green roof	 Semi-intensive green roof	 Intensive green roof
Maintenance	Low	Periodically	High
Irrigation	No	Periodically	Regularly
Plant communities	Moss-Sedum-Herbs and Grasses	Grass-Herbs and Shrubs	Lawn or Perennials, Shrubs, and Trees
System build-up height	60–200 mm	120–250 mm	150–400 mm on underground garages > 1000 mm
Weight	60–150 kg/m^2	120–200 kg/m^2	180–500 kg/m^2
Costs	Low	Middle	High
Use	Ecological protection layer	Designed Green Roof	Park-like garden

and cooling load. The green roof can also benefit building energy saving indirectly, for example, possibly reducing the capacity of mechanical heating or cooling system.

Green wall has gained increasing interests as a strategy to improve indoor thermal performance and reduce building total energy use. Greening the building facades is an alternative strategy to compensate the area limits for the building roof to incorporate a vegetation system. Building facades have much more surface area than the building roof, especially for high-rise buildings. Similarly, according to the difference in construction technology and operation, green walls can be broadly divided into two groups: green facades and living walls. On green façades, plants embedded in the ground or at some intervals climb on the walls by themselves, which does not require extensive maintenance. Green facades can be irrigated either naturally by rain or manually. A living wall is a much more complicated and sophisticated system that consists of numerous pre-cultivated modular panels. Each panel comprises a series of elements: plants, growing medium (substrate) and container. All panels are connected with an irrigation system and attached to a trellis that is used to adhere the system to the façade. The plants on living wall are irrigated automatically through the pipes distributed along the panels. Living walls require periodically examination and maintenance to check the functionality of the system. Appropriate plants for green walls should be selected in response to the prevailing conditions of air temperature, humidity, solar radiation, and ventilation, in order to maximize the performance of the system.

Green walls contribute to energy saving by controlling the heat transfer between indoor and outdoor environments, through the evapotranspiration process and shading effect of plants, as well as the thermal insulation of growing substrates. Green walls installed on different sides of the building facades perform differently due to the varied exposure levels to solar radiation.

Energy Efficient Service Systems

After optimizing the passive design of a building, the internal thermal load of the building has been minimized. For residential buildings, the residual building energy load largely depends on the building service equipment and how the occupants use them. Building service systems supplement the building performance by providing thermal comfort, fresh air, hot water and artificial lighting, consuming the energy resources in the form of electricity, natural gas, oil, etc. Energy efficient building service systems are able to engender significant energy savings. In terms of energy end uses in residential buildings, air conditioning system, water heating system, and lighting system are the major energy consumers.

Air Conditioning System

An air conditioning system usually accounts for the primary part of the total building energy consumption, especially for residential buildings. Coefficient Of Performance (COP) of the chillers in the air conditioning system is a measurement of cooling efficiency, defined as the ratio of cooling power provided by the chiller to the input power. It has no unit. Higher COP figure means a higher level of performance the equipment can provide. Generally, water-cooled rotary screw or scroll chillers are the most efficient type of chiller, with a high COP over 6 (Kibert, 2013). The determination of the COP level of chillers should be made based on comprehensive analysis of building thermal load and prevailing climatic conditions. The efficiency of chillers is also affected by the operation status. The highest efficiency occurs when the chiller is operated at full load. Correctly sizing the chiller offers great potential to operate the equipment to meet the thermal load both effectively and efficiently.

There are basically two types of HVAC (heating, ventilation and air conditioning) systems, central and decentralized systems. Central systems produce cooling or heating in one location and distribute it to multiple places. They are mainly adopted in large non-residential buildings. Among the decentralized systems, the split-type air conditioners are most commonly utilized in residential buildings. They consist of two units, indoor and outdoor units. The outdoor unit includes the compressor and a fan, while the indoor unit (referred as air-handling unit, AHU) includes the evaporator coil and a fan. Compared to central systems, decentralized systems have various advantages when applied in residential units: require a much lower initial cost on the equipment and installation; easy to repair when a malfunction occurs; do not require extra building height as the central system; more suitable for a small individual room and can be operated efficiently at full load since they have much smaller size than central systems; much less energy would be wasted compared to the energy loss in central systems due to the long distribution pipeline.

Water Heating System

According to the energy breakdown for residential buildings, water heating constitutes a large fraction of the total energy demand. In order to reduce energy consumption by water heaters in residential units, priority should be given to reducing hot water usage, through utilizing water-efficient appliances, such as low-flow showerheads and faucets. Consequently, the energy consumed in water heating could be decreased due to less hot water demand.

Compared to conventional water heating equipment (water heater, boilers, and furnaces) which consumes electricity or natural gas, Solar Water Heating System (SWHS) is much more cost-effective and energy conservative. SWHS has been widely regarded as an energy-efficient technology due to energy deficiency and environmental concern. In warm or hot climates, SWHS can provide sufficient hot water for use, while for other climates, hot water supplied by the SWHS could constitute a large part of the total hot water demand. SWHS has gained remarkable popularity in residential buildings and various types are made available in the market. The same basic technology is using the solar thermal collector to capture solar radiation and then transfer to thermal energy to produce hot water. SWHS normally consists of solar collectors, a storage tank, and interconnecting pipeline. Solar collectors are often mounted on the building roof to maximize the exposure to sunlight and minimize the obstruction from building components or other buildings. They can be installed at an inclined angle to the roof ground, facing south (for the northern hemisphere) to receive maximum annual solar energy. The angle varies according to the geographical latitude of the project site. Generally, SWHS requires relatively low maintenance. It can normally last for 15–20 years.

Lighting System

Lighting is also an important consumer of electrical energy in residential buildings. Energy efficient design of lighting system is to maximize the natural daylighting use and minimize the artificial lighting use, without compromising the lighting quality. Considerable advancement of energy efficient lighting device and lighting control technology have been made to achieve major energy savings.

The general approach to regulate lighting energy use in building energy standard is to impose maximum lighting power density for a specific space. In addition to lighting power allowance, lighting control technology has been addressed in building energy code. Lighting control systems could detect the occupancy and respond by turning the lights on or off when occupants are present or absent. They can also increase the lighting level when natural lighting is unable to provide sufficient illuminance, and dim the lights when the desired lighting level can be achieved by daylighting. Lighting control technologies are primarily required in the public area. Table 11.3 presents the

Table 11.3. Lighting power density and automatic lighting control for various types of space (EMSD, 2015).

Type of space	Maximum allowable LPD (W/m^2)	Automatic lighting control required (Yes/No)
Atrium/Foyer with headroom over 5m	17	Yes
Corridor	8	Yes
Entrance Lobby	14	Yes
Kitchen	13	No
Lift Lobby	11	Yes
Public Circulation Area	13	Yes

requirements set in the Building Energy Code (Code of Practice for Energy Efficiency of Building Services Installation) in Hong Kong. It applies to space within which the total electricity used by the complete fixed lighting installations exceeds 70 W.

With respect to lighting equipment, some governments impose standards to grade the product according to the performance, in order to facilitate the application of energy-efficient products. Efficacy is used to measure the efficiency of lighting sources, defined as the ratio of luminous flux emitted by the lamp to the electricity used by the lamp. Its unit is lumens per watt (l m/W). Higher efficacy represents more efficient lighting source. Fluorescent lighting and Light-Emitting Diodes (LED) lighting are most commonly used in energy efficient design. The efficacies of fluorescent lamps range from 80 to 93 lm/W, while the maximum efficacy of LED light could be up to 130 lm/W (Kibert, 2013).

Renewable Energy Solutions

Renewable energy refers to the energy that is produced from renewable resources, such as sunlight, wind, tide, geothermal heat, biomass, etc. In contrast, conventional fossil fuels, such as coal, natural gas, and oil, are non-renewable resources, since they cannot be replenished in the short term. Exploitation and application of renewable energy play an essential role in pursuing environmental sustainability. Using renewable energy as a substitute for conventional energy in building operations has gained growing popularity in green building design. Onsite renewable energy generation can cover a large fraction of the total building energy consumption, with supplementary power from the utility grid to cover the rest. If the renewable energy generated on site is more than the building requires, the excessive power can be fed into the utility grid, or stored on site using batteries. There are three main types of renewable energy that can be generated on the project site, namely photovoltaics, wind energy, and biomass, which are also most often adopted in residential buildings.

PV and BIPV

Photovoltaic (PV) system consists of a series of components: PV panels to absorb and convert sunlight into electricity by semiconductor materials; an inverter to convert the electricity from Direct Current (DC) to Alternating Current (AC) before the electricity is fed into the utility grid; mounting, wiring, and other accessories. If the generated electricity cannot be connected to the grid, battery storage solution can be integrated into the system. PV panels are modular and composed of multiple solar cells. An array of PV panels forms a PV system. There have been various configurations of PV systems available, within which rooftop mounted and building-integrated systems are most commonly adopted in building projects. Building-integrated photovoltaic (BIPV) are PV panels formed directly as building materials to be used as part of building envelope. They can be used to replace conventional building materials, such as roofing, skylight, window glazing, curtain wall, shading device, and façade cladding. There have been numerous types of BIPV products that can be adopted on a building envelope, which increases the sunlight collection area and covers a larger proportion of building energy use. Part of the cost of BIPV system can be offset since the cost of conventional building components has been saved.

The efficiency of solar panels of the PV system depends on the material of the semiconductor (Keeler and Vaidya, 2016). Monocrystalline solar panels have the highest efficiency and are most expensive. Polycrystalline solar panels perform moderately worse than monocrystalline solar panels at high temperatures. Among all types, thin-film solar cells have the least efficiency. However, they are more flexible to support a wider range of application. They can be applied directly to skylights and window glazing.

The annual power generation of PV system depends on the latitude, climatic condition, orientation and quantity of PV panels, and obstruction from surrounding buildings or facilities. For the rooftop mounted systems, PV panels should be tilted to maximize the intensity and period of solar collection. For BIPV system, PV materials should be arranged on the facades that receive the most annual solar radiation to improve the system performance.

Wind Energy

Wind turbines are used to convert the kinetic energy of wind flow into electrical power. There are basically two kinds of wind turbines classified by the rotating direction of the blade, namely Vertical-Axis Wind Turbine (VAWT) and Horizontal-Axis Wind Turbine (HAWT) as shown in Fig. 11.6. HAWTs are generally more common than vertical-axis type. HAWTs are often applied in large scale with bigger blades, such as utility-scale, whereby large amounts of electrical power can be generated and fed into the power plants. VAMTs are usually small in scale and less efficient, which are the main products that are applied at building scale.

The capacity of a wind turbine is measured in Watts (W), kilowatt (kW), megawatt (MW), and gigawatt (GW). Electrical power generation and consumption are often measured in kilowatt-hours (kWh). The capacity levels vary within a quite large range, with the largest capacity being around 9 MW. The performance of wind turbine systems depends essentially on wind speeds. Higher wind speed could lead to more generated electrical power. Cut-in speed means the minimum wind speed at which the turbine blade first begins to rotate and generate electricity. The cut-in speed for most wind turbines is approximately 11 km/h, while some VAWTs are more sensitive

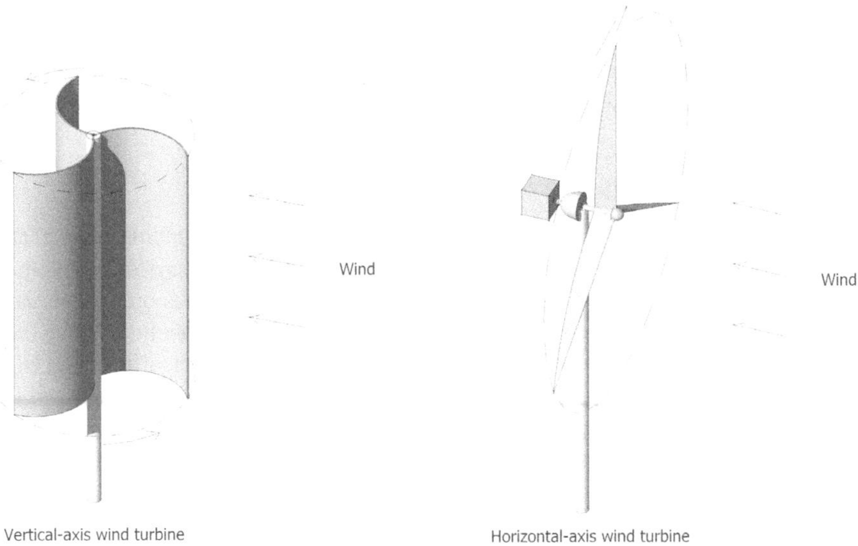

Figure 11.6. Vertical-axis wind turbine and horizontal-axis wind turbine.

and could cut in at lower than 8 km/h (Keeler and Vaidya, 2016). The wind speed and direction at a specific urban site vary substantially due to the complex urban environment. Therefore, an onsite wind measurement is suggested to be conducted to explore the potential of exploiting wind power before wind turbine system is applied to the building project. When applied to buildings, wind turbine systems are normally installed at the highest level to face the highest wind speed. The effectiveness of wind turbines also depends on wind conditions.The desirable wind condition for optimal performance of wind turbines is smooth wind flow with low turbulence as in wind farm. It means wind turbines installed in a building site located in the high-density area may not perform efficiently.

Biomass Energy

Biomass energy technology refers to the technologies that produce energy from organic matters, such as plants, agricultural crop, forest residues, and animal wastes, by burning or converting into biofuels. Advances in this technology have made it possible to convert biomass into different forms of energy, such as electricity, gaseous and liquid fuels, etc. Biomass energy is considered as renewable energy since the organic material can be replenished shortly and the stored energy of biomass comes from the sun. Compared to solar and wind energy, biomass energy is not constrained by the dynamic climate conditions, as long as there is sufficient supply of biomass source. Biomass energy can be generated at any time of the day. However, some concerns have arisen regarding the life-cycle carbon emission analysis. When biomass is burned, the stored CO_2 would be released back into the air, which seems to be carbon neutral. The extra CO_2 emission resulted from the process of growing, harvesting, storing, transporting and burning biomass, is often neglected. It may lead to almost equal CO_2

emission as fossil fuels. Therefore, bio-waste, as an important source of biomass, not only contributes to less greenhouse gas emission but also reduces landfill waste and generates clean energy. For residential communities, it is suggested to use kitchen waste to generate bioenergy by converting it into gaseous fuels, solid fuels, and biodiesel.

Economic Feasibility and Social Benefits

Installation of renewable energy generation systems inevitably entails an increased initial cost. The economic feasibility and life cycle cost of these systems have been extensively investigated. However, there has not been a consensus on the economic feasibility of renewable energy systems. Some researchers concluded that these small-scale systems are economically viable. In a study conducted in India, BIPV systems installed in different orientations were compared. The results showed that the energy payback time periods range from 7.3 years to 16.9 years, which means the additional cost can be fully recouped in the form of lower energy bills within the expected effective duration (30 years) of the system (Tripathy et al., 2017). Moreover, with the progressive development of renewable energy technologies, the costs have been declining considerably. Some other researchers claimed that the cost cannot be redeemed by the energy generated within the systems' lifetime (Hammond et al., 2012). A comprehensive economic analysis of small wind turbines in 88 regions in Iran was conducted and the results showed that these systems can be cost-effective in around 30% of the investigated area. It was found that the cost-effectiveness of the wind turbine systems depended on both financial issues (feed-in tariff, initial cost, and interest rate) and climate background (wind speed) (Hosseinalizadeh et al., 2017). Therefore, a case by case integrated environmental and economic evaluation should be performed at the design phase.

Except for the potential direct economic benefit, installation of these visible renewable energy systems plays a prominent role in increasing the occupants' awareness on energy conservation and sustainability (Bahaj and James, 2007). It leads to a radical change in user behavior in terms of how to use the energy in the daily life, which has been acknowledged to be an important factor for building energy conservation (Delzendeh et al., 2017).

Discussion and Conclusion

Compared to traditional housing estates, constructing energy efficient housing would inevitably increase the construction cost. The problem is that occupants benefit from the energy efficient technologies with a lower energy bill while the developers have to pay for the extra cost initially. For private housing developers, except for the various economic incentives provided by the government such as tax discounts, lower development application costs, and related funding, etc. (Olubunmi et al., 2016), the burden of additional cost can also be transitioned to the home buyers in the form of higher home prices. However, affordable housing developers do not have the same ways to repay the financial investment since their target customers are low-income renters and the increase of rents is usually strictly restricted. Aside from regular government incentives, the main government financial incentive for affordable housing development would be lower land prices (McManus et al., 2010). Affordable housing developments are often in charge by the non-profit organizations that are partly or fully funded by the local government, such as Registered Social Landlords in the UK and Hong Kong Housing Authority in Hong Kong. It may be a paradox to increase the supply of affordable housing and improve the energy performance level of the estates simultaneously with respect to the tight financial budget. The government needs to weigh the negative impacts of the deficiency of housing supply at present and environmental degradation in the future (McManus et al., 2010).

This chapter introduces and analyzes three main approaches to deliver energy efficient affordable housing, namely passive design strategies, energy efficient building service systems, and renewable energy solutions. With respect to the cost-effectiveness of the strategies for affordable housing, passive design is addressed as the main focus of this chapter, wherein the effects of each strategy in different urban and climate contexts are compared and demonstrated. Related assessment factors are introduced to facilitate the performance evaluation. For building service systems, the major energy consumers in residential buildings are addressed, including air conditioning system, water heating system, and lighting system. In terms of alternative renewable energy resources, PV, wind energy, and biomass energy are considered to be the most feasible to be implemented in affordable housing. The economic viability and social benefits of these advanced systems are also discussed. It is indicated that the effects of these strategies vary with the urban environment and climate backgrounds. An integrated energy performance evaluation, environmental and economic life-cycle assessment should be conducted at the design phase to determine the optimal design. This chapter would be helpful for professionals who make design decisions. Energy efficient affordable housing could benefit the low-income families with a more thermally comfort, economical, and healthier living environment in the long term.

References

Aflaki, A., N. Mahyuddin, Z. Al-Cheikh Mahmoud and M. R. Baharum. 2015. A review on natural ventilation applications through building facade components and ventilation openings in tropical climates. Energy and Buildings 101: 153–162. https://doi.org/10.1016/j.enbuild.2015.04.033.

Aksoy, U. T. and M. Inalli. 2006. Impacts of some building passive design parameters on heating demand for a cold region. Building and Environment 41(12): 1742–1754. https://doi.org/10.1016/j.buildenv.2005.07.011.

ANSI/ASHRAE. 2013. ANSI/ASHRAE 55-2013: Thermal Environmental Conditions for Human Occupancy. Atlanta, GA.

ASHRAE. 2013. ANSI/ASHRAE Standard 90.1-2013. energy standard for buildings except low-rise residential buildings. Atlanta, GA.

Bahaj, A. S. and P. A. B. James 2007. Urban energy generation: The added value of photovoltaics in social housing. Renewable and Sustainable Energy Reviews 11(9): 2121–2136. https://doi.org/10.1016/j.rser.2006.03.007.

Bojić, M. and F. Yik. 2005. Cooling energy evaluation for high-rise residential buildings in Hong Kong. Energy and Buildings 37(4): 345–351. https://doi.org/10.1016/j.enbuild.2004.07.003.

Bojic, M., F. Yik, and W. Leung. 2002. Thermal insulation of cooled spaces in high rise residential buildings in Hong Kong. Energy Conversion and Management 43(2): 165–183. https://doi.org/10.1016/S0196-8904(01)00018-8.

BRE. 2011. Passivhaus. Retrieved August 3, 2017, from http://www.passivhaus.org.uk/index.jsp.

BRE Global. 2016. BREEAM International New Construction 2016.

Coxon, R. 2013. Research into the effect of improving airtightness in a typical UK dwelling. REHVA Journal (January), 24–27.

D&R International Ltd. 2012. Buildings energy data book. United States Department of Energy. Washington DC. Retrieved from http://buildingsdatabook.eren.doe.gov/DataBooks.aspx.

DCLG. 2017. Affordable housing supply: April 2016 to March 2017 England. UK. Retrieved from www.nationalarchives.gov.uk/doc/open-government-licence/.

Delzendeh, E., S. Wu, A. Lee and Y. Zhou. 2017. The impact of occupants' behaviours on building energy analysis: A research review. Renewable and Sustainable Energy Reviews 80(August): 1061–1071. https://doi.org/10.1016/j.rser.2017.05.264.

Dixit, M. K., J. L. Fernández-Solís, S. Lavy and C. H. Culp. 2010. Identification of parameters for embodied energy measurement: A literature review. Energy and Buildings 42(8): 1238–1247. https://doi.org/10.1016/j.enbuild.2010.02.016.

Doan, D. T., A. Ghaffarianhoseini, N. Naismith, T. Zhang, A. Ghaffarianhoseini and J. Tookey. 2017. A critical comparison of green building rating systems. Building and Environment 123: 243–260. https://doi.org/10.1016/j.buildenv.2017.07.007.

EMSD. Code of Practice for Energy Efficiency of Building Services Installation. 2015. Hong Kong.

EMSD. 2017. Natural ventilation. Retrieved August 9, 2017, from http://gbtech.emsd.gov.hk/english/utilize/natural.html.

Etheridge, D. 2015. A perspective on fifty years of natural ventilation research. Building and Environment 91: 51–60. https://doi.org/10.1016/j.buildenv.2015.02.033.

Hammond, G. P., H. A. Harajli, C. I. Jones and A. B. Winnett. 2012. Whole systems appraisal of a UK Building Integrated Photovoltaic (BIPV) system: Energy, environmental, and economic evaluations. Energy Policy 40(1): 219–230. https://doi.org/10.1016/j.enpol.2011.09.048.

HKEMSD. 2016. Hong Kong Energy End-use Data 2015. HKEMSD. Hong Kong.

HKHA. 2017. Housing in Figures 2017. Hong Kong.

Hosseinalizadeh, R., E. S. Rafiei, A. S.Alavijeh and S. F. Ghaderi. 2017. Economic analysis of small wind turbines in residential energy sector in Iran. Sustainable Energy Technologies and Assessments 20: 58–71. https://doi.org/10.1016/j.seta.2017.02.018.

IGRA. 2017. Green roof types. Retrieved August 11, 2017, from http://www.igra-world.com/types_of_green_roofs/index.php.

Keeler, M. and P. Vaidya. 2016. Fundamentals of integrated design for sustainable building (second edi). New Jersey: John Wiley & Sons, Inc.

Kibert, C. J. 2013. Sustainable Construction: Green Building Design and Delivery. Hoboken, N.J.: John Wiley & Sons.

Kubba, S. 2017. Handbook of Green Building Design and Construction (Second edi). Kidlington, United Kingdom: Joe Hayton.

McManus, A., M. R. Gaterell and L. E. Coates. 2010. The potential of the Code for Sustainable Homes to deliver genuine "sustainable energy" in the UK social housing sector. Energy Policy 38(4): 2013–2019. https://doi.org/10.1016/j.enpol.2009.12.002.

NHBC Foundation. 2016. The challenge of shape and form: Understanding the benefits of efficient design. Milton Keynes, UK. Retrieved from https://www.nhbcfoundation.org/wp-content/uploads/2016/10/NF-72-NHBC-Foundation_Shape-and-Form.pdf.

Okeil, A. 2010. A holistic approach to energy efficient building forms. Energy and Buildings 42(9): 1437–1444. https://doi.org/10.1016/j.enbuild.2010.03.013.

Olubunmi, O. A., P. B. Xia and M. Skitmore. 2016. Green building incentives: A review. Renewable and Sustainable Energy Reviews 59: 1611–1621. https://doi.org/10.1016/j.rser.2016.01.028.

Pacheco, R., J. Ordóñez and G. Martínez. 2012. Energy efficient design of building: A review. Renewable and Sustainable Energy Reviews 16(6): 3559–3573. https://doi.org/10.1016/j.rser.2012.03.045.

Raji, B., M. J. Tenpierik and A.van den Dobbelsteen, 2015. The impact of greening systems on building energy performance: A literature review. Renewable and Sustainable Energy Reviews 45: 610–623. https://doi.org/10.1016/j.rser.2015.02.011.

Ratti, C., N. Baker and K. Steemers. 2005. Energy consumption and urban texture. Energy and Buildings 37(7): 762–776. https://doi.org/10.1016/j.enbuild.2004.10.010.

Tang, M. 2002. Solar control for buildings. Building and Environment 37(7): 659–664. https://doi.org/10.1016/S0360-1323(01)00063-4.

Testa, J. and M. Krarti. 2017. A review of benefits and limitations of static and switchable cool roof systems. Renewable and Sustainable Energy Reviews 77(March): 451–460. https://doi.org/10.1016/j.rser.2017.04.030.

Tripathy, M., H. Joshi and S. K. Panda. 2017. Energy payback time and life-cycle cost analysis of building integrated photovoltaic thermal system influenced by adverse effect of shadow. Applied Energy 208(October): 376–389. https://doi.org/10.1016/j.apenergy.2017.10.025.

U.S. Green Building Council. 2013. LEED Reference Guide for Building Design and Construction.

CHAPTER 12

Passive Design Strategies for Affordable Housing in Indian Tropical Regions

Manjari Chakraborty

Introduction

The threats to the very existence of our world are real now. Out of the many components of danger, pollution problems stemming from the aspects of human shelter are huge in proportion. The fallacy is, because a vast number of the world's population is in varying states of homelessness and the numbers are growing as world population burgeons (The last time a global survey was attempted—by the United Nations in 2005—an estimated 100 million people were homeless worldwide. As many as 1.6 billion people lacked adequate housing: Habitat, 2015), we need to provide more and more shelters and that would cyclically create more and more pollution. The number of homeless people is additionally increasing because of manmade factors like discord and unrest and natural calamities like increasing trends of typhoons and dry spells, ironically, both being end results of uncontrolled population explosion and resultant activities.

So it is critical to seek less polluting ways of providing human shelters. We need to focus analytically on the issue, trying to define its core elements.

Firstly, housing means primarily shelters for humans, with a predominant trend of preference for quantity over quality. Housing can still be of various standards, ranging from temporary or makeshift to permanent and lavish, involving varying time, technology, cost, land and comfort elements. This chapter mainly dwells on targeting environmental sustainability and economical sustainability in Affordable Housing. Out of these two broad areas of Sustainability, it has further focused to explore the

Birla Institute of Technology (Deemed University), Ranchi, India.
Email: profmanjari@gmail.com

various passive design objectives needed to be followed to achieve the thermal and visual comfort at three major tropical climatic regions in India.

As per the studies done on the existing affordable housing and ongoing housing projects, it is found that there are missing policies to make climate resilient housing by incorporating the passive strategies. There are also missing policies to make affordable housing green. Awareness on green initiatives with passive design features need to be integrated in affordable housing in India. There is also a need for financial policies and incentives to promote passive housing. But, actually, there is no or limited incentives for borrowers from Government or financial institutions, for climate responsive houses.

Thus, this chapter intends to explore the opportunities of acquiring environmental and on the long run to acquire economical sustainability by incorporating passive design strategies in Affordable Housing. Three case studies have been discussed related to three major tropical climates in India—warm humid, hot dry and composite. The methodology adopted here starts with an introduction of the importance of the topic in its domain. Then the scope of the current discussion is clarified. After this, the various climatic zones of India are introduced and explained in brief. It is followed with a background containing a comprehendible definition of the term 'Affordable' from all applicable angles, in order to clarify the interrelation of the various elements of the entire discussion. Thereafter several dimensions, aspects and characteristics of affordable housing are discussed. Once these basic issues are clarified in clear detail, the main focus of the chapter, i.e., passive design strategies, its applicability in affordable housing types, its various components, relevance and justification are discussed and elucidated. The levels of affordable housing and the applicability and scope of passive design in such cases have been discussed, citing appropriate and relevant case studies. The discussion ends with a conclusion which is a gist of the discourse and an affirmation of its relevance and affectivity.

Background

Affordable Housing in India

Affordable housing is defined in many different ways in different countries. It is basically shelter for very low to moderate income households and priced in such a manner that they are also able to meet their other basic survival requirements, which covers food, clothing, transport, medical care and education. As a rule of thumb, housing is usually considered as affordable if it costs less than 30% of gross household income. Affordability in India is guided by several socio-economic variables. Here, 'affordable housing' refers to residences that have been especially designed for the Economically Weaker Section (EWS) and Lower Income Group (LIG) who also desires the same comfort, security of a self-owned property/home.

In urban India, there currently exists a wide gap between the demand and supply of housing (both in terms of quantity and quality. The reason behind this is rapid urbanization which is forcing people to increasingly live in slums and squatter settlements and thus developing poor housing conditions of the economically weaker sections of the society. According to Ministry of Housing and Urban Poverty Alleviation (MHUPA), the urban housing shortage in the country at the end of the 10th

Five-Year Plan was estimated to be 24.71 million for 66.30 million households. And it covers a shortage of 99.9% for EWS and 10.5% for LIG people (Source : Making Affordable Housing Work in India, November 2010). So, it is very clear that the housing requirements of the economically weaker sections and lower income groups are mainly neglected, and there exists a huge dearth in the supply of affordable houses primarily demanded by this income group in India. The Central Government desires to provide housing for all by 2022 and as per KPMG report 70% of this housing need is for Affordable Housing. According to the KPMG Report on 'Affordable Housing—A Key Growth Driver in the Real Estate Sector', affordable housing is defined in terms of three main parameters, namely income level, size of dwelling unit and affordability. Where, the third is a dependent parameter that can be correlated to income and property prices.

	Income Level	Size of Dwelling Unit	Affordabillity
EWS	< INR 1.5 Lakhs per annum	Upto 300 sq ft	**EMI to monthly income:** 30% to 40%
LIG	< INR 1.5–3 Lakhs per annum	300–600 sq ft	**House price to annual income ratio:** Less than 5:1 (Task Force headed by Deepak Parekh)
MIG	< INR 3–10 Lakhs per annum	600–1200 sq ft	

Figure 12.1. Definition of Affordable Housing. Source: KPMG.

	Size	EMI or Rent
EWS	• minimum of 300 sq ft super built-up area • minimum of 269 sq ft (25 sq m) carpet area	not exceeding 30–40% of gross monthly income of buyer
LIG	• minimum of 500 sq ft super built-up area • minimum of 517 sq ft (48 sq m) carpet area	
MIG	• 600–1,200 sq ft super built-up area • maximum of 861 sq ft (80 sq m) carpet area	

Figure 12.2. Affordable Housing as per JNNURM. Source: Guidelines for Affordable Housing in Partnership (Amended), MHUPA, 2011.

The **Jawaharlal Nehru National Urban Renewal Mission (JNNURM)** Directorate of MHUPA has also defined affordable housing in its amended Guidelines for Affordable Housing in Partnership released in December 2011 (Fig. 12.2 below).

Climate Zones and Their Characteristics in India

Bansal et al., 1994 had carried out detailed studies and reported that India can be divided into six climatic zones, namely, hot and dry, warm and humid, moderate, cold and cloudy, cold and sunny, and composite. According to a code of the Bureau

Table 12.1. Classification of climates in India.

As per Bansal et al.1994			**As per Bureau of Indian standards (BIS)**		
Climate	**Meanmonthly temperature (°C)**	**Relative humidity (%)**	**Climate**	**Meanmonthly maximum temp (°C)**	**Relative humidity (%)**
Hot and dry	> 30	< 55	Hot and dry	> 30	< 55
Warm and humid	> 30	> 55	Warm and humid	> 30 > 25	> 55 > 75
Moderate	25–30	< 75	Temperate	25–30	< 75
Cold and cloudy	< 25	> 55	Cold	< 25	All values
Coldand sunny	< 25	< 55			
Composite	This applies,when six months or more do not fall within any of the above categories		Composite	This applies,when six months or more do not fall within any of the above categories	

of Indian Standards(BIS), the country may be divided into five major climatic zones. Table 12.1 presents the criteria of this classification as well. In this chapter, the three tropical classification has been taken into consideration, i.e. Warm Humid Climate, Hot Dry Climate and Composite Climate, which have covered major parts of India.

The characteristics of each climate differ and accordingly the comfort requirements vary from one climatic zone to another. As per ASHRAE, thermal comfort is, "that condition of mind which expresses satisfaction with the thermal environment." It is also, "the range of climatic conditions within which a majority of the people would not feel discomfort either of heat or cold". Such a zone in still air corresponds to a range of 20–30 degree C dry bulb temperature with 30–60% relative humidity. Besides, various climatic elements such as wind speed, vapor pressure and radiation also affect the comfort conditions.

Various Dimensions of Affordable Housing

Affordable housing doesnot consider only area, price and affordability of the occupier as per the basic definition; it does possess a few more dimensions also. Affordable

	Size	**EMI or Rent**
EWS	• minimum of 300 sq ft super built-up area • minimum of 269 sq ft (25 sq m) carpet area	not exceeding 30–40% of gross monthly income of buyer
LIG	• minimum of 500 sq ft super built-up area • minimum of 517 sq ft (48 sq m) carpet area	
MIG	• 600–1,200 sq ft super built-up area • maximum of 861 sq ft (80 sq m) carpet area	

Figure 12.3. Various dimensions of affordable housing. Source: Guidelines for Affordable Housing in Partnership (Amended), MHUPA, 2011.

housing needs to be supported with required amenities and facilities for users to sustain and also the optimum level of comfort. Besides defining a minimum house (30 s qm minimum dwelling unit or 10 s qm per capita), a housing cluster should also provide a minimum space of 5 sq. m per capita each for greens/open space/play area, social infrastructure and transport and utilities (Jain, 2016).

Affordable Housing also should be interlinked with local community, health, livelihood and culture.

Other basic facilities required for the Affordable Housing are as follows:

- Infrastructural facilities: roads, drains, water, sanitation, street and outdoor lighting
- Health service facilities
- Educational facilities
- Infrastructural facilities for Community development: Community centers, institution building, gender awareness, vocational training and economic support;
- Small scale occupations entrepreneurship
- Promoting micro-credit facilities and Community based participatory planning
- Community Networking

Thus, when we consider passive ways for energy, natural resources and cost savings, it is required to consider not only housing units, but also all supporting facilities and amenities required by the users.

As per Jones Lang LaSalle's Report, 2012, Affordable Housing should possess the following characteristics:

Minimum Volume of Habitation

With the increasing pressure on urban land, all structures should go vertical, even in case of affordable housing also. This provides flexibility to architects to work on vertical planning of a dwelling unit as well.

Provision of Basic Amenities

Though the definition of affordability covers the minimum area and cost considerations, provision of basic facilities such as sanitation, water supply and other services to the dwelling unit is a must. Also, amenities like community spaces, parks, schools and healthcare centers, either within the project or in the neighborhoods, are desirable with the housing project.

Cost of the House

Affordability of the buyer should not only cover the initial cost of buying the property, it should take care the maintenance and operational cost also. Passive design strategies related to sustainable features is key to any affordable housing project mainly to reduce the operational/lifetime cost of the building.

Location of the House

An affordable housing project should be located within reasonable distances from workplaces and should be connected adequately through public transport. If housing is developed very far away from major workplace hubs in a city affordability will be greatly affected.

Housing development in all countries should aim to improve the quality of life of people by better citizen services, governance and urban mobility. It should also transform the existing housing areas, including slums, into better planned ones with the application of different types of retrofitting, thereby improving livability, as well as physical and mental health.

Though as per Jones Lang LaSalle's Report, 2012, Affordable Housing needs to go vertical. But, high rise living has the following negative environment consequences, such as heat island effect, air pollution, increase in carbon foot print and danger prone, high amount of energy is spent on vertical transportation, water pumps and other basic services.

So, compact, high density, low/medium height housing can give a range of environmental benefits, reduce travel distances and transmission losses and reduce the pressure on land, public transport and services.

Passive Design Strategies in Affordable Housing

Passive Design

Passive Design strategies are the most cost effective means of making buildings energy efficient and thus saving operational cost to a great extent. Passive strategies, when included in initial building design, adds little or nothing to the cost of a building, yet has an effect of realizing a reduction in operational costs and reduced equipment demand. It is reliable, mechanically simple, and is a viable asset to a building.

In passive design the system is integrated into the building elements and materials —the windows, walls, floors, and roof are used as the heat collecting, storing, releasing, and distributing system. These very same elements also act as major element in passive cooling design but in a very different manner. At the same time, passive design does not necessarily mean the elimination of standard mechanical systems.

Passive solar design can reduce 75% energy bills with an added construction cost of only 5–10% (sol power people). This surely adds to affordability of LIG and EWS people of society to a great extent.

Designers can achieve a passive building by following the steps below:

1. Modulating microclimate of the site by proper landscaping
2. Optimizing the building orientation and building configuration
3. Optimizing the building envelope and fenestration to reduce heating/cooling demand
4. To assure maximum possible daylight use in the building
5. Adopting passive heating/cooling strategies to conserve energy

Technological aspects of Passive Design:

- Site Selection, Site Analysis, Site Engineering
- Settlement patterns and Site planning
- Orientation of the Building
- Housing Forms and Building Shapes
- Shading Calculations to optimize radiation and daylighting
- Size and Position of Openings: related to wind flow, daylight and shading
- Indoor temperature balance: careful use of materials for improved indoor conditions.

Relevance of Passive Strategies in Affordable Housing:

Passive design is a process which directly uses natural energy such as solar radiation, daylighting, wind, air temperature, humidity to provide comfort without artificial energy. Passive designs are often valued for their simplicity and inexpensive ways. They also tend to have zero operational costs. As they mostly contain no moving parts, passive designs potentially last for centuries.

Passive design results when a building is created and simply works "on its own". The plan, elevation, section, materials selections and siting create a positive energy flow through the building and "save energy".

Thus, Passive Design strategies are justified by incorporating in affordable housing with much lower operational cost, better environment and also overall improvement of health. Though Passive Design considerations need to be intrinsic to all architecture, but can make special contributions to define affordable housing.

Justification of Passive Building Systems in affordable housing:

- Cost investment is one time , during the construction of the building
- Techniques are incorporated with building elements only
- No operational cost because there is no need of power and control
- Generally requires no or very little maintenance
- Most of the parts are static
- Functional 24 hours per day

Few limitations of Passive Building Systems:

- Some additional initial cost to be taken care
- Building elements are mostly fixed
- There are also limitations in upgradation

Passive Design refers to a series of Architectural Design Strategies to be considered in the design process in order to respond adequately to climatic requirements along with all other contextual necessities. It has been proven that it reduces the energy use of the building by even more than 50%, during its functioning.

In case of Affordable Housing, the basic objective of maintaining thermal comfort by heating, cooling, humidification and dehumidification is required to be achieved mostly by natural means, avoiding as much as possible non-renewable energy sources. Similarly, visual comfort may be maintained with adequate number

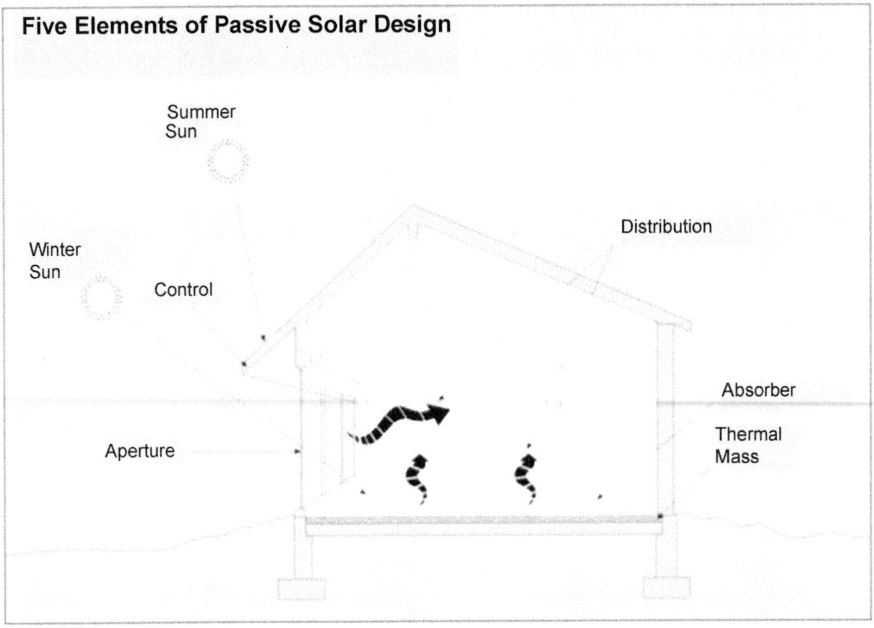

Figure 12.4. Elements of passive solar design. Source: BARN again Solar-Powered Home.

and sized window openings, which will provide enough daylighting. The shape and orientation of the buildings, the arrangements of openings, shading devices and all other design elements are required to be given very careful consideration, in order to take maximum advantage of the climate.

Solar energy has been proven as the most common renewable energy source to be incorporated in passive building design in India.

The distinctive design features for Solar Passive House is as follows:

1. South facing windows (in the northern hemisphere) to trap solar radiation during daytime and to re-radiate that heat energy during night time to keep the indoor environment comfortable.
2. To provide high density thermal mass to store heat energy.
3. West and south-west windows are a source of heat gain during summer should be shaded with sun-shading devices as per solar angle.
4. Generally, the house plan with longer east-west axis and optimized south-facing fenestration will be the best passive solar house in northern hemisphere.
5. The houses need to be well insulated with minimum infiltration.
6. Open floor plans are desired in warm humid climate and compact floor plans are desired in hot dry climate.
7. Passive solar housing should take advantage of the changes of angles of the sun throughout the year, allowing low angle winter sun penetration through the glazing during winter and blocking hot summer sun with shading devices.

Two levels of passive design strategies in Affordable Housing:

Passive strategies need to be integrated in two different levels of Affordable Housing –

1) Settlement level
2) Building level

Settlement level

As the provisions are made on minimum area and cost consideration issues, the provision of basic amenities and facilities, such as water supply, drainage, sanitation, open spaces, schools, and health care facilities in the neighborhood is desirable to provided at settlement level.

1. Passive features at settlement level to be covered are—Preservation of natural resources—existing trees, contours, topography, top soil
   Preservation of existing tree and top soil is affordable because it is cheaper than the conventional practice. It indirectly takes care of environmental sustainability by preserving the eco system, improving micro climatic environment, minimizes soil erosion and manages storm water drainage.
2. Landscaping: (a) Soft landscaping—Native species and (b) Hard landscaping—Pervious pavement: It is affordable and can be made using tiles and stones left from homes. Water run-off can be used for ground water recharge.
3. Water preservation, treatment and reuse
4. Rainwater harvesting
5. Renewable energy based outdoor lighting
6. Maintaining micro-climate
7. Waste Segregation and Treatment
8. Low Embodied Energy and Local Materials Usage
9. Renewable Energy Based Outdoor Lighting

Increasing Temperatures and Heat Stress

In urban areas, one of the reasons for increase in temperature is urban heat island effect. Thus, reduced impervious pavements, increase in vegetation and shaded as well as light colored building surfaces help maintain cooler microclimate.

Building Level

Housing forms and building shapes: Of all geometrical shapes, the lowest surface-volume ratio is that in case of a circular building. The circular form of the building also enhances natural ventilation inside the building. The lesser the Surface-Volume Ratio of a dwelling unit lesser is the heat gained by the building. But since functionally, a circular shape is not ideal, alternative similar alternatives can be hexagonal or octagonal shaped dwelling units.

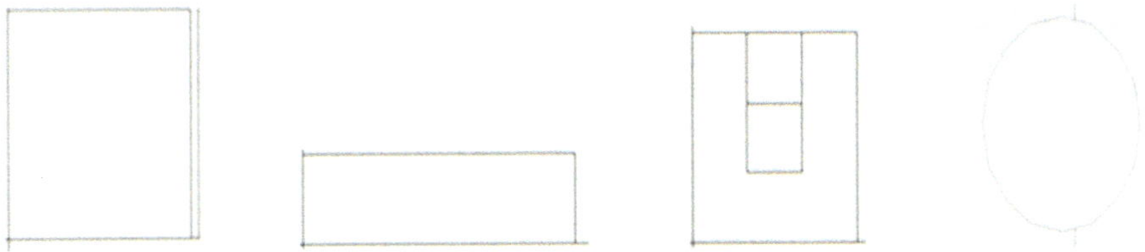

Figure 12.5. Housing Forms. Source: The Energy and Resource Institute (TERI) guidelines, Solar Passive Design for buildings.

Efficient Space Utilization: Minimum and Standard Room Sizes can reduce material wastage and real estate cost with minimum corridor width. Window area should be optimized and wherever possible to provide opening areas, more than by law requirement which will reduce the load on artificial light, enhances ventilation especially for warm humid climate.

Other passive features at settlement level to be covered are use of locally available materials with low embodied energy. Alternate low cost construction techniques may be followed. Indoor temperature may be balanced with comfort ventilation at warm humid climate and with thermal mass at hot dry climate.

Applications of relevant passive technologies at major Indian tropical regions: warm humid, hot dry and composite

Passive building design techniques are solely climate specific. Based on the characteristics of climate, the comfort requirements for each climatic zone are different and the techniques to be adopted in buildings are also different. Following are the guidelines for different climatic conditions in tropical climate for **naturally conditioned affordable housing.**

Table 12.2. Passive design strategies at warm and humid region.

DESIGN OBJECTIVES: To reduce heat gain	**PASSIVE DESIGN SOLUTIONS**
1. Decrease exposed surface area	• East-west orientation • Shape of the building
2. Increase thermal resistance	• Providing insulation in building envelope • Reflective surface of roof
Increase shading	• Building envelope may be protected by verandahs, balconies, overhangs, fins, trees.
3. Increase buffer space	• Verandahs and balconies surrounding the external walls
DESIGN OBJECTIVES: To promote heat loss	
1. Adequate ventilation	• Big size openings • Cross ventilation
2. Increase air exchange rate	• Ventilated roof construction • Courtyards • Wind Towers
DESIGN OBJECTIVES: To decrease humidity	
1. Dehumidification	• To avoid surface water and fountains on windward side

WARM HUMID REGION

In India, the North Eastern region and coastal region (16 states) are under this type of climate.

It is characterized by high temperature and high humidity, causing discomfort for people. The sky is mostly covered by cloud blankets. Daytime and night time temperature variation is negligible.

So, the housing needs to be protected from heat gain and the heat loss also needs to be promoted.

Ventilation and air movement is desirable. The building should have enough provision for cross ventilation.

Design Guidelines for Incorporating Passive Features

For warm and humid climates, where ventilation is considered an asset, desirable conditions for thermal comfort and ambient air cooling can be achieved by following design considerations:

Housing Site: As ventilation and wind flow only can bring comfort naturally, the building should be oriented towards windward side and shaped to be benefited by cool breezes. Surface water to be avoided, as it further increases the humidity.

Following guidelines need to be followed for best utilization of natural ventilation:

a) At least one window should be provided in the windward wall and the other on the leeward side to achieve cross ventilation.
b) Maximum air movement in a room at a particular plane is achieved by keeping the sill height at 85% of the height of the plane. Incidentally, this also agrees closely with the recommended sill height for optimum daylight.

Building Orientation and Building Configuration

Open linear planning is preferred to encourage ventilation. Housing units should be positioned in a staggered arrangement to avail prevailing wind. In warm humid regions, buildings should be oriented to minimize solar heat gain. In the northern hemisphere, an east-west orientation should be preferred.

Arrangement of buildings and urban furniture should permit horizontal winds to enter all living spaces. The choice of urban forms that present minimal obstacles to prevailing winds and building designs for ventilation will maximize enhancement of the same.

In the housing units an unobstructed air path should be made to make sufficient airflow within the building. Semi open spaces like balconies, verandahs and porches may be added for outdoor activities, as the air temperature in the shade is not very high.

Buildings should be designed and oriented in such a way that they prevent indirect and direct solar radiation from entering living spaces and avoid heat storage in building materials. Glazed surfaces exposed to the sun should be avoided, because of the resulting 'green house' effect. Even some heat absorbing glasses should not be used because they heat up when exposed to the sun and re-radiate heat into building interiors.

Building Envelope

The major components of a building envelope are walls, roof, fenestration and shading devices and finishes.

WALLS: Walls must be designed to have maximum airflow. Techniques like baffle wall, fins should be incorporated to divert the airflow inside the building. As the rainfall is more in this climate, external surface of the wall should be protected. Insulation does not significantly improve the performance of a non-air conditioned building. Walling materials should be light-weight.

ROOF: The design of the roof should also promote airflow within the building. Vents at the roof top may be provided for induced ventilation. Separate roof and ceiling, if possible, with an air-gap in between the two, allows air-flow in between the ceiling and roof. Also light-weight roofs should be used as far as possible.

FENESTRATION: In this climate, cross ventilation is the best solution for comfort. So, there should be adequate number of openings in both the windward and leeward side. Openings should be shaded by external overhangs. Stack effect and venturi effect can be used for efficient ventilation. Fenestration area should be 20 to 30% of the floor area to attain the average wind velocity outside. Further increase in the window sizes increases available velocity only marginally.

Materials and Construction Techniques

Building material should be selected with low embodied energy, efficient structural design and construction techniques to save energy and natural resources. Mostly lightweight building materials are used and at the site of heavy rainfall stilted structures are made.

Buildings, urban surface and materials should have low thermal admittance and heat storage capacity, in order to reduce urban 'heat island' intensity. In this case also large open spaces and parks can be beneficial; insofar as they create a cooler environment and may allow fresh air to enter the adjacent built up areas while permitting the incoming wind to descent and penetrate these areas.

Colors, Textures and Finishes

Light colored surface are preferred for less heat gain. The roof surface may be finished with glazed tiles to gain more reflectivity of solar radiation.

Daylighting

Direct sunlight is not desirable. Provisions like light shelves may be used for indirect and diffused lighting.

CASE STUDY: BELAPUR HOUSING IN NAVI MUMBAI, INDIA BY CHARLES CORREA

Charles Correa's housing at Belapur, Mumbai at coastal warm humid region in India, uses and re-interprets traditional Indian urban spatial syntax. It holds lessons for housing design and production in both developed and developing countries.

The housing has been designed as an traditional Indian settlement with one and two storyed high structure, plot sizes are between 45 sqm–75 sqm. There is flexibility and modularity in spaces.

Open to sky space in every housing unit, which again connects to community open space.

Passive Strategies followed: Open planning to encourage ventilation. Slope roof for rainwater drainage, as well as to provide shade with eave projection. Outdoor activity areas at the building level, also at community level and traditional building materials with white plastering. Traditional building method followed with local masons and craftsmanship. Roof structures covered with shingles to provide thermal insulation. Grass grid pavers provided outdoors.

HOT DRY REGION

In India, North western regions are under this type of climate.

Characterized by very high solar radiation and ambient temperature and very low humidity. Sudden temperature falls during night time.

Table 12.3. Passive design strategies at hot and dry region.

DESIGN OBJECTIVES: To resist heat gain	**PASSIVE DESIGN SOLUTIONS**
1. Decrease exposed surface area	• Compact inward looking plan • Orientation and shape of building
2. Increase thermal resistance	• Insulation of building envelope
3. Increase thermal capacity (Time lag)	• Massive structure
4. Increase buffer spaces	• Air locks/lobbies/balconies/verandahs
5. Decreases air exchange rate (Ventilation during day-time)	• Weather stripping and scheduling air change
6. Increasing shading	• High density of settlements for mutual shading • External surfaces protected by overhangs, fins and trees
7. Increase surface reflectivity	• Pale color, glossy surface to reflect solar radiation
DESIGN OBJECTIVES: Promote heat loss at Night	
1. Ventilation of appliances	• Provide windows/exhausts
2. Increase air exchange rate (Ventilation during night–time)	• Courtyards/wind towers/arrangement of openings
DESIGN OBJECTIVES: To increase humidity	
Humidification	• Trees, water ponds, evaporative cooling

So, the housing needs to be protected from heat gain and the heat loss also needs to be promoted.

Design guidelines for incorporating Passive features:

Housing Site: In this zone, land is mostly flat, allowing surrounding areas to heat up uniformly. In case of contour or an undulating site, the north facing slope is desirable to protect from direct solar radiation and dusty winds.

Water bodies are desired for evaporative cooling, as well as to increase humidity. Water is a good modifier of microclimatic.

Open spaces as courtyards or atriums are preferred to encourage ventilation.

Building Orientation and Building Configuration

Building orientation is a very significant design consideration, with regard to two major climatic elements, sun and wind. In hot regions, buildings should be oriented to minimize solar gain, as well as dusty hot air in. In the northern hemisphere, an east-west orientation should be preferred.

The s/v (surface to volume) ratio is determined by the building form, which effects building heat gain and loss to a great extent. For hot and dry area, compact form with low s/v ratio is desired.

Courtyards: An effective passive-design strategy in hot-dry and composite climates with proper placement.

Building Envelope

The key determinants of the amount of heat gain and loss in a building is the building envelope. Flat roofs in housing units are preferable for night time activity and sleeping for hot dry climate because the diurnal temperature range being large, there is sudden fall in night temperature that can assure a comfortable outdoor environment during the night. The material of the roof should be massive. In hot regions, the roof should have enough insulating properties to minimize heat gains.

Among all the elements in the building envelope, windows and other fenestrations are most vulnerable to heat gain or loss. Proper location, sizes, distribution, attachments and detailing on windows, ventilators, skylights, help to keep the sun and wind flow, as desired. In hot dry climate, window openings should be less in number and small in size. All openings should be protected from the sun with shading devices. A wind tower is ideal for cooling effect, but may be an expensive way for bringing comfort in affordable housing. While the use of brick, stone and RCC grill may be used for stopping solar radiation, while allowing ventilation.

Materials and Construction Techniques

Building material should be selected with low embodied energy, efficient structural design and construction techniques to save energy and natural resources. Locally available materials may be largely used to save transportation energy.

Thermal Insulation

Insulation is of great value for controlling of heat flow inward, as well as outward. In hot climates, insulation is placed on the outer face (facing exterior) of the wall so that thermal mass of the wall is weakly coupled with the external source and strongly coupled with interior (Bansal et al., 1994). But, in affordable housing adding insulation may lead to some extra cost.

Colors, Textures and Finishes

Colors with low absorptivity should be used in external finishes. Darker shades also should be avoided to reduce heat gain. Surface may be finished with glazed tiles to gain more reflectivity of solar radiation.

Daylighting

In hot and dry climates due to clear skies, radiation is strong and direct. It creates glare and overheating. So, direct daylight is not preferred and advisable. Accordingly, small openings at higher levels is the design solution for diffused lighting.

Case Study: HUDCO Housing in Jodhpur, Rajasthan, India by Charles Correa

The Central Housing Agency HUDCO commissioned Correa to add 176 houses as an expansion of medium sized district. Correa grouped the houses around a hierarchy of open spaces.

Due to the hot dry desert arid climate, each unit is built around an enclosed courtyard. Locally available construction material, local stones are used as basic building materials in a traditional manner. The unit themselves are massed in single and double storyed blocks.

The residential types provided are two, which are added to as many (of which Correa does not deal). Destined for two distinct social classes, for both designs a space built around a main courtyard. All functional spaces are facing towards the courtyard. The architect assumes as module a square of side 3 meters as a unit for all housing.

COMPOSITE REGION

In India, the central part is under this type of climate.

For designers, it is a challenge to provide design solution for climate with changing seasons, characterized by high air temperature during summer and low air temperature during winter. Composite climate is basically a combination of hot dry and warm humid climate. At some places there is a third season known as cool dry climate.

So, it is a difficult task for architects to design a satisfactory housing for users. Though some constructional features may serve equally well in both the seasons.

Table 12.4. Passive design strategies at composite region.

DESIGN OBJECTIVES: To resist heat gain in summer and resist heat loss in winter	PASSIVE DESIGN SOLUTIONS
1. Decrease exposed surface area	Orientation and shape of building
2. Increase thermal resistance	Roof insulation and wall insulation
3. Increase thermal capacity (Time lag)	Thicker walls
4. Increase buffer spaces	Verandahs, Air locks, Balconies
5. Decreases air exchange rate	Weather stripping
6. Increase shading	Walls, glass surfaces protected by overhangs, fins and trees
7. Increase surface reflectivity	Pale color, glazed reflective tiles
DESIGN OBJECTIVES: Promote heat loss in summer	
1. Ventilation of appliances	Provide exhausts
2. Increase air exchange rate (Ventilation)	Courtyards/wind towers/arrangement of openings
DESIGN OBJECTIVES: Humidification in Dry season and dehumidification in wet season	
1. Increase humidity levels in dry summer	Trees and water ponds for evaporative cooling
2. Decrease humidity in monsoon	Dehumidifiers/desiccant cooling

Housing Site: In this zone, land is mostly flat, allowing surrounding areas to heat up uniformly. In case of contour or an undulating site, the north facing slope is desirable to protect from direct solar radiation and facing buildings towards windward direction during warm humid season.

Building Orientation and building configuration: Moderately compact planning of housing is desirable. Open spaces such as courtyards or atriums are preferred to encourage ventilation, also as shaded outdoor area. A moderately dense, low rise housing is desired.

Building Envelope

The key determinants of the amount of heat gain and loss in a building is the building envelope.

Walls and the roof should be constructed of solid masonry to achieve a 9–12 hour time-lag in heat transmission, which will serve in both hot and cold season. Low rise construction is preferable to have ground to act as better thermal mass.

Materials and construction techniques

Building material should be selected with low embodied energy, efficient structural design and construction techniques to save energy and natural resources.

Colors, textures and finishes

The colors with low absorbance should be used as external finishes on the roof and walls.

Case Study: Aranya Housing in Indore, India by B. V. Doshi

B. V. Doshi's Aranya Housing in Indore, at a composite region in India, is a low cost vernacular incremental, adaptable housing project.

The objective of the project was slum development. The plan was informal with mixed and multiple land uses. The creation of small neighborhoods and houses extending to the outdoors. Houses were clustered in a group of 10. At settlement level it took care of all type of neighborhood facilities. Public spaces provide a place for social gathering in each cluster.

Plot size started from 35 sqm. Plots were mostly small in size and clustered in low rise blocks. Each house had minimum exposure to wall surface and a common wall. Streets were mostly in shade. Openings on the north and south allowed cross ventilation and enough daylighting. Building materials used were locally available, primary materials being brick and stone. Bright colors were used as exterior finish. All building services at building level and at settlement level have been taken care in a very cost effective manner.

Conclusion

Providing shelter for the millions at affordable cost is a formidable task in itself.

The needs for balancing land economics, strategic location, political influences, legalities and social and other economic factors and above all the need for finally making the settlements an overall success, also taking care of the environmental aspects, becomes a really tough task. Beside durability, dweller density, serviceability, another huge technical challenge faced by the designers and providers of affordable housing is the degree of human comfort level that these cost-controlled shelters can offer, because creating cheaper housing does not justify unmanageable discomfort to be borne by the dwellers. The dual angles of environmental sustainability and human comfort level can be addressed to a significant degree by prudent adaption of passive solar techniques in the design of such affordable housings. The proven techniques and materials that facilitate a higher degree of environmental and human comfort, are identified broadly in this chapter. The cases of affordable housings vary among low-rise, mid-rise and high-rise configurations, each having its pros and cons, challenges and opportunities. On a case-specific and particular location-specific way, if these guidelines of solar passive techniques as mentioned here are incorporated and further researched upon for enhanced results, it is surely expected that the provision of affordable housing shall be able to add to itself the elements of better human comfort and environmental sustainability.

References

Anderson, N. M. 2009. Approaches to Affordable Housing Design: Science, Art, Communication and Strategy Iowa State University.

Arvind, K., B. Nick and Y. S. V. Simos 2001. McGraw Hill Education Private Limited. India.

Bansal, A. K. H. N., Jagadeesh and H. Guruvareddy. 2001. Bamboo based housing system. New Building Materials and Construction World 7(6): 33–36.

Bansal, N. K., G. Hauser and G. Minke. 1994. Passive Building Design, Elsevier Science B.V.

Brown, D. K. 1994. Solar Energy and Housing Design. ETSU India.

Bhikhoo, N., A. Hashemi and H. Cruickshank. 2017. Improving Thermal Comfort of Low Income Housing in Thailand through Passive Design Strategies. MDPI.

CatalunyaGeneralitat de. 2004. Sustainable Building Design Manual Volume 1. The Energy and resource Institute, New Delhi.

CatalunyaGeneralitat de. 2004. Sustainable Building Design Manual Volume 2. The Energy and resource Institute, New Delhi.

Fathy, H. 1973. Architecture for the poor: an experiment in rural Egypt Chicago. University of Chicago Press.

Goulding John R., Owen J. Lewis and Theo C. Steemers. 1994. Energy in architecture. The European passive solar handbook, Readwood books, Wiltshire.

Guidelines for Affordable Housing in Partnership (Amended), MHUPA, 2011.

Habitat. 2015. Global Homelessness Statistics, https://homelessworldcup.org/homelessness-statistics/.

Jain, A. K. 2016. Housing for All, pp. 2–9, HUDCO Shelter, vol 17, no 1.

Janmejoy Gupta, Manjari Chakraborty, Arnab Paul and Vamsi Korrapatti. 2017. A comparative study of thermal performances of three mud dwelling units with courtyards in composite climate. Journal of Architecture and Urbanism, 41(3): 184–198.

Jones Lang LaSalle's Report. 2012. RICS Research Report, Making Affordable Housing Work in India, November 2010.

Kabre, C., U. Pottgiesser and J. P. Sharma. 2012. Energy Efficient design of Buildings & cities. Ram University of science and Technology, Murthal India.

Karuppanna, S. and A. Sivam. 2009. Sustainable Development and Housing Affordability. Institute of Sustainable System and Technology, Australia.

Koenigsberger, O.H., T.G.I Ingersol, Mayhew Alan and S.V. Szokolay. 1998. Manual for Tropical housing and building; Orient Longman.

McGee, C. 2013. http://www.yourhome.gov.au/passive-design.

Nayak, J. K., R. Hazra and J. Prajapati. 1999. Manual of solar passive architecture. Solar energy center, MNES, Government of India.

Palit, D. 2004.Green Buildings. World Energy Efficiency Association.

Roaf, S. C. 2001. Natural ventilation of buildings in India. pp. 157. *In*: Krishan, A., N. Baker, S. Yannas and S. V. Szokolay (eds.). Climate responsive Architecture: A Design Handbook for Energy Efficient Buildings. New Delhi: Tata McGraw-Hill Publishing Company Limited.

Shinde, S. S. and A. B. Karankal. 2013. Affordable housing Material & Techniques for Urban Poor's. North Maharashtra university, India.

SP 7: 2005. National Building Code of India 2005, BIS, New Delhi, 2005\.

The Energy and Resource Institute (TERI) guidelines, Solar Passive Design for building

Yannas, S. 2001. Passive heating and cooling design dtrategies. pp. 82–83. *In*: Krishan, A., N. Baker, S. Yannas and S. V. Szokolay (eds.). Climate Responsive Architecture: A Design Handbook for Energy Efficient Buildings. Tata McGraw-Hill Publishing Company Limited, New Delhi.

filters-now.com

CHAPTER 13

Tiny Houses

An Innovative and Sustainable Densification Solution

Heather Shearer

Introduction

The tiny house trend has grown to such an extent that it is no longer on the fringes of the housing market, and is regularly reported in the mainstream media. Of course, tiny dwellings are not unusual and are the norm in most countries in Asia, Africa, South America, and even Europe. Nor are very large dwellings particularly unique. Nonetheless, the trend for 'average' people to build very large houses, sometimes known as McMansions, is mostly confined to countries (mostly ex colonies of the UK) such as Australia, Canada, and New Zealand. Perhaps unsurprisingly, these are also the countries where the tiny house movement is most apparent. Yet, even in Australia, very small houses are common, and include granny flats, converted sheds or containers, relocatable houses, prefabricated houses, studio units, beach shacks, caravans and even houseboats. So what is it that defines a tiny house, and moreover, how can a tiny house be a sustainable and affordable option for the urban environment? This chapter briefly details the method used, then explores a working definition of tiny houses. It then describes some of the economic and environmental sustainability benefits that tiny house living can bring to the city; as well as a range of barriers, that pertain to tiny houses in the Australian context, and uses case studies to show how these might be overcome through modifications of planning schemes. It concludes that tiny houses can be a sustainable option for urban infill, and as a housing form, can address a number of housing and environmental needs.

Griffith University, Southport, Queensland 4222.
Email: h.shearer@griffith.edu.au

Methodology

First, given that the term 'tiny house' is so unclear, methods used to come up with a working definition included a textual search and content analysis of Facebook pages as well as interviews. Analysis of the sparse academic literature on tiny houses, as well as popular tiny house books and websites was also conducted. This resulted in a broad description of the some general commonalities of tiny houses, and their placement within the actual tiny house movement. These include size, mobility, design, environmental sustainability, and affordability.

Second, a survey was created in SurveyMonkey™ and posted online to a number of Tiny House Facebook groups, as well as sent by email to respondents known to be interested in tiny houses. The survey consisted of 18 questions, including whether the respondent had built or was intending to build a tiny house, type of tiny house, location, cost and demographics. Questions were also asked on perceived drivers and barriers to building tiny houses. Finally, the author of this paper attended a number of tiny house specific meetings (organized by tiny house Facebook Groups and other organizations), as well as held informal interviews with tiny house advocates.

What is a Tiny House?

Despite the popularity of tiny houses on Facebook, Instagram and other social media, the definition of tiny houses is somewhat amorphous. This is partly because the tiny house movement is largely under-researched, and information is primarily from non-academic sources such as social and popular media. The only common criteria on tiny houses is size and tiny house books and the like state that a tiny house is generally under 400 square feet (about 37 square metrs). But if the sole criteria is size, then surely any type of dwelling, even an apartment, could be considered a tiny house? In practice, yes, but the majority of tiny house advocates would say that there is more to it than just size. So obviously in the context of the tiny house movement, tiny houses are something different, over and above a mere small dwelling.

Other definitions vary widely, and range from the specific; "A tiny house is a house built to fit within the New Zealand Transport Agency's light trailer (TB class) specifications" (Langston, nd) or a house built on a trailer "that conforms to the maximum trailer sizes that govern shipping containers and RVs" (Murphy, 2014) to the general: "There are many different kinds of houses and tiny houses are no exception. Because tiny houses are built individually, not just moulded and assembled, there are many different styles to choose between. From log cabins with wood stoves to modern designs with integrated solar roofs" (RPMNational, 2015) and, "people have tried to define exactly what a tiny house is, and these definitions have brought much debate" (Mitchell, 2014). Some tiny house advocates also emphasize the importance of a differentiation between a tiny house and a Recreational Vehicle (RV) or caravan; yet others regularly share pictures of their converted buses, caravans and trucks.

Size

In practice, the size of houses identified as tiny houses ranged from the very small (under 10 square metres) to over 40 square metres. The size limit of 400 square feet (37 square metres) however, is primarily because the original tiny house advocates in the USA built their tiny houses on trailers; and 340 square feet is the maximum possible size that could be built on a trailer base in the USA (Mitchell, 2014); thus the size was rounded up to 400 square feet. It is also important to note however, that size is relative, and may be considered absolute (a total size), or per capita. In addition, many tiny houses may also include additional space, such as outbuildings, outdoor kitchens, storage sheds, large decks, etc.

Mobility

Another characteristic of tiny houses, partly related to the size, was mobility. Tiny houses generally tended to be fully mobile (on a trailer base), partly mobile (on a trailer base, skids, or otherwise moveable) or permanent. Generally, tiny houses, no matter on wheels or not, are moved infrequently; and are not designed to be fully mobile, such as RVs or caravans. If on wheels, they are (usually) compliant with trailer regulations, but are not particularly suited to frequent movement, as they are constructed as houses, and are often extremely heavy (greater than 3 tonnes, for example) and have poor aerodynamics for towing. Nonetheless, a subset of tiny house advocates are more mobile, and live in caravans, RVs, converted buses or trucks; these include the Grey Nomads in Australia, snowbirds in the USA, itinerant workers, travellers, etc. Unlike tiny house dwellers, there is substantial research on such people, and they are not usually considered as part of the tiny house movement (Mitchell, 2014; Murphy, 2014; Onyx and Leonard, 2005).

Economic Freedom

The desire for economic freedom (in various guises) is a strong characteristic of tiny house advocates. Research by Shearer (2015a,b) shown in Fig. 13.1, showed that of the drivers for tiny house living, the desire for 'freedom' was the top ranked motivation for wanting to build a tiny house, and other economic variables, such as reducing costs, transitioning to part time work, reduction of debt including mortgage debt and housing affordability. This is supported by other studies, showing that the "...biggest incentive for living tiny is saving money" (Kilman, 2016).

Environmental

Green building principles are highly relevant to the tiny house movement; and these are generally ranked similarly high as economic drivers. The environmental drivers that motivate tiny house living is complex, and include not only the building material but also the desire to live more sustainably, general environmental values, wanting to downsize and minimize possessions and to live off the grid (Fig. 13.1). Of course,

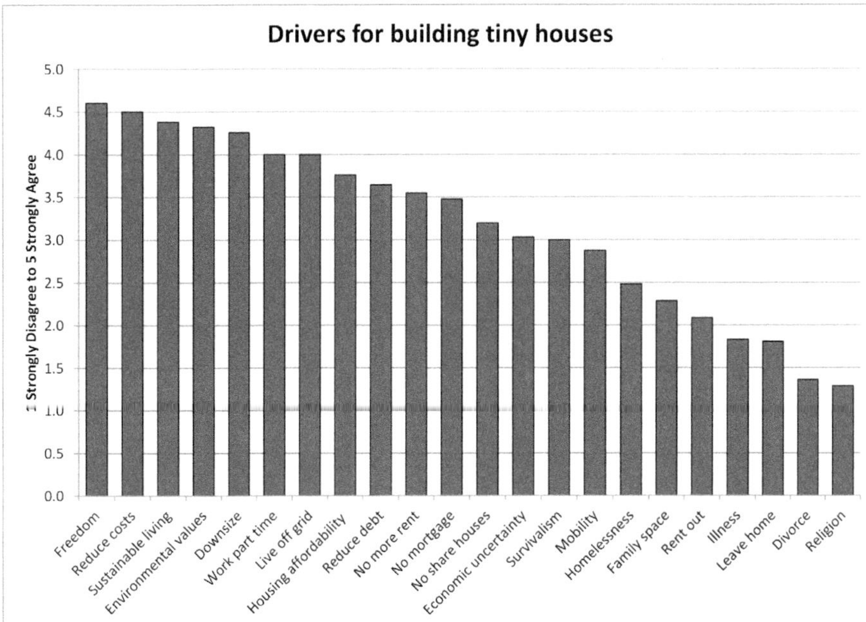

Figure 13.1. Main drivers for building a tiny house (summed Likert scale scores in descending order). Source: Shearer (2017).

both economic and environmental drivers are to some extent interrelated, as there is a strong ethos of wanting to escape the consumerist culture for environmental and social reasons. This is discussed further in the Environmental Sustainability section.

Design

A strong emphasis on architecture and design of tiny houses is evident. Design is also fundamental to the concept of sustainability, in that it informs the function and lived experience of the dwelling as it interacts with the built and natural environment (Ford and Gomez-Lanier, 2017). Tiny houses can be considered as 'intentional dwellings' with a deliberate "use of space, materials, light and function" (Mitchell, 2014). Tiny houses are specifically designed for the people who will live in them, and not for resale purposes. Thus they are often unique, creative, and indeed, beautiful. Tiny houses also epitomize an idea of 'home', as a cosy refuge or the adult equivalent of the childhood 'fort'.

Related to design is a strong Do It Yourself (DIY) mentality, in that tiny house design and construction is commonly undertaken by the individual/s intending to live in the house, and often friends and family. This gives the owner/builder almost total control over almost all aspects of the house, including construction material, plans, interior finish, furniture, etc. Interestingly, more women than men build their own tiny houses thus they can also be considered as empowering (Mitchell, 2104). Tiny houses are also frequently built for quality, and over a much longer period than a project house.

Table 13.1. Types of tiny houses.

Typology	Description	
Type 1. Small and mobile		
1a	**Iconic tiny house on a trailer.** • Mobile, on a trailer (i.e., Tumbleweed) • House fully owned, trailer and vehicle might be under finance • Owner/friends/family built (*in the USA, you can buy tiny houses*) • Size: < maximum dimensions allowable on a trailer • Inhabitants 1–2 • Moves from construction site to permanent or semi-permanent site in rural land, free campsites, friends/family land, national parks, caravan parks	
1b	**Relocatable tiny house, moved to site then fixed or semi-permanent** • Relocatable, cabin, shipping container, 'tiny house', mine hut, kit home, etc. • House fully owned; land owned, mortgaged or rented • Owner/friends/family build • Size –larger than 1a • Cost (likely greater than 1a, but not huge) • Moved infrequently (< 6 months *selected this as the general min lease time*) • Inhabitants (1 or 2, sometimes children) • Semi to permanently placed on land (either owned or rented)	
1c	**Fully mobile dwellings: Caravans, boats, bus, trucks, tents, tepees** • Environmental ethos as per tiny house movement • Very small (smaller than 1a and 1b) • Cost; varies wildly (< A$1000 to A$1,000,000+) • Not generally built from scratch, purchased new/used or repurposed (e.g., bus/truck) • Generally very mobile and often temporary structures • Caravan parks, free camping, state forests, friends/family property,	
Type 2. Small and permanent		
2a	**Purpose built tiny house/cottage** • Size, under 60 square meters (could be bigger if more inhabitants) • Detached • Permanent • Own land or rented • Owner/friends/family built, or by a registered builder • Generally in a rural/rural residential area • Cost (varies widely, likely $50 k plus, depending on material and builders)	

Table 13.1 contd. ...

... *Table 13.1 contd.*

Typology	Description	
2b	**Converted non-residential building (shed, garage, barn)** • Under 70 square meters (could be bigger if more inhabitants) • Detached or semi-detached (attached to a larger property) • Any area, though generally suburban to rural (suburban legality depends on LGA) • Temporary (at least, in theory) • Often rented from family or built on own/family/friend's land • Cost?	
2c	**Apartment/cottage/townhouse in community (intentional or co-housing)** • Shared facilities (i.e., kitchen, garden, tools) • Rented or owned (or alternate tenure such as tenants in common) • Cost, variable, depends on tenure and location • Location, can be CBD to rural • Often have a guiding philosophy (i.e., eco-friendly)	

Image 1, by Lynn Cargill; Images 2–4, by Heather Shearer; Images 5 & 6, Country Living Magazine, 2017.

History of Tiny Houses

In the USA, a response to similar problems resulted in the birth of the tiny house movement. The tiny house movement emerged in western USA in the late 1990s in response to affordability problems, and the desire to live more sustainably and later, to the Global Financial Crisis (Mutter, 2013). Significant early adopters included the writers Sarah Susanka and Lloyd Kahn, as well as Jay Schafer who founded the Tumbleweed Tiny House Company, and espoused his own philosophy by living in a tiny house himself. Although the tiny house on wheels trope originated after the 1970s, the movement to smaller dwellings has its antecedents in much earlier times, even as far back as the 1850s, with the publication of Walden by Henry David Thoreau (Anson, 2014; Diguette, 2017; Ford and Gomez-Lanier, 2017) in which he espoused the values of self-sufficiency, elimination of debt and building one's own tiny house. In the 20th century, Jean Prouve's 64 square metres prefabs were built to ease post war housing shortages in the 1940s (Fig. 13.2); and tiny houses also had a resurgence concurrently with the environmental movement of the 1960s.

The original tiny houses in the USA were very small, and nearly all were built on wheels (trailer bases). This was less to do with wanting to be permanently mobile, as say, in a caravan, but because the planning laws of the towns were highly restrictive, in some areas mandating a minimum size of house, mandatory connection to utilities and of course, ownership of land, which many advocates of tiny house living could not and still cannot afford (Kilman, 2016). There was also a strong desire to be environmentally sustainable, to be off the grid, and for freedom and independence. This latter point

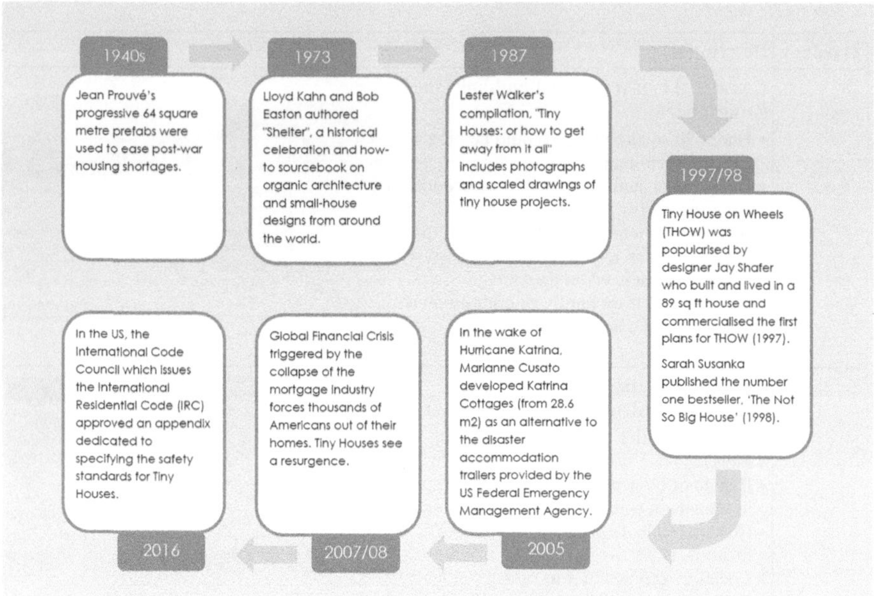

Figure 13.2. A Short History of the Tiny House Movement (from Bares et al., 2017).

is particularly relevant in the birthplace of the tiny house movement, the USA. The USA has a strong anti-establishment, even libertarian worldview which meshes well with the tiny house movement. Indeed, Thoreau and another contemporary, Ralph Waldo Emerson, were strong advocates of individualism and minimal government interference in the lives of citizens (Wikipedia, 2017).

Economic Sustainability of Tiny Houses

Our big, Expensive McMansions

Since European settlement, Australian housing has largely been characterized by the archetypal detached house with a big backyard. This is similar to other ex-colonies of the UK, such as the USA, where land was and is still seen as an abundant resource. More recently however, all three levels of Australian governments promote urban densification policies, which are aimed to consolidate urban settlements into a smaller footprint, reduce resource use and encourage public transport use rather than private motor vehicle uses. In response to these strategies, and also to the cost of land and changing cultural makeup, a wider variety and number of house types has emerged, including low and high rise apartments, semi-detached properties and master planned projects such as urban villages. Nonetheless, the dominant urban form in Australian capital cities is a small CBD and inner city area, dominated by high rise commercial and residential buildings, surrounded by large areas of mostly detached houses. Interspersed around the city are areas of semi-detached properties, either as semi-detached townhouses or duplexes, or retirement village type developments.

These detached houses are generally very large by world standards, with Australia having some of the largest houses in the Organisation for Economic Co-operation and Development (OECD), and new houses currently averaging around 241 square metres (Cox and Pavlevitch, 2014; Worthington, 2012; Yates et al., 2006). As is shown in Fig. 13.3, mean house sizes have increased from around 100 square metres in 1950 to over 240 square metres in the 21st century, while at the same time, household sizes have decreased (from approximately 3.6 to 2.6 persons per household). Thus, contemporary households now have over three times the space per person (93.5 square metres) than those in the 1950s (27.7 square metres).

In addition, the vast majority of detached houses are three bedrooms or larger (see Fig. 13.4), but in contrast, the majority of apartments are two bedrooms or smaller, and where larger apartments exist, are usually found either in expensive and desirable areas, or as townhouse complexes in master planned estates relatively far from the city. This housing type also includes retirement villages, which are limited to certain demographics (usually over 55s), and do not appeal to all.

As is obvious in Fig. 13.4, there is little choice available for either large apartments or small detached houses. There remains a strong desire for one's own space, as in the Great Australian/American Dream, for a detached house with a backyard (or at least, to have private space around one's dwelling) (Worthington, 2012). This desire **not** to live in apartments is one of the major drivers of the tiny house movement, and one reason why they are usually not considered as tiny houses (although many are, of course, extremely small, even down to the 'coffin' apartments in Hong Kong and Japan where people pay about US$300 per month for a room measuring 3x6 feet (18 square feet) (Taylor, 2017). Indeed, a recent post about this trend on to a tiny homes Facebook page (Tiny Houses Brisbane, 2017) met with extreme negativity from members of the group. This response emphasized that, although very small, such apartments are *not* considered tiny houses, at least by tiny house advocates in developed, western countries.

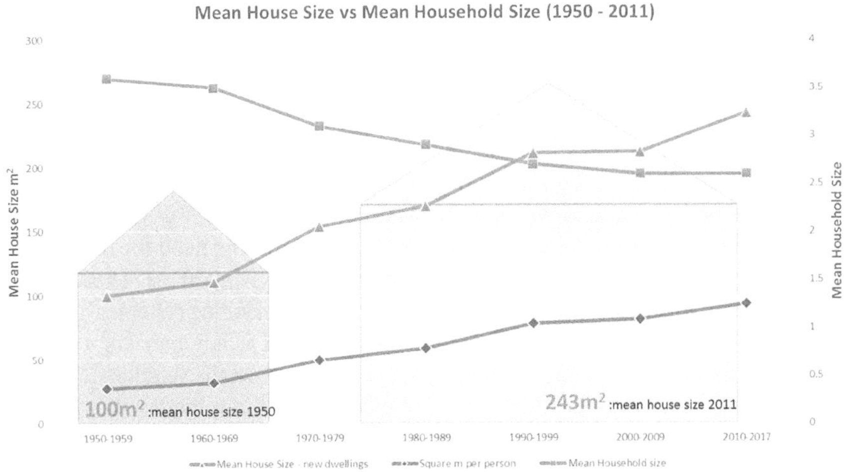

Figure 13.3. Mean house and household sizes over time (Australia).

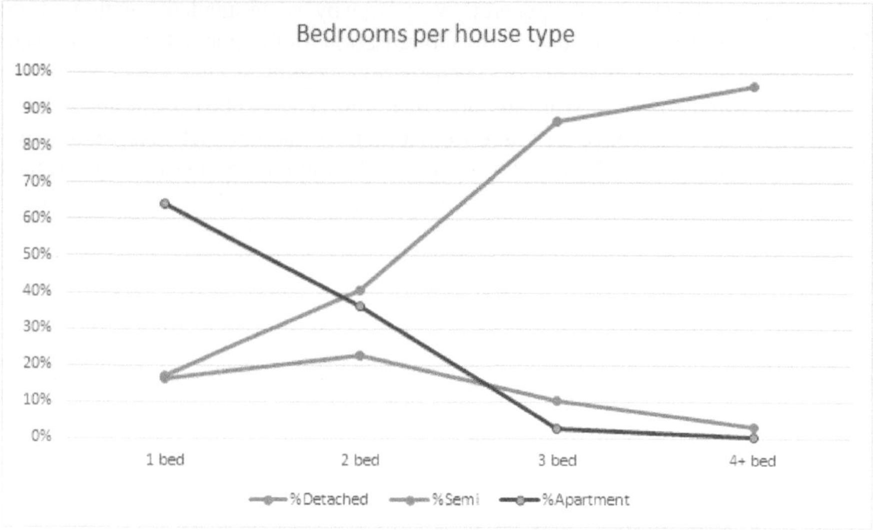

Figure 13.4. Bedrooms per dwelling type (based on ABS, 2012).

Moreover, there is a locational aspect to the conundrum of house sizes and whether they are fit for purpose, and this is related to affordability of housing. Australian houses are not only the second largest in the OECD, but also amongst the most unaffordable, with all major housing markets in Australia ranked as severely unaffordable, a trend that shows no signs of slowing (Cox and Pavlevitch, 2014; Worthington, 2012). Similar to Australia, the decline in housing affordability is also common to nations such as Canada, New Zealand and the USA (Berry et al., 2006; Cox and Pavlevitch, 2014; Yates, 2008). The continued rise in house prices is a complex, multi-faceted issue, with reasons including, on the demand-side, population growth and in migration to urban areas. Other demand-side drivers of house price rises include easily accessible housing finance, tax incentives such as negative gearing and lack of capital gains tax on the family home, and a 'strong cultural preference for owner-occupied detached houses' (Worthington, 2012; Pawson et al., 2015). These are important points; the demand for housing, including for tiny houses, is for detached dwellings in the inner and middle areas of the capital cities, and not for the more affordable regional areas of Australia, where houses can be bought relatively cheaply, even under A$2 00,000.

On the supply side, the rise in house prices has been attributed to inefficient planning processes and development assessment, mostly by state and local governments, resulting in inflexible and slow responses to the need for new housing stock, and lack of or poor infrastructure (Cox and Pavlevitch, 2014). These are also directly related to the tiny house movement, as inflexible planning schemes and overly onerous development assessment requirements are noted as key barriers to building tiny houses in most urban areas. Generally speaking however, the scholarly consensus on house price rises attributes it to a combination of all these factors, but primarily demand-side issues (Pawson et al., 2015). It must also be strongly emphasized that unaffordable housing is not just an issue related to purchasing property; indeed, housing stress is often more marked in renters than purchasers. The rental market is largely uncontrolled with respect to the minimum proportion of income to rent. Property purchaser buyers generally have to demonstrate that they have a deposit and the mortgage will not cost more than 30% of their gross income, but renters, particularly low income renters, will often spend 60% or of their income on housing. The rental market too is very insecure, and rental tenures are rarely more than 12 months. Moreover, leases frequently prohibit pets and will rarely allow (or pay for) any modifications by the tenant, for example, to improve the sustainability of the dwelling.

Unaffordable housing is not only a problem for those who cannot afford to live in their desired area; it can have significant negative consequences for the city itself; with implications for general economic growth. It can also lead to increased polarization of the population, with affluent people living in the inner cities, and the lower-socio-economic classes living further and further out of the city. This exacerbates issues such as traffic congestion, GHG emissions, land clearing, etc. Clearly, reducing both the size and the cost of houses can save significant amount of resources and energy.

Cost of Tiny Houses

A desire to achieve economic independence is a strong driver for building a tiny house. Research on the drivers to building tiny houses showed that proponents viewed them both as a vehicle to own one's own house, but at the same time, reduce debt such as mortgages and other debt such as credit and store card debt (Shearer, 2015a,b). Related to this is the concept that 'debt consumes time', and getting rid of debt will give more disposable income, provide a buffer against property and market crashes, and reduce future risk of energy, water and food shortages, as well as escape natural disasters. Tiny houses are seen as a means to 'get out of the rat race', give more flexibility in working, enable working from home, and even provide an income by renting out the tiny house, such as on Airbnb. Some property owners wanted to build a tiny house so that their adult children or aged parents could live on the same property, yet still be independent. Anecdotally, others wished to build a tiny house and live in it themselves, and rent out their larger dwellings.

A common theme running through all tiny house publications is that of 'freedom', which can be unpacked to mean both economic freedom, and freedom from government interference. If one's income is not being spent on what is usually the largest cost in a household, the mortgage or rent then this can give freedom from the responsibility and time commitments of a full time job and commuting; it also means freedom to travel, reduce chores, to follow one's passion or to find adventure. Another strong motivation is freedom from planning restrictions and building codes (including the cost of compliance). This of course, reflects the origin of the Tiny House Movement in the USA, with its 'first world' Libertarian and frontier world views (Wikipedia, 2017).

Of course, economic sustainability also refers to the cost of building a tiny house. Broadly speaking, are more affordable than standard houses, sometimes costing under A$20,000 (Kilman, 2016), due to the cost of purchasing fewer construction materials, the frequent use of recycled or upcycled materials, and because they are usually owner built. Of note, the majority of tiny house costings available do not include the cost of the owner/builder's time. However, full cost accounting of building a tiny house, and the time spent building it, has shown that tiny houses can and probably do cost *more* per square metre than standard houses (Nobel, pers. comm. 2016), as the larger the house, the more that economies of scale reduce the price per square meter.

Another difficulty with estimating building cost of tiny houses is that these are often built for cash, or using scavenged or recycled building materials. Mortgage finance is not available for temporary dwellings such as tiny houses; and other than cash, these are financed by short term loans (i.e., personal or vehicle loans), mortgages on land, credit cards, family loans, microloans, gifts, crowdsourcing, joint purchases or even proceeds of an annuity.

In the USA, where the tiny house market is much more mature, these can be bought direct from manufacturers or builders, in a multitude of designs, fixtures and fittings, costing from around US$50 to US$100,000 (Kilman, 2016). Many tiny houses are also interchangeable with 'trailer' type houses or other relocatable dwellings, which come in a myriad of styles. In Australia, the nascent tiny house market is far less developed, and despite a number of tiny house builder start-ups, is still only on the very fringes of the large construction industry. Australia does have however, a well-developed granny flat/relocatable dwelling market, but as most of these have been designed for mining sites, remote communities and holiday parks, they are not particularly attractive or sturdy, and are also expensive (in comparison to the USA market).

The real way in which tiny houses can be much more economically sustainable however, is not due to their building cost, but due to the location of the tiny houses. As discussed above, Australian capital cities are ranked as highly unaffordable. This affordability issue is not related to construction cost, but to the cost of urban land. Various forms of tiny houses, such as backyard leases, tiny house communities, and tiny house 'granny flats' are ways to site such properties in areas of high demand (i.e., the inner and middle suburbs of the capital cities) yet not incur the same type of cost as buying an existing property, or buying a lot and building. This will be discussed further in the section on how tiny houses can be provided for in local government planning schemes.

As well as economic sustainability, a major driver of tiny houses is environmental sustainability; with almost all tiny house Facebook group members stating that they want a tiny houses so they can live more sustainably, off the grid, and minimize their footprint on the earth (Shearer, 2015b). Thus, can tiny houses be environmentally as well as economically sustainable?

Environmental Sustainability

Environmental Impacts of Large Houses

Very large houses are unsustainable; the larger the house, the greater the amount of resources used in its construction, and its ongoing operation and maintenance (Kilman, 2016; Wentz and Gober, 2007; Wilson and Boehland, 2005). Traditional building construction uses a significant proportion of non-renewable resources, such as fossil fuels, hardwood timber, concrete, plastic and metal products. It also creates huge amounts of waste material, either from new construction, or renovation of old housing stock. Heating and cooling buildings also uses energy, as well as water, thus contributing to greenhouse gas (GHG) emissions. Housing ranks second only to transportation in its contribution to the GHG emissions (Jones and Kanmen, 2011; Carlin, 2014).

The majority of suburban houses (as well as high rise apartments and duplexes) are frequently built by large development companies or building contractors (Shearer et al., 2016). These firms generally use standard sized and sourced building materials, with purchasing based on cost and not environmental factors. Many building materials such as carpeting, finishes like paint and varnishes, cabinetry, etc., release toxins such as Volatile Organic Compounds, often for years, and these can pose significant health hazards (Ford and Gomez-Lanier, 2017). Their manufacture too uses large amounts of energy, and produces GHGs and other pollutants, further impacting the natural environment.

In addition to the environmental impact of the construction and ongoing operation of large houses, these are often poorly designed, requiring heating and/or cooling. In many contemporary housing developments in Australia, the native vegetation is almost completely cleared, and the lots are insufficiently shaded. Many houses too are built without features that help ameliorate climate extremes, such as eaves or awnings, have dark coloured roofs, and are often built without taking aspect or prevailing winds into account. These houses then require heating in winter and cooling in summer, all of which use significant amounts of power, and contribute to GHG emissions. The trend to smaller lots in such developments has not improved the situation, as houses tend to be larger, with smaller gaps between them, further preventing cooling breezes; and large expanses of hard surfaces increase the heat island effect. Moreover, should a house buyer wish to build a smaller house, the sales people at the display villages will often subtly (and not so subtly) dissuade them from doing so, saying that they will lose resale value by doing this.

Urban Density and Environmental Impact

In addition, when housing is unaffordable (economically unsustainable), it can also lead to indirect environmental impacts. In search of affordable housing, many people move to the outer suburbs or regional areas that are often an hour or more from employment, schools and even retail. These areas too are often characterized by inadequate public transport, so a high proportion of people use private motor vehicles to access work and services. Private vehicle use is a major contributor to GHG emissions, and it also results in fragmentation of the landscape, and pollution impacts. Moreover, such outer suburbs have increased vulnerability to the impacts of climate change (Dodson and Sipe, 2008).

The majority of city managers and State agencies in Australia promote ideals of urban densification, but in practice, the cities are characterized by a small dense CBD, and large expanses of suburban detached houses. But large residential high rises can be even more unsustainable than detached houses, often using significantly greater amounts of water and energy per capita in day to day operations (Pullen et al., 2006). This is largely due to the closed in nature of such dwellings, necessitating high energy uses such as clothes dryers and air-conditioning; as well as high levels of water use from water features such as large swimming pools, which have to be emptied and refilled often weekly (Pullen et al., 2006). These buildings also have extremely high embodied energy, because their construction involves the extensive use of steel, concrete and other high strength materials. While GHG emissions per household are higher in detached dwellings, mostly because of increased use of cars, but per capita, they are higher in apartments, because of high embodied and operational energy used in these dwelling types.

It is beyond the scope of this chapter to give an exhaustive description of the environmental impacts of building construction, urban sprawl and the like, but it is clear that neither the horizontal sprawl of the large detached houses of the suburbs, or the vertical sprawl of the inner city high rise apartment buildings are particularly sustainable. So, what about tiny houses?

Sustainability of Tiny Houses

Environmental sustainability and the related desire to minimize possessions and consumerism are key drivers of the tiny house movement. Tiny houses have very low energy and water requirements, and are frequently built 'off-grid' (not connected to urban utilities and generating their own solar power, collecting rainwater and using composting or chemical toilets) (Kilman, 2016). Their much smaller footprint means that they use proportionately less resources than larger houses, and those resources tend to be more sustainably sourced. They are often built from found or recycled building materials, natural products such as stone, rammed earth or sustainably harvested timber, and even industrial waste such as containers or old tyres (as in 'Earthships'[1]).

Figure 13.5 (below) shows some data from the United States about how tiny houses can be more sustainable than standard houses. These data are generally applicable to other countries, including Australia (for example, the USA and Australia have similar sized houses). As the infographic shows, the construction of tiny houses (even if they are built using standard building products and not recycled or reused products, which is more common in tiny house construction) uses less timber, and other construction material. Tiny houses also create far less carbon dioxide emissions, merely due to their size, but also because they frequently are powered by solar sources and 12V power/ batteries instead of electricity from the grid. They also produce less waste, including sewerage, as often use composting or other sustainable toilets. Water too is obtained from rainwater or other sustainable sources, and even if reticulated water is used, the small size of the household (often singles or couples) and the small appliance size means that much less water is used than a standard house, and virtually none, if any, water is used for outdoor purposes such as watering lawns or water features. Moreover, tiny houses use less land, and the sharing ethos of many in the tiny house community means that people living in tiny houses or tiny house communities will share things such as cars, appliances, power tools and other consumer goods.

On top of using less non-renewable resources, tiny houses also have secondary environmental benefits. Tiny house advocates express a strong desire to minimize

[1] "An Earthship is a type of passive solar house that is made of both natural and upcycled materials such as earth-packed tires, pioneered by architect Michael Reynolds." https://en.wikipedia.org/wiki/Earthship.

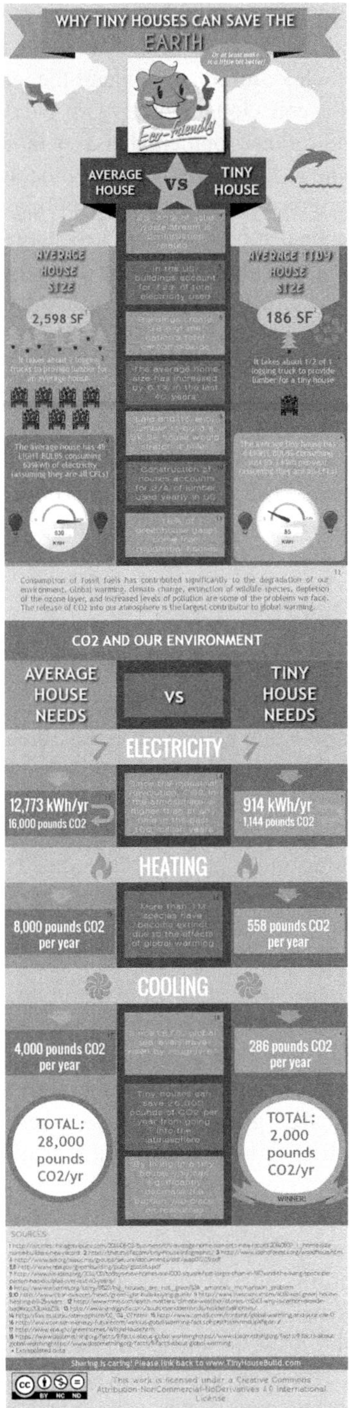

Figure 13.5. How tiny houses can save the earth (Source, Tiny House Build, 2014).

possessions and reduce consumerism (although of course, this is not unique to the tiny house movement, and has been a popular theme in environmental movements for many decades). Nonetheless, minimization and the lack of space in a tiny house means that consuming is a conscious choice, with every new purchase evaluated whether it is necessary, and for environmental and social harm (Kilman, 2016). Tiny house advocates often opine that stuff is making people unhappy; it ties them down, costs money and is largely unused. Conscious consuming on the other hand is aimed to reduce debt and possessions, and prevent environmental degradation and waste.

Living in a tiny house, particularly if the house is off-grid, forces the inhabitants to directly confront the use and waste of resources that is generally invisible in contemporary society (Anson, 2014; Kilman, 2016). The inhabitants of the off-grid tiny house have to ensure that they have sufficient sunshine (or battery backup) to power their house; that their water tank is full, whether this is from rain or from purchasing water; and to directly influence how their own waste is reused or disposed of (Anson, 2014; Kilman, 2016). Moreover, the limited space in the tiny house means that people are almost forced to interact with the outdoor environment, instead of, as in the contemporary McMansion, seeking solitude in one more empty room (Kilman, 2016).

Barriers to TIny Houses

Planning Provisions and Tiny Houses (where do you put the tiny house)

As we saw in the history of the tiny house movement, the impetus behind the genesis of the tiny house on wheels that became popular in the late 1990s, was largely planning restrictions in the USA. Some have even mooted that legal and planning issues pose the greatest barrier to tiny houses (Ford and Gomez-Lanier, 2017). For example, in many USA municipalities, planning and zoning rules mandated *minimum* house sizes; with house sizes based on the International Residential Code, which mandates that houses should have one room that is no smaller than 120 square feet, and other rooms must be at least 70 square feet (6.5 square metres) (Kilman, 2016; Nonko, 2016). People wishing to build a small house were stymied by minimum size rules, thus built their tiny houses on trailers to circumvent planning scheme restrictions (Nonko, 2016).

In Australia, maximum sizes in Local Government Authority (LGA) planning schemes are more common; although these are not aimed at the primary residence (which can be almost any size, given site coverage restrictions), but for ancillary (secondary) dwellings such as ‘granny flats’. These planning scheme rules differ by LGA, by State and by zone and lot characteristics within each zone, within each LGA, so are highly complex to interpret. For example, one LGA planning scheme permits up to 65 square metres granny flats no more than 25 m from the main residence and the neighbouring LGA permits up to 100 square metres granny flats no further than 20 m from the main residence.

Ancillary dwellings built by the owner of the land are, however, a relatively simple, and well established property type in most Australian urban areas. The planning problems arise when a non-homeowner wishes to ‘park’ their tiny house on someone else’s property and pay them rent; or find a tiny house community. Most if not all

Australian planning schemes consider a tiny house on wheels (or one placed on a property, but able to be moved) as a temporary dwelling, equivalent to a caravan, and restrict living in these to no more than two weeks.

Moreover, there is no legislation pertaining to the type of rental situation where a person 'brings their own property' and parks it in someone's backyard. This not only causes issues with rental legislation, but also with LGAs which planning schemes mandate additional parking for new dwellings, connections to utilities, and even infrastructure charges for increasing the population of a LGA. There are also related issues with building regulations, as tiny houses are currently ill defined and do not fall under the Building Code of Australia (MB Qld nd.), which means that there are no real controls on their construction. While this may appeal to the more libertarian of tiny house advocates, it has potential and possibly serious implications for health and safety. In addition, by not falling under the BCA, the tiny house builder leaves themselves open to complaints by neighbours, especially if the tiny house is owner built to a lesser standard.

Even well-built tiny houses have fallen foul of the Not In My Back Yard (NIMBY) syndrome. For example, an attractive and well-built tiny house on wheels (THOW) as situated on a back yard in Brisbane, Australia with rent paid to the owners of the property. However, after a neighbour complained, the Brisbane City Council issued the owners of the tiny house with an infringement notice, stating that "…that the siting of the THOW on the subject site constituted building work which is assessable development and that no development permit for building work had been obtained for the THOW" (BDDRC, 2016). After the court case, the Building and Development Dispute Resolution Committee ruled that the THOW was neither a building nor a structure (as defined) thus did not constitute building works, and was considered rather a road registered moveable dwelling.

This raises another issue with tiny houses, in that if they are on wheels, then even if they do not have to comply with LGA building rules, then they do have to comply with transport rules regarding trailer sizes, weight, width and height of the THOW. This constrains the size of tiny houses, in that the THOW must be under the maximum weight, width and height for a specific State and it has to conform to another, complex set of legislation. Nonetheless, even if road legal, THOWs can be extremely heavy and require towing by a high powered 4WD or even a truck. This is one reason why THOWs are not designed to be 'permanently' mobile, such as a caravan or 5th wheeler.

So, where can you put them?

As discussed earlier, the issue in Australia is not with housing in rural or remote areas; it is with housing affordability in the urban areas. Eight seven per cent of Australians live in urban areas with over two thirds living in the Capital Cities (ABS, 2017). At the same time, all Australian capital cities are considered severely unaffordable. To address these issues, Bares et al. (2017) have identified a range of planning options whereby tiny houses can, with minor modifications, be situated in urban areas in Australia (Table 13.2). The options in Table 13.2 can all be achieved with minor modifications to most planning schemes in Australia. Indeed, granny flats and residential (tiny house)

Table 13.2. Tiny house planning options in Australia (based on Bares et al., 2017).

Option	Benefits	Barriers
Granny flat (ancillary dwelling to existing house)	• Permitted by planning schemes. • Can buy do-it-yourself or 'off the shelf' kits • Adds to land value (permanent) • Encouraged as densification option by many LGAs • Can be larger than a tiny house	• Requires ownership of the land. • Potential legal issues if house owner wants to relocate it • Requires legal deed if other person than landowner builds it • May incur substantial development costs, infrastructure charges
Tiny lots (small freehold subdivided lots)	• Allows the ownership of land • Can build and rent out a tiny house on the lot • Allows infill development without high rises • Communal open space and facilities	• Currently not supported by most LGAs in Australia • Not a given that tiny lots would be cheaper or have smaller houses • Potential issues with car parking, needs sensitive planning of design, layout, access, etc.
Tiny villages (a small number of houses on a lot)	• Resident owns property under a strata title arrangement • Can be rented out to others • Entry level property ownership • Communal benefits	• Not currently supported in most LGAs in Australia • Needs sensitive planning of design, layout, access, etc. • Land prices in urban areas
Tiny backyard leases (permitting THOW in backyards)	• Property owner can get rent from the tiny house owner • Tiny house is mobile • Privacy for both parties (unlike group housing/sharing) • Good for THOW	• Unclear in most local laws • May need connection to infrastructure and services • Potential for conflict (NIMBY, property owner, etc.) • Car parking in some areas
Tiny house parks (Similar to a caravan/trailer park)	• Could be suitable for ageing caravan parks in Australia • LGA could zone specifically for tiny houses • Good for communal living	• Stigma about living in trailer parks • Some caravan parks in desirable areas under development pressure • These already exist, why are THOW any different?

parks already exist. Adapting the planning schemes to allow the other options would be relatively simple, although it would require some change in various legislations, to protect the tiny house owner and the landowner. The real issue with all of these options is that even if the law is changed, the price of urban land in Australia is still extremely high, and the closer the land is to the CBD, the higher priced (and the higher the development pressure) it is under.

The highest and best use of inner city urban land is not for tiny houses, in whatever guise, but for high density building. Thus, these planning options are likely to eventuate in the areas of the city that are already zoned for single residential or medium density dwellings. Most Australian cities are very low density, and in Brisbane for example, 74% of houses are detached dwellings (.idcommunity, 2017), with mean lot size of around 600 square metres. Brisbane has substantial areas that could be utilized at least for granny flats and tiny backyard leases; for example Fig. 13.6 shows land in relatively close proximity to Brisbane, zoned for low and medium density dwellings, and with lot size over 1000 square metres.[2] In other areas, the State and Local Governments often

[2] This is indicative only, and does not indicate whether or not the property is developed, and whether it already has an existing granny flat or medium density buildings.

Figure 13.6. Low and medium density residential lots in Brisbane > 1000 square metres.

own large tracts of land, which potentially could be developed as tiny house villages. Because tiny houses are frequently mobile, the government could retain ownership of the land, and lease it to a tiny house developer.

Financial Considerations (How do You Pay for the Tiny House)

Despite the possible economic benefits in building a tiny house compared to a standard dwelling, this still poses a problem. In Australia, unlike a normal house, it is impossible to get a mortgage for a tiny house, unless the owner owns the land on which it will be built. Dwellings are considered as depreciating in value, unlike the land on which they are sited. Generally speaking, banks loan money not on the house, but on the land. This is slightly different in the USA, where there are a small number of tiny house financing options available, albeit at a higher interest than a standard mortgage (Kilman, 2016).

This means that tiny house builders either have to pay cash to build their tiny house (from whatever source) or have to take out a short term loan with high interest (in essence, a personal loan, such as for a car or boat) (Kilman, 2016; Shearer, 2015). Given that many want to build a tiny house for financial reasons, which may include inability to qualify for a home loan or to save up sufficient deposit for one, this poses a serious issue. How do they pay for their tiny house? Arguably, this a major reason why many do not build a tiny house, or else, settle for the option of a converted bus or caravan, which can be had for a relatively small outlay.

Research also indicated other financial and related considerations to building tiny houses and these include lack of capital growth from a depreciating asset, inability to insure the tiny house, no legacy for children, inability to use the tiny house as collateral for other loans, and long term insecure tenure similar to that experienced by renters. This is particularly pertinent to one demographic for which tiny houses have significant appeal, retirees or pre-retirees. As discussed in the following paragraphs, tiny houses are mostly not suited to people with mobility problems, so once the tiny house owner becomes incapable of towing or moving around the tiny house, then they have no asset to be used if buying into an aged care facility.

Design Restrictions

Tiny houses are generally suited to specific demographics only, such as young singles or couples, older singles and couples with small children. They are probably not suited to the archetypal nuclear family especially those with teenagers, and nor are they suited to extended families or to group living (unless they are built on to an existing dwelling or are part of a tiny house community). A number of international case studies have also explored tiny houses as options for the homeless.

The very small space of a tiny house can pose specific problems, especially to the mobility impaired. For example, a common feature of tiny houses is a loft bedroom which is accessible by ladder. For many older or disabled people, this design is impractical if not impossible. Tiny houses also do not have much space for those who are reliant on a wheelchair or other walking aids, to move around, cook or use the bathroom.

In subtropical or similar climates, such as Australia, tiny houses can include outside areas such as decks to extend the space. Although heavy (decks may have to be moved separately to the tiny house when moving the structure), they allow indoor-outdoor living and more options for entertainment. However, in very cool climates, tiny house dwellers may be restricted to the indoor areas in inclement weather, which could give rise to conflict, lack of privacy and 'cabin fever'.

No matter what demographic or climate, tiny houses can pose challenges to those living within; and ideally should be designed using principles taken from boats, caravans or tiny dwellings in other parts of the world, such as Japan. Living in a very small space requires sacrificing a lot of things which are taken for granted in countries such as Australia, such as individual private spaces, entertainment areas and storage areas for things like sporting equipment or tools. Anecdotally, a large proportion of those who have abandoned tiny house living have done so for this reason; although most still live in small dwellings, the problems of family living in a tiny house proved too difficult.

Conclusion: So, Why Tiny Houses?

There is of course, no simple panacea for all the problems of the urban society; but tiny houses can fill a valuable niche in our future cities. Tiny houses are above all sustainable; environmentally, economically and socially. They combine green building principles with economic sustainability, and can be fitted into the physical and legal infrastructure of contemporary cities. Our cities are largely path dependent; planning for future infrastructure and densification requires working around the existing built environment, as well as pushback from residents. Because tiny houses are attractively designed, and well built, these can be slotted into the existing city fabric with relative ease.

For example, in most Australian capital cities, the older suburbs within approximately 10 km from the CBD are characterized by large old houses, and small ageing households. Many of these people have lived in such suburbs for decades, and are unwilling to sell the family home, even if it is too large for them. This is often due to the lack of affordable smaller housing to be found within the same area. Permitting tiny houses in such areas would enable densification of these areas, without high rises, and even allow the residents of larger houses to build and move into a tiny house on their own property whilst getting an additional income from the large main house. This would also allow families to move closer to the city and employment, and such shared living spaces facilitate community, and intergenerational mixing.

Environmentally, the earth's population continues to increase, with increasing impacts on finite resources such as water, food, soil and non-renewable power sources. Housing is a significant source of GHG emissions, environmental degradation, deforestation, mining impacts and the like. Reducing the amount of materials used in construction and recycling other materials, will have a long term positive benefit for the environment. Moreover, the mobility of many tiny houses has the potential to allow these to be situated in areas that are threatened by climate change. The houses in a tiny house park for example, could be relocated if a cyclone or bushfire were to threaten the area. They could also be sited as temporary dwellings without degrading a site, either from building or from having to construct infrastructure for connection to utilities.

Economically, tiny houses are much more affordable than standard houses, because they use far less building materials, and much of this can be sourced from recycled or discarded building products. Because they are mostly owner built, further economies can be gained by saving on labour costs of building, and fees and taxes charged by governments, such as infrastructure charges. The USA has a strong tiny house building sector, and in other areas, such as Australia, this has the potential to be a future growth industry, adding to employment and training of the workforce. Tiny houses also have the potential to be built and sited in remote areas, thus improve employment outcomes.

Tiny houses are of course, not suited to all demographics, but perhaps their true significance lies in their disruptive value; their ability to precipitate a paradigm shift in awareness of the waste of resources in modern construction (Kilman, 2016). They fall at one end of the continuum, of which the suburban McMansion and 50 storey high rise fall at the opposite end. The tiny house movement with its emphasis on quality, community, sustainability and minimalism is the philosophical antithesis of the McMansions and shopping malls of consumer society (Ford and Gomez-Lanier, 2017). Consuming (and discarding) in a tiny house is a conscious choice; not made because something is on sale or because it is the latest gadget, but because it is a necessary product.

While it is unlikely that tiny houses will ever be a major part of the housing market, they have the potential to precipitate a new environmental and sustainability ethic (Kilman, 2016). Their benefit may well lie in their power to disrupt the neo-liberal consumerist culture, and change social norms so that the McMansion goes the way of the huge petrol-guzzling cars of the past. They truly epitomize sustainability, in that they can motivate beneficial social, environmental and economic changes over the long term to benefit both current and future generations (Ford and Gomez-Lanier, 2017).

Research into the tiny house movement is still in its infancy, and future studies, by this author and others, will no doubt build on the information in this and other articles.

References

Anson, A. 2014. The World is my Backyard: Romanticization, Thoreauvian Rhetoric, and Constructive Confrontation in the Tiny House Movement. From Sustainable to Resilient Cities: Global Concerns and Urban Efforts. 2014: 289–313.

Australian Bureau of Statistics (ABS). 2012. 1301.0—Year Book Australia, 2012. Types of dwellings. http://www.abs.gov.au/ausstats/abs@.nsf/Lookup/by%20Subject/1301.0~2012~Main%20 Features~Types%20of%20Dwellings~127.

Australian Bureau of Statistics (ABS). 2013. 4130.0 - Housing Occupancy and Costs, 2011-12.

Australian Bureau of Statistics (ABS). 2017. 2071.0 - Census of Population and Housing: Reflecting Australia - Stories from the Census, 2016.

Bares, V., R. Pieters, L. Nobel, B. Winkle, K. Meathrel, H. Shearer and F. Caniglia. 2017. A Place for Tiny Houses. Exploring the possibilities: Tiny House Planning Resource for Australia, 2017.

Berry, M. 2003, 'Why it is important to boost the supply of affordable housing in Australia—and how can we do it?' Urban Policy and Research 21: 413–35.

Berry, M., C. Whitehead, P. Williams and J. Yates. 2006. Involving the Private Sector in Affordable housing provision: can Australia learn from the United Kingdom? Urban Policy and Research 24: 3.

Building and Development Dispute Resolution Committees (BDDRC). 2016. Decision Appeal no. 39-16. Online http://www.hpw.qld.gov.au/SiteCollectionDocuments/FinalDecision39-16.pdf.

Burke, T., S. Pinnegar and P. Phibbs. 2007. Experiencing the housing affordability problem: blocked aspirations, trade-offs and financial hardships. AHURI.

Burnett, K., E. Holt, L. Fisman and G. Chang. 2010. Research on factors relating to density and climate change.

Buxton, M. and E. Taylor. 2011. Urban land supply, governance and the pricing of land. Urban Policy and Research 29(01): 5–22.

Carlin, T. M. 2014. Tiny homes: Improving carbon footprint and the American lifestyle on a large scale. Celebrating Scholarship and Creativity Day. Paper 35. http://digitalcommons.csbsju.edu/elce_cscday/35.

Carlyle, E. 2014. 12 Tricked-Out Tiny Houses, And Why They Cost So Much. Forbes, online. http://www.forbes.com/sites/erincarlyle/2014/04/28/12-tricked-out-tiny-houses-and-why-they-cost-so-much/.

Chia, G. 2015. Building a tiny house. Online. http://www.resilience.org/resource-detail/2544932-building-a-tiny-house.
Coiacetto, E. 2006. Real estate development industry structure: Consequences for urban planning and development. Planning, Practice and Research 21(4): 423–441.
Coiacetto, E. 2007. The role of the development industry in shaping urban social space: a conceptual model. Geographical Research 45(4): 340–347.
Country Living Magazine. 2017. 69 Impressive Tiny Houses That Maximize Function and Style. Online: http://www.countryliving.com/home-design/g1887/tiny-house/?thumbnails&slide=1.
Cox, W. and H. Pavlevitch. 2014. Cox and Pavlevitch international housing affordability survey.
Diguette, R. 2017. Lessons from the first 'tiny house' evangelist, Henry David Thoreau. Washington Post, March 2017. Online https://www.washingtonpost.com/lifestyle/home/lessons-from-the-first-tiny-house-evangelist-henry-david-thoreau/2017/03/13/3dddc69e-02d0-11e7-b1e9-a05d3c21f7cf_story.html?utm_term=.08d80b37891f.
Dodson, J. and N. Sipe. 2008. Shocking the suburbs: urban location, homeownership and oil vulnerability in the Australian city. Housing Studies 23(3): 377–401.
Ford, J. and L. Gomez-Lanier. 2017. Are tiny homes here to stay? A review of literature on the tiny house movement. Family and Consumer Sciences Research Journal 45(4).
Florida, R. 2000. Competing in the Age of Talent: Quality of Place and the New Economy. Report Prepared for The R. K. Mellon Foundation, Heinz Endowments, and Sustainable Pittsburgh.
Glaeser, E. L. and J. Gyourko. 2002. The impact of zoning on housing affordability (No. w8835). National Bureau of Economic Research.
Google Trend Search. 2015. Tiny Houses. https://www.google.com/trends/explore#q=tiny%20houses.
.idcommunity. 2017. Dwelling type: Greater Brisbane. Online http://profile.id.com.au/australia/dwellings?WebID=270.
Jones, C. M. and D. M. Kammen. 2011. Quantifying carbon footprint reduction opportunities for U.S. households and communities. Environmental Science and Technology 45(9).
Kilman, C. 2016. Small House, Big Impact: The Effect of Tiny Houses on Community and Environment. Undergraduate Journal of Humanistic Studies. Winter 2016. Vol 2.
Lewis, D. and R. Wright-Summerton. 2014. Communes and intentional communities. The Routledge Companion to Alternative Organization, 89.
Langston, B. 2015. Living Big in a Tiny House: What is a tiny house? Online. http://www.livingbiginatinyhouse.com/tiny-house/.
Langston, B. (nd) Living Big in a Tiny House: Tiny House Trailer FAQ. Online http://www.livingbiginatinyhouse.com/tiny-house-trailer-faq/.
Lloyd, R. and T. N. Clark. 2001. The CIty as an Entertainment Machine. Critical Perspectives on Urban Redevelopment 6(3): 357–78.
Master Builders Queensland (MBQld). nd. Building Code of Australia. Online https://www.mbqld.com.au/laws-codes-and-regulations/building-act/building-code-of-australia.
Mitchell, R. 2014. Tiny House Living: Ideas for Building and Living Well in Less than 400 Square Feet. Betterway Home: Cincinnati, Ohio.
Murphy, M. 2014. Tiny Houses as Appropriate Technology. Communities. Winter 2014. Pg. 54.
Mutter, A. 2013. Growing Tiny Houses: Motivations and Opportunities for Expansion Through Niche Markets. Thesis for the fulfilment of the Master of Science in Environmental Sciences, Policy and Management Lund, Sweden, June 2013.
Nobel, L. 2016. Personal Communication.
Nonko, E. 20 https://www.curbed.com/2016/9/22/13002832/tiny-house-zoning-laws-regulations 16. Tiny house zoning regulations: What you need to know. Curbed, 22 September 2016. Online.
Onyx, Jenny and R. Leonard. 2005. Australian grey nomads and American snowbirds. Journal of Tourism Studies 16, no. 1 (2005): 61.
Organisation of Economic Cooperation and Development (OECD). 2015. Focus on house prices. Internet website accessed 10 July, 2015. http://www.oecd.org/eco/outlook/focusonhouseprices.htm.
Pawson, H., B. Randolph, J. Yates, M. Darcy, N. Gurran, P. Phipps and V. Milligan. 2015. Tackling housing unaffordability: a 10-point national plan. The Conversation. Online https://theconversation.com/tackling-housing-unaffordability-a-10-point-national-plan-43628. June, 2015.
Pullen, S., D. Holloway, B. Randolph and P. Troy. 2006. Energy profiles of selected residential developments in Sydney with special reference to embodied energy. pp. 22–25. *In*: Proceedings of the Australian & New Zealand Architectural Science Association (ANZAScA) 40th Annual Conference.

Petersen, M. and C. Parsell. 2014. Homeless for the First Time in Later Life: An Australian Study. Housing Studies (ahead-of-print), 1–24.

RPMNational. 2015. What is a tiny house? Online http://www.tinymountainhouses.com/what-is-a-tiny-house/.

Rugg, J., D. Rhodes and A. Jones. 2002. Studying a niche market: UK students and the private rented sector. Housing Studies 17(2): 289–303.

Rugg, J. and D. Rhodes. 2008. The Private Rented Sector: Its Contribution and Potential.

Schelkle, W. 2012. A Crisis of What? Mortgage Credit Markets and the Social Policy of Promoting homeownership in the United States and in Europe. Politics and Society 40(1): 59–80.

Senate Economics References Committee. 2015. Out of reach? The Australian Housing Affordability Challenge.

Seyfang, G. and A. Smith. 2007. Grassroots innovations for sustainable development: Towards a new research and policy agenda. Environmental Politics 16(4): 584–603.

Shearer, H. 2014. Move over McMansions, the tiny house movement is here. The Conversation. Online. https://theconversation.com/move-over-mcmansions-the-tiny-house-movement-is-here-32225.

Shearer, H. 2015a. Australians love tiny houses, so why aren't more of us living in them? The Conversation. Online. https://theconversation.com/australians-love-tiny-houses-so-why-arent-more-of-us-living-in-them-44230.

Shearer, H. 2015b. Tiny houses: a radical new solution for addressing urban housing affordability, or a just another niche market? State of Australian Cities Conference, Gold Coast Australia. 9 to 11 December 2015.

Shearer, H., E. Coiacetto, J. Dodson and P. Taygfeld. 2016. How the structure of the Australian housing development industry influences climate change adaptation. Housing Studies 31(7): 809–828.

Shearer, H. 2017. Tiny House Take 2. Online Survey. SurveyMonkey™.

Tasman Ecovillage website. 2015. http://tasmanecovillage.org.au/.

The Salvation Army. 2014. The Salvation Army Submission to the Senate Economics Reference Committee on Affordable Housing. http://www.salvationarmy.org.au/en/Who-We-Are/Publications-reports-submissions/Reports--Submissions/Latest-Reports/Submission-to-2014-Senate-Economics-Reference-Committee-on-Affordable-Housing/.

Tiny Houses Australia. THA. 2015. Online Facebook Forum. https://www.facebook.com/groups/TinyHousesAustralia/.

Tiny House Build. 2014. How tiny houses can save the earth: infographic. Online, https://tinyhousebuild.com/tiny-houses-infographic/.

Taylor, A. 2017. The 'Coffin Homes' of Hong Kong. The Washington Post, 16 May 2017. Online https://www.theatlantic.com/photo/2017/05/the-coffin-homes-of-hong-kong/526881/.

Walsh, L. 2014. By the masses: the emergence of crowdfunded research in Australia. The Conversation. Internet website, accessed 10 September 2014. http://theconversation.com/by-the-masses-the-emergence-of-crowdfunded-research-in-australia-22837.

Wentz, Elizabeth A and Patricia Gober. 2007. Determinants of small-area water consumption for the city of Phoenix, Arizona, Water Resources Management 21(11): 1849–1863.

Wikipedia. 2017. Libertarianism in the United States. Online https://en.wikipedia.org/wiki/Libertarianism_in_the_United_States.

Wilson, A. and J. Boehland. 2005. Small is beautiful U.S. house size, resource use, and the environment. Journal of Industrial Ecology 9: 277–287. doi:10.1162/1088198054084680.

Worthington, A. C. 2012. The quarter century record on housing affordability, affordability drivers, and government policy responses in Australia. International Journal of Housing Markets and Analysis 5(3): 235–252.

Yates, J., B. Randolph and D. Holloway. 2006. Housing Affordability, Occupation and Location in Australian Cities and Regions—Final Report, Australian Housing and Urban Research Institute.

Yates, J. 2008. Australia's housing affordability crisis. Australian Economic Review 41(2): 200–214.

CHAPTER 14

Identification and Establishment of Weightage for Critical Success Factors in Sustainable Affordable Housing

AbdulLateef Olanrewaju,[1,*] *Seon Yeow Tan*,[1] *Jia En Lee*[1] and *Naoto Mine*[2]

Introduction

This research seeks to advance the sustainability theory to affordable housing supply. This study draws on, and combines, two different theoretical perspectives: affordable housing and sustainability. The pieces of literature on affordable housing examine the nature and indicators of affordable housing, while the literature on sustainability examines the relevance of the theory to affordable housing. Putting these two concepts together, the research moves further to evaluate what frameworks are needed in order to deliver sustainable affordable housing. Since Our 'Common Future' (Keeble, 1988), or the 'Brundtland Report' that introduces the concept of sustainable development, there has been the proliferation of similar terms including sustainable construction, sustainable management, sustainable maintenance, sustainable government, sustainable culture, and sustainable business or sustainability. Each of these concepts has similar objectives as sustainable development. Sustainable development is very broad in terms of applications and may involve the government at national levels or cross boundary intergovernmental arrangements because the concerns of sustainable development have global implications. However, the other concepts that were developed as nodes on the sustainable development paradigm are more specific and focused on a particular area

[1] Universiti Tunku Abdul Rahman, Malaysia.
[2] 13-19, Enoki-cho, Tokorozawa, Saitama, 359-1104 Japan.
* Corresponding author

of practice or research. There is a proliferation of literature toward the achievement of sustainable development, however, addressing it through different pieces would likely be easy to control and achieve. For instance, as explained by Cole (2011), climate change required individual solutions, engagement and action played out on a local scale. This is truism because the sustainable development goal cannot be achieved simultaneously. Sustainable development involves a multi-attributes decision-making process. Therefore, to develop realistic solutions to achieve sustainable development problems, it is practical to break the problem into series of a simpler entity. Hence, a solution to each of the simple entity is then linked together to provide solutions to the sustainable development goal. Housing supply is one of the major strategic aspects by which sustainable development goal can be achieved considering the size of wastes that are generated and the volume of resource consumptions during design, construction and operation of housing. Housing is classified differently by virtue of accessibility (e.g., high-income, medium and lower income housing), physical height (e.g., low, high rise, and skyscrapers) as well as differentiated level of luxury (e.g., flats, terraced, condominium and apartment buildings). Financially, low and medium income housing are considered as affordable housing. Terms including mass housing, low-cost housing, and middle-income housing are sometimes used to denote affordable housing.

Studies on affordable housing are aplenty (Glaeser and Gyourko, 2003; Nguyen, 2005; Assaf et al., 2010; Olanrewaju et al., 2016). Paradoxically, only a few have made a logical inquiry to examine the compatibility of affordable housing with sustainability (Susilawati and Miller, 2013; Olanrewaju et al., 2016b). In 2016, Olanrewaju et al. reported in a study that the housing industry impression is that affordable housing and sustainability are not compatible, extensive researches revealed that sustainable is expensive. The major reason for this is because of the capital cost of constructing a house. Green projects are often between 5 and 15% higher as compared to projects that follow conventional standards design, specifications and regulations (Spiegel and Meadows, 2010) however, the operation costs of sustainable buildings are low and more importantly the building occupants' comfort, convenience, satisfaction will increase, the building will enhance social integrations, reduce negative externalities, reduces pollutions and waste generations. Ideally, sustainable housing should be cheaper because they use recycled materials, are adaptable and are planned for recycling at the end of their life span (Spiegel and Meodows, 2010). Presently, there is a stable source for sustainable materials and many manufacturers are now retooling their plants to use recycled materials. Analyzing the benefits of sustainable buildings will require whole life appraisal analyses, where both the capital and operating costs and value added to housing is considered over its lifespan for meaningful comparisons. This is to ensure that the capital cost of the housing is not the major decision making determinant, especially in material and components selections. In fact, as the market for sustainable materials and components grow, manufacturers and contractors will gain more experience and this will consequently reduce the capital cost of the sustainable materials and components in future. A case in point, smartphones, flat screen television, HIV drugs, malarial drugs, hybrid cars were initially beyond the reach of many but as the market developed the prices began reducing correspondingly. So it is a matter of time before the capital cost of recycling materials lowers.

Recent researches on sustainable affordable housing (Olanrewaju et al., 2016b; 2017), Ganiyu et al. (2015), Ganiyu et al. (2017) and Ang et al. (2016) specify a necessity to analyze the sustainability criteria that can be implemented in affordable housing. While Olanrewaju's studies are theoretical, Ang et al.'s (2016) and Ganiyu et al.'s (2017) studies are empirical. Ganiyu et al.'s works emphasized on construction management and construction procurement related techniques including lean construction, concurrent engineering, modular construction, health and safety principle and waste reduction to increase the performance of the housing projects. Other sustainable management practices such as quality management, value management, and maintenance management could facilitate achieving sustainability. However, sustainability has no direct relationship with any of these techniques. To illustrate, value management might not improve the holistic sustainability of projects except sustainability is defined as a factor in the client value system. Similarly, lean construction, concurrent engineering quality management may be introduced to reduce project duration or/cost without giving priority to environmental or social sustainability. In fact, these techniques are broad and only large contractors/developers could actually implement them in practice. However, a major challenge facing the Malaysian affordable housing sector is in meeting the call for environmentally sustainable design (Government of Malaysia, 2010). For this purpose, roles of the federal government will be rationalized, Green Guidelines and a Green Rating System be introduced, and sustainable housing industry will be emphasized (Government of Malaysia, 2010). Although, in order to achieve the government aim of incorporating green technology into the country's construction industry and to embark on the green building practices, and to a reduction of 40% of the CO_2 emission, green rating tools have been developed. The tools are Green Performance Assessment System (Green PASS), Skim Penilaian Penarafan Hijau JKR (PH JKR), Malaysian Carbon Reduction and Environmental Sustainability Tool (MyCREST), Green Real Estate (GreenRE) and Green Building Index (GBI). However, because of the methods of assessment, and the stage at which each of the tools is applied in the building cycle, there are complications among the stakeholders (Abd-Hamid et al., 2014), thereby discouraging their applications. The low awareness among the Malaysian developers of the concept of sustainable development (Zainul Abidin, 2010), moderate readiness in applying sustainability concept (Ibrahim et al., 20013) and low awareness of the building rating tools (Hamid et al., 2014) implies the examination of the current state of knowledge of the housing industry of implementation of sustainability 'ingredient' in affordable housing. However, this current research aims to investigate the possibility of incorporating sustainable factors into affordable housing design, construction, and operation. This is compelling, because, solutions to sustainable affordable solution need engagement at a more practical level. Moving forward, research on sustainable housing has largely been conceptual, qualitative and descriptive (Hayles et al., 2010; Dempsey et al., 2012; Miller and Buys, 2013; Olanrewaju et al., 2016). Therefore, there is the need to develop quantitative sustainable indicators for sustainable affordable housing. The quantitative sustainability indicators are a strategic tool in benchmarking of progress towards sustainable housing. They are useful for monitoring and measuring the performance of affordable housing by considering a measurable number of variables.

Conceptual Justification

Sustainable development or sustainability has many interpretations. It is often defined as "the development that meets the needs of the present without compromising the ability of future generations to meet their own needs" (Keeble, 1988). The topical issues in the efforts towards sustainable development are the need to integrate economic, environmental and social aspects in decision making to increase productivity, quality of life, enhance integration and reduce wastes and pollutions. The bottom line is that at the current global scenario, we are currently consuming more than what the earth could productively support and producing and discharging wastes that are far beyond what the earth can accommodate. At a practical level, sustainable development is a practice that integrates various criteria including energy efficiency, durability, waste minimization, social impacts, good indoor environment, durability, pollution control, life-cost, user-friendliness, user comfort and others. The impact of building construction and operation on sustainability issues is huge. For instance, buildings consume more than 40% of the world energy, release 1/3 of CO_2, use about 25% of harvested woods, release about 50% of fluorocarbons, produce 40% landfill materials, use 45% of energy in operations, emit 40% of Green House Gas (GHG) emissions and 15% of world's usable water (See Wood, 2009; Killip, 2006; Sherwin, 2000; UNEP-SBCI, 2014). Building interiors contain five times more pollutants than the air outdoors. There is a projected 56% upsurge in building CO_2 emissions and buildings are expected to use 12% of global fresh water, and generate 30% of total waste in the European Union by 2030 (World Economic Forum, 2016) When other CO_2 emissions attributable to buildings are considered- such as the emissions from the manufacturing and transportation of building construction and demolition materials and transportation associated with urban sprawl—the result is an even greater impact on the climate. The energy that buildings consumed is even higher if we take into account other energy use attributable to buildings. For example, the embodied energy in a single building's envelope is around 8–10 times the annual energy used to heat and cool the building. Buildings occupants generate more waste than in other sectors. From these statistics, buildings are part of the threat to sustainable development goal and it must form part of the solution as well. In order to keep the global warming level below 2°C threshold, a 36% reduction in total CO_2 emissions in the real estate sector is required by 2030 (World Economic Forum, 2016).

Affordable housing stock has to be a significant part of this profile simply because it has to cater to an ever increasing middle class as globalized world economy created more wealth and enables more people to afford formal housing with access to banking credit. Therefore, if affordable housings are supplied as a major part of the housing stock, and operate sustainably to reduce energy and water consumptions and curtail carbon emission, enhance occupants' productivity, this will impact greatly in saving millions of metric tons of carbon emissions, while increase well-being and total user values systems. Imagine the impact of innovation in the manufacturing of just one of the primary global construction resources such as cement, for instance, which can significantly help reduce water requirements and increase the compressive strength of the concrete (Sanchez and Sobolev, 2010). The production [brief, design, and construction] and use of the sustainable housing influence a variety of criteria especially materials, components, design, layout and the delivery process. The relevance of its design is essential to continuity of use and avoiding from premature obsolescence of use. It must be durable not only physically but also socially in order to be sustainable. Collectively, both in professional and academic literature, these strategies for selecting and implementing these criteria explain whether a sustainable building is delivered or not.

Unfortunately, affordable housing supply is not taken advantage of where innovative approaches to design, construction, and maintenance are used. It is as if only luxurious and unaffordable projects can fulfill all the trappings of 'green' such as photovoltaic, solar, energy and water saving technologies. These higher-end buildings make up relatively only a small percentage to impact significantly on global warming. Affordable housing is still largely supplied using the traditional procurement methods and traditional method of construction as the housing industry believes that sustainability is expensive, hence incompatible with affordable housing, as in the case of UK experience (Pitt, 2013). The design and construction team are reluctant to explore the innovative methods and there are uncertainties in objective achievement of sustainability in affordable housing unless the issue of cost savings is the primary and measurable concern. Sustainable innovative housing production should basically emphasis on the use of localized materials, components and labor, efficiency components sufficient to the housing requirement even during the peak periods and reduces energy use. Current evolution of literature outlined that to deliver sustainable housing it has to be energy efficient and able to generate its own energy by applying the building as an energy producing infrastructure, in order to contribute to decarbonizing the built environment (Patterson, 2007).

However, sustainable housing is more than this, is it about total quality of life of occupants where satisfaction leads to its continual usage for a longer lifespan without the occupant feeling the need to 'upgrade' to another property, thus reducing the upward demand for housing 'wants' instead of 'needs'. For instance, a question that would probably come to mind is, is it energy cost that is important or occupants' comfort? Certainly, the two are not correlated. In Malaysia, the green rating systems, however, only accord less importance to thermal comfort. For instance, PH JKR, GBI, and GreenRE allocate 22, 21, and 4% to indoor environmental quality. While sustainable development requires government supports through laws and regulations, developers may require incentives to promote and implement such measures. Academic and professional literature also led to the conclusion that land use and efficient use of brownfield sites, lands with irregular shapes, creating access to inaccessible lands, modular construction, prefab construction, housing maintainable, highly flexible and adaptable (Mora, 2007; Olanrewaju et al., 2017), safety and security (Tankard, 2016) energy conservation are some of the sustainability factors. The consequences of unsustainable resource use are leading the world towards climate change, pollution, resource depletion, environmental degradation, and social disorder. The measurable environmental impacts from the lack of affordable housing supply are uncertain, but with the increase in the number of homes and poor housing operations, the probability increases. This poses risks for the future of urbanity. Perhaps the question should be posed; can the multi facet drive towards sustainable development be partly achieved through the implementations of sustainability in affordable housing? After all, many resources are required and wastes are generated during affordable housing construction and operations. The land, materials, water, energy and other resources required to build and operate houses are mined from the natural environment. It is also the environment that will 'receive', 'store' and/or 'process' the waste generated in the production and operation of the house.

However noble the environmental stewardship and is essential for humankind's survival, it's uncertain future benefits struggles with the reality of everyday life at a personal level. Investing in a 'green' home often conflicts with other needs of homebuyers who are allocating more of their disposable income on transportation, on food, children's education and health due to urban inflation, many urban housing has become unaffordable. However, while developers in the UK, Australia, and Hong Kong have been delivering sustainable low-cost housing, in Malaysia this is yet to commence. For instance, affordable housing is still measured in terms of housing financial-ability (BNM, 2017). The main obstacle to the slow uptake in implementing sustainability measures in the affordable housing delivery stems from lack of understanding on the constituent of sustainable affordable housing, especially at an applicable level. Therefore it is essential that the availability of data that support the possible implementations of sustainability in affordable housing will enhance its credibility and help educate industry stakeholders to trigger its mass implementations. While, most studies on sustainable building focused towards commercial buildings (i.e., hotels, corporate offices, hospital) and high-end residential buildings, empirical investigations into the compatibility of housing affordability and sustainability are lacking. This research aims to fill this gap.

Overview of Sustainable Affordable Housing in Malaysia

Consistent with commitments by major countries, the Malaysian government aims to reduce energy intensity ratios to less than 1.0 from the current 1.3 to be more in line with the global average of 0.73 (Government of Malaysia, 2010). The government also aims to lower carbon emissions to 40% by 2020 and reduce energy consumptions to 40% by 2050, relative to 2005 levels. It has also set a goal of 22% household recycling rate, reduce non-revenue water to 25% by 2020 (Government of Malaysia, 2015). Government of Malaysia (2010), while increasing solar photovoltaic usage form 1 MW in 2011 to 21.4 MW in 2050 (SEDA, 2012). In 1990 to 2012, electricity consumption for buildings increased more than 300% Energy Commission (2015). As the housing stock in Malaysia is accountable to some 90% of the total building stock and more than 90% were neither designed nor constructed for sustainability compliance, this may post a challenge in meeting the stated target. Residential building accounts for some 21% of the electricity consumption and the average water consumption per day of Malaysians was 212 liters compared to the recommended 165 liters (Energy Commission, 2013). The implications of these statistics are that if the government commitment and targets for sustainable development goals are to be achievable, the supply and operation of affordable housing stock need to align with sustainability requirements.

Framing Indicators for Sustainable Affordable Housing

Critical Success Factors (CSFs) and Critical Success Criteria (CSCs) are two common terms in project performance research and practice. Critical success factors of projects are the key factors that when applied during the planning, design, and construction/maintenance of the projects will ensure the attainment of the project's design objectives. Critical success criteria, on the other hand, are the criteria of determinants when the design objectives are attained or not. The critical success factors of a project are the key factors that are required to make the projects as part of its designed functions to meet stakeholders' "checklist" of requirements. The term 'factor' in the phrase implies, factor, activity, event, process or action. CSFs analysis could enable the identification of key important areas on a project that required considerable interests of the stakeholders to achieve project's goal. However, the indicators could be technical or behavioral. Technical indicators for sustainable housing are the indicators that by virtue of their presence in the design/scheme, they are sustainable. For example, using green material in a building or installing solar panel in building regardless of whether or not the behavior of providers/users is sustainable. Authors and bodies (i.e., BEEAM, LEED, and GBI) provided critical success factors for buildings. Behavioral sustainable indicators depend on the behavior of the users for the element to be sustainable. But, there are many deficiencies connected with the previous factors to allow practical applications into affordable housing. First, the previous CSFs are not for affordable housing. Second, the previous factors are not comprehensive enough to accommodate major aspects of affordable housing such as distance from work opportunity, recreation, relevance of design for a growing family, etc. Finally, there are separate factors for measuring housing affordability and separate for sustainable housing. CSFs for sustainable affordable housing are the major components of the structure that govern the evaluation and improvement on the sustainable affordable housing process. Therefore, in an attempt to develop a framework for sustainable affordable housing, the sustainability CSFs need to be adequately discovered, documented and analyzed to define the set of standard, measures, and procedure required when implemented on sustainable. Ultimately the house will generate less waste, pollution, emit less carbon, consume less water, energy, enable energy conservation, be inexpensive to maintain, operate and own, comfortable, convenient, accessible for those with restricted mobility, and its lifecycle end, its components be recyclable as future resource.

While there is no conclusive list on what are the basic elements that form sustainable housing, Table 14.1 contains factors collated from literature on construction industry (Shen et al., 2007), housing (Priemus, 2004; Zavr et al., 2009; Tam and Zeng, 2013) of various types of buildings including hotels, industry, hospitals, and office buildings (Spiegel and Meadow, 2010; Shad et al., 2017). The table was designed to contain factors in sustainable buildings that can be implemented **in** affordable housing. This list is not meant to be comprehensive but indicative.

Table 14.1. Summary of previous studies on sustainable factors in buildings.

Factor	Author
Install energy efficiency equipment in the house	Miller and Buys (2013); Sayce et al. (2010); Kamali and Hewage (2015); UNEP (2015); Al-Jebouri et al. (2017); Shad et al. (2017); Tam and Zeng (2013); Akadiri et al. (2012); Bragança et al. (2010); Zavr et al. (2009)
Install thermal insulation in the house	Miller and Buys (2013); Sayce et al. (2010); Kelly and Hunter (2010); Kamali and Hewage (2015); Shad et al. (2017); Tam and Zeng (2013); (2010); Zavr et al. (2009)
Install gray water use system for flushing, etc.	Miller and Buys (2013); Kelly and Hunter (2010); Devitofrancesco et al. 2010; Kamali and Hewage (2015); UNEP (2015); Al-Jebouri et al. (2017); Shad et al. (2017); Tam and Zeng (2013)
Install thermal comfort equipment in the housing	Miller and Buys (2013); Sayce et al. (2010); Kelly and Hunter (2010); Kamali and Hewage (2015); UNEP (2015); Al-Jebouri et al. (2017); Shad et al. (2017); Tam and Zeng (2013); Bragança et al. (2010); Zavr et al. (2009)
Install rainwater treatment equipment	Miller and Buys (2013); Sayce et al. (2010); Kelly and Hunter (2010); Kamali and Hewage (2015); Al-Jebouri et al. (2017); Shad et al. (2017); Tam and Zeng (2013)
Design to provide natural lighting	Miller and Buys (2013); Kelly and Hunter (2010); Kamali and Hewage (2015); UNEP (2015); Al-Jebouri et al. (2017); Shad et al. (2017); Tam and Zeng (2013); Zavr et al. (2009)
Design for the use of rainwater	Miller and Buys (2013); Kelly and Hunter (2010); Kamali and Hewage (2015); Tam and Zeng (2013); Zavr et al. (2009)
Use recycle materials construction &maintenance	Miller and Buys (2013); Kelly and Hunter (2010); Devitofrancesco et al. (2010); Kamali and Hewage (2015); UNEP (2015); Al-Jebouri et al. (2017); Shad et al. (2017); Tam and Zeng (2013); Akadiri et al. (2012); Bragança et al. (2010); Zavr et al. (2009)
Install storm water discharge system	Miller and Buys 2013; Kelly and Hunter (2010); Kamali and Hewage (2015); UNEP (2015); Al-Jebouri et al. (2017); Shad et al. (2017); Tam and Zeng (2013)
Community center provided or planned	Kamali and Hewage (2015); UNEP (2015) ; Akadiri et al. (2012)
Housing proximity to workplace	Miller and Buys (2013); Sayce et al. (2010); Kelly and Hunter (2010); Devitofrancesco et al. (2010); Kamali and Hewage (2015); UNEP (2015); Akadiri et al. (2012); Zavr et al. (2009)
Housing proximity to public transport service	Miller and Buys (2013); Sayce et al. (2010); Kelly and Hunter (2010); Devitofrancesco et al. (2010); Kamali and Hewage (2015); UNEP (2015); Akadiri et al. (2012); Zavr et al. (2009)
Design for reduction of energy consumption and embodied energy	Miller and Buys (2013); Kelly and Hunter (2010); Devitofrancesco et al. 2010; UNEP (2015); Shad et al. (2017); Akadiri et al. (2012)
Housing proximity to open green public area	Miller and Buys (2013); Kamali and Hewage (2015); UNEP (2015)
Desirability of neighborhoods	Miller and Buys (2013); Kamali and Hewage (2015); UNEP (2015)
Ensure ecological value of the site	Kamali and Hewage (2015); Tam and Zeng (2013); Zavr et al. (2009)
Design plan for adaptability-for future changes	Sayce et al. (2010); UNEP (2015); Tam and Zeng (2013); Bragança et al. (2010); Zavr et al. (2009)

Table 14.1 contd. ...

Table 14.1 contd. ...

Factor	Author
Low house price in relation to income	Kamali and Hewage (2015); UNEP (2015); Shad et al. (2017) Zavr et al. (2009)
Design plan to ensure flexibility	Sayce et al. (2010); Kamali and Hewage (2015); UNEP (2015); Tam and Zeng (2013); Bragança et al. (2010)
Design plan to reduce waste disposal bills	Sayce et al. (2010); Shad et al. (2017); Tam and Zeng (2013); Akadiri et al. (2012); Bragança et al. (2010)
Application of artificial sustainable space plan	Miller and Buys (2013); Devitofrancesco et al. (2010); UNEP (2015); Al-Jebouri et al. (2017); Tam and Zeng (2013); Akadiri et al. (2012)
Design to provide natural ventilation	Miller and Buys (2013); Sayce et al. (2010); Kelly and Hunter (2010); Devitofrancesco et al. 2010; Kamali and Hewage (2015); UNEP (2015); Shad et al. (2017); Tam and Zeng (2013); Akadiri et al. (2012)
Design to reduce housing maintenance costs	Kamali and Hewage (2015); UNEP (2015); Tam and Zeng (2013); Akadiri et al. (2012)
Provides bicycle path and bicycle parking lot	UNEP (2015); Shad et al. (2017); Tam and Zeng (2013); Akadiri et al. (2012)
Use of low-toxicity materials	Miller and Buys (2013); UNEP (2015); Al-Jebouri et al. (2017); Tam and Zeng (2013); Akadiri et al. (2012); Zavr et al. (2009)
Connection of pedestrian pathway to the outer pedestrian network	Miller and Buys (2013); Shad et al. (2017); Tam and Zeng (2013); Akadiri et al. (2012); Zavr et al. (2009)
Use eco-friendly methods of construction or new technology	Miller and Buys (2013); UNEP (2015); Al-Jebouri et al. (2017); Shad et al. (2017); Tam and Zeng (2013); Akadiri et al. (2012); Zavr et al. (2009)
Install skylight for illumination	Sayce et al. (2010); Al-Jebouri et al. (2017); Tam and Zeng (2013); Akadiri et al. (2012); Zavr et al. (2009)
Design plan for reduction of CO_2 emissions	Miller and Buys (2013); Sayce et al. (2010); Kelly and Hunter (2010); UNEP (2015); Tam and Zeng (2013); Akadiri et al. (2012); Bragança et al. (2010); Zavr et al. (2009)
Separate living waste collection for recycling	Tam and Zeng (2013); Akadiri et al. (2012)
Rational site environmental management plan	Kelly and Hunter (2010); Tam and Zeng (2013); Akadiri et al. (2012); Bragança et al. (2010); Zavr et al. (2009)
High green space area ratio	Kelly and Hunter (2010); UNEP (2015); Tam and Zeng (2013); Akadiri et al. (2012); Zavr et al. (2009)
Use solar energy systems	Miller and Buys (2013); Sayce et al. (2010); Devitofrancesco et al. 2010; Kamali and Hewage (2015); UNEP (2015); Shad et al. (2017); Tam and Zeng (2013); Akadiri et al. (2012); Zavr et al. (2009)

Other likely factors include densities providing social mix, educate homebuyers on energy saving behavior and design plan for noise emission and control, use of alternative energy, variety in housing types accommodation, minimize noise and nuisance, senior citizen and handicap accessibility, recreational areas as well enabling the minimal use of furniture. Based on the professional and academic literature, the following proposition or hypothesis is that it is possible to implement sustainability factors into affordable housing supply.

Outline of the Research Method

Research can be inductive or deductive. In this research, the deductive approach is used to develop sustainability indicators for affordable housing. The primary assumption in deductive research or reasoning is that logically and valid conclusions can be made from general to specific (Fig. 14.1). Deductive research commences with general theory and applies this theory to a specific case for testing hypothesis. The sustainability factors in buildings were collated from extensive professional and academic literature and were administered to the respondents. While there is much research that discusses the possibility of implementing sustainability in affordable housing, these studies are based on case studies, thereby making a comparison, measurements and improvement difficulty. It is difficult to determine the precise factors that are common to a different building and if they can be implemented in affordable housing. To fill this gap, this research first provides a summary of the indicators for sustainable buildings of various types and goes ahead to measure the possibility of implementing them in affordable housing (Table 14.1). The main goal of a theory of sustainable affordable housing indicators is to derive a framework for implementing sustainability in the affordable housing delivery. The practical implication of this study is to evaluate if those implemented sustainable factors in affordable housing can contribute to a reduction in climate change, pollution, waste generation, and ozone depletion and to apply the knowledge learned in making the decisions that will enable affordable housing delivery to serve as a tool for sustainability.

The primary data is collected based on convenience sampling. Convenience sampling is a data collection method where the survey is administered to those that are available, accessible and willing to be respondents. It is an appropriate method where information on population size is not available. However, its findings may not be generalizable, but as Sekaran and Bougie (2010) explained, with large respondents the findings can be representative. The survey was conducted in two phases, through hand delivery and an online survey. The first phase was administered to respondents that attended the ARCHIDEX (International Architecture, Interior Design & Building Exhibition, 2017) in the Kuala Lumpur Convention Centre. ARCHIDEX is held annually and attended by architects and other stakeholders in the construction sector (i.e., engineers, clients, developers, quantity surveyors) in Malaysia and other South East Asian countries. The ARCHIDEX 2017 was held between 19 July 2017 and 22 July 2017. The survey was conducted on 22 (Saturday) July 2017 and 37 completed

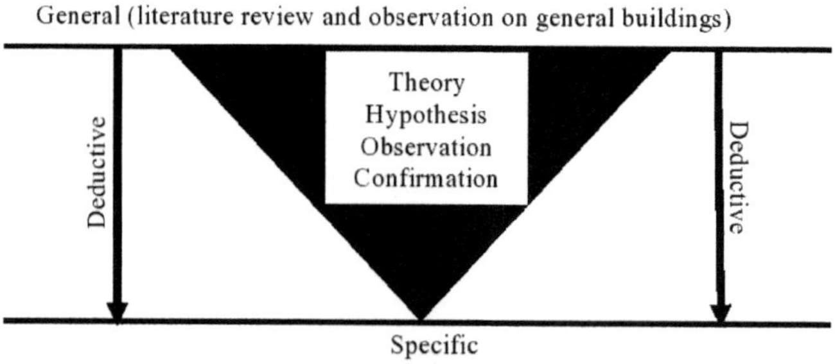

Figure 14.1. Deductive research (Towards a theoretical framework sustainable affordable housing).

survey forms were returned. The second survey based on purposive sampling was administered to the expert professionals in the housing sectors including architects, quantity surveyors, and engineers. The second survey commenced on 18 of September and ended on 28 October 2017. Altogether 85 completed surveys forms were received. The sustainability factors implementable of affordable housing is included in the survey form were obtained from extensive professionals and academic literature (Table 14.1) and the author's experiences. The questionnaire went through two pilot surveys that comprised of clients, developers and developer organizations. Respondents were asked to select the degree of the possibility of implementing sustainability factors into affordable housing on a five-continuum scale; where 1 denoted extremely possible, 5 denoted least possible, 3 denoted possible and 2 and 4 were located in between. The constructs were positively worded. Low scores indicated a higher possibility and a high score indicated a lower possibility. The degree possibility was calculated by a weighted mean score. To interpret, a mean score/possibility index of 1.00–1.79 denotes extremely possible; 1.80–2.59 very possible; 2.60–3.39 possible, 3.40–4.19 less possible and 4.20–5.00 denotes least possible. There is a common difference of 0.8 between each of the scales. To ensure the results were not influenced, missing data were not replaced with a mean or mode of a valid response. The mode technique was used to analyze the demography of the respondents. The mode was also used to determine the distribution of respondents with respect to the scales. All data gathered adopted IBM SPSS Statistics Data Editor for analysis. The statistical tests computed are the one-way test, crunch alpha reliability tests, convergent validity and factor analysis.

Analyzing the Results of the Survey

Altogether 122 completed responses were received during the survey period. The results are presented in tables and figures and discussed in the following sections.

Demographic Profiles of the Households

Some 86% of the respondents obtained a minimum of degree level and most (36%) of the respondents have their degrees in quantity surveying (Table 14.2). Majority of those with Ph.D. obtained degrees in engineering and most of those with MBA/MSc obtained degrees in architecture. Forty percent had degrees in architecture or engineering. Most of the respondents held strategic positions and about 90% have

Table 14.2. Cross-tabulation between respondent's academic qualification and academic background.

Academic qualification	Architecture	Quantity surveying	Engineering	Estate management	Town planning	Others	Total
Diploma	0	7	1	2	0	2	12
BSc	3	28	8	6	2	6	53
MSc/MBA	18	9	7	5	0	2	41
Ph.D.	3	0	7	1	0	0	11
Other	0	0	0	0	0	4	4
Total	24	44	23	14	2	14	121

more than five years working (Fig. 14.2). Approximately 12% have more than 20 years working experiences (Table 14.3). Most of those with more than 15 years working experience were architects and most of those with more 20 years were managing directors.

Majority of the respondents work with housing developers and from the contracting organizations (Table 14.4). On this basis, it is inferred that the respondents have adequate knowledge and skill to have the capabilities and abilities to provide unbiased and valid information reflect the housing industry.

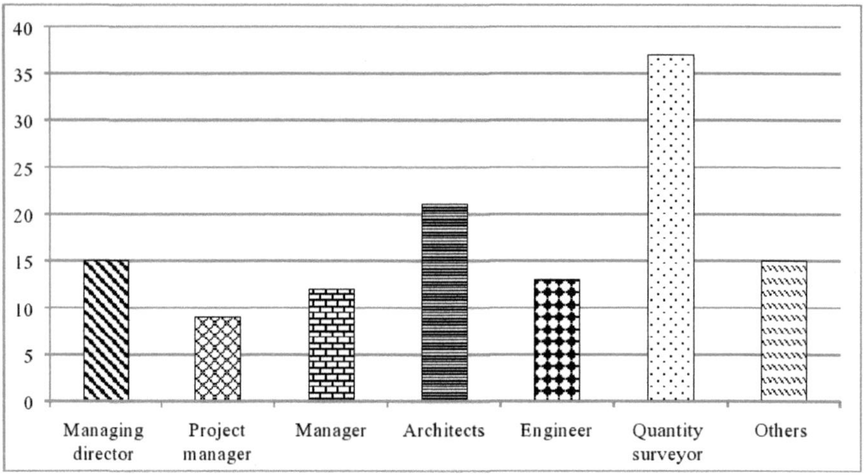

Figure 14.2. Respondent's position.

Table 14.3. Cross-tabulation between respondent's position and working experience (year).

Your current position	Not more than 5 years	5 years–10 years	10 years–15 years	15 years–20 years	20 years and above	Total
Managing director	0	1	1	5	8	15
Manager	0	1	1	3	0	5
Project manager	0	0	7	1	1	9
Principal partner	0	1	3	4	0	8
Manager	0	1	1	3	2	7
Architects	0	2	8	11	0	21
Engineer	2	4	3	4	0	13
Quantity surveyor	12	10	8	6	1	37
Partners	0	0	1	2	0	3
Other	1	1	0	0	2	4
Total	15	21	33	39	14	122

Table 14.4. Category of respondent's organisation.

Organisation	Frequency	Percent
Government	2	1.639
Developers	46	37.705
Contractors	23	18.852
Private client	16	13.115
Private quantity surveying firm	18	14.754
Private architectural firm	12	9.836
Others	5	4.098

Analysis of Sustainable Affordable Housing's Indicators

Firstly, reliability and validity tests were performed to determine the goodness of fit of the data. The reliability test results indicate that the Cronbach's alpha for all the determinants is very high at 0.930. The validity test, using the 'commonalities', produces values that range from 0.461 (Design plan to reduce waste disposal bills) to 0.792 (use recycle materials construction and maintenance) (Table 14.6). Table 14.6 also contains t-test statistics conducted to test the first research hypothesis and further examine the measurements of the population with respect to the each of the factors. For this reason, the null hypothesis was that it is not possible to implement the sustainability factor into affordable housing (H0: U = U0) and the research hypothesis was that it is possible to implement the sustainability factor into affordable housing (Hr: U > U0). U0 is the population mean. 1.5 was set as the t-test level because the 1 denotes extremely possible to implement and 2 denotes very possible to implement. As shown in Table 14.5, the significance (i.e., Pr > |t|) of each of the factors (Hr: U > U0) show that the entire factor is statistically significant. The degrees of freedom for all the factors are the same at 121. Furthermore, small standard errors, in this case, close to zero suggest that the measurements of the respondents with respect to the factors are reflections of the population. The standard error measures the accuracy of the extent to which the sample mean is close to the population mean. A small standard error is an indication that the sample mean is a more accurate replication of the actual population mean. More specifically, standard errors show that the sample mean is plus or minus 1.96 standard errors from the population mean. The standard deviation for the factors, being less than I.0 is an indication of consistency. These statistics suggest that the factor measure what they are intended for. Consequently, all the sustainability factors can be implemented in affordable housing. While the Kolmogorov-Smirnov test could not confirm the normality of the sample, the KMO and Bartlett's test is high and the result is significant (X2 (595) =1620, p = 0.000).

The average mean for the factor is 2.667 and the average standard deviation is 0.661 respectively. The general relationship between mean and standard deviation assume that 68% of the respondents' estimations will fall mean ±1 standard deviation, 95 will fall between mean ± 2 standard deviation and all the respondents will estimations must fall between mean ± 3 standard deviation. The plain interpretations of these statistics are that all the professionals in the housing industry measured that the factors can be implemented in the affordable housing. In particular, 25.64% measured

Table 14.5. One-sample test.

	t	Sig. (2-tailed)	Mean Difference	Lower	Upper	Std. Error Mean
Install energy efficiency equipment in the house	17.076	0.000	1.180	1.044	1.317	0.069
Install gray water use system for flushing, etc.	24.26	0.000	1.566	1.438	1.693	0.065
Install rainwater treatment equipment	17.786	0.000	1.164	1.034	1.294	0.065
Design to provide natural lighting	11.413	0.000	0.836	0.691	0.981	0.073
Design for the use of rainwater	16.265	0.000	1.098	0.965	1.232	0.068
Use recycle materials construction & maintenance	22.026	0.000	1.516	1.380	1.653	0.069
Install storm water discharge system	26.064	0.000	1.516	1.401	1.632	0.058
Community center provided or planned	22.433	0.000	1.500	1.368	1.632	0.067
Housing proximity to workplace	27.637	0.000	1.615	1.499	1.730	0.058
Housing proximity to public transport service	18.805	0.000	1.107	0.990	1.223	0.059
Design for reduction of energy consumption and embodied energy	23.122	0.000	1.393	1.274	1.513	0.060
Housing proximity to open green public area	12.756	0.000	0.926	0.783	1.070	0.073
Desirability of neighborhoods	27.174	0.000	1.516	1.406	1.627	0.056
Design plan for senior citizens	23.706	0.000	1.361	1.247	1.474	0.057
Maintain ecological value of the site	21.61	0.000	1.566	1.422	1.709	0.072
Design plan for adaptability-for future changes	24.417	0.000	1.541	1.416	1.666	0.063
Use minimal furniture	25.125	0.000	1.672	1.540	1.804	0.067
Low house price in relation to income	21.948	0.000	1.426	1.298	1.555	0.065
Design plan to ensure flexibility	22.154	0.000	1.443	1.314	1.572	0.065
Design plan to reduce waste disposal bills	23.599	0.000	1.492	1.367	1.617	0.063
Application of artificial sustainable space plan	24.551	0.000	1.525	1.402	1.648	0.062
Design to provide natural ventilation	18.378	0.000	1.230	1.097	1.362	0.067
Design to reduce housing maintenance costs	19.815	0.000	1.311	1.180	1.443	0.066
Provides bicycle path and bicycle parking lot	21.568	0.000	1.393	1.266	1.521	0.065
Increase floor space index	28.963	0.000	1.664	1.550	1.778	0.057
Use of low-toxicity materials	29.9	0.000	1.713	1.600	1.827	0.057
Connection of pedestrian pathway to the outer pedestrian network	18.758	0.000	1.246	1.114	1.377	0.066
Use eco-friendly methods of construction or new technology	19.661	0.000	1.492	1.342	1.642	0.076

Table 14.5 contd. ...

... *Table 14.5 contd.*

	t	Sig. (2-tailed)	Mean Difference	Lower	Upper	Std. Error Mean
Install skylight for illumination	17.949	0.000	1.246	1.109	1.383	0.069
Design plan for reduction of CO2 emissions	24.573	0.000	1.582	1.455	1.709	0.064
Rational site environmental management plan	24.083	0.000	1.418	1.302	1.535	0.059
High green space area ratio	22.52	0.000	1.557	1.421	1.694	0.069
Use solar energy systems	13.468	0.000	1.074	0.916	1.232	0.080

that the factors are extremely or very possible to implement in affordable housing. Approximately 56% know it is possible to implement the factors however about 18% recognized that the implementations of the sustainability factors are less or least possible. The research finds that the possibility of implementing the sustainability can be grouped into two. The first group comprises of four factors that are very possible to implement and the reminder 30 are measured as possible to implement. Specifically, none of the factors was measured as less or least possible to implement. In general, the surveyed results are consistent with our undisclosed hypothesis. Because of space constraint, only seven of the factors will be explained further.

The research found that to design affordable housing in providing nature lighting is the easiest sustainability factors to implement in affordable housing. Natural energy and ventilation are some of the hallmarks of sustainable buildings. The costs of energy are increasing, and in order to reduce energy cost and production, making use of the nature lighting has been recognized as an essential part of the sustainable design. Energy generation is the major source of greenhouse gas because energy generation involves burning substance including petroleum, coal, wood, natural gas, or fuels. While carbon dioxide emission from energy generation is not exclusive to electricity generation, it takes up the major portion. Harnessing affordable housing location near to open green public areas seems to be very possible to implement in affordable housing. This finding is not easy to interpret in the context of sustainability. For instance, this could mean that, affordable housing is not located in cities because of land restrictions. In general, most of the affordable housing because of the location factor are located at the outskirts of the cities and isolated and surrounded by virgin lands. Will this compromise on the accessibility to work and education remains to be seen. Using solar energy is a major characteristic of sustainable buildings. However, because of capital cost of implementing solar energy like HVAC systems and, photovoltaic panels, it is seldom installed for affordable housing. However, advancement in technology is bringing the cost of implementing solar energy system down. Therefore, it is interesting to find that, the responding housing professional measured it is very possible to implement the solar systems in affordable housing. However, instead of using the solar panels, harnessing the natural lighting remains the most greening option because the solar panels in Malaysia are imported. Using imported materials instead of using localized material vitiate the principle of sustainability itself. While housing needs to be energy efficient, it is important to stress that, the homeowners comfort is paramount to energy

Table 14.6. Distribution of statistics.

Factor	Extraction	Total Correlation	Cronbach's Alpha	Std. Deviation	Mean
Design to provide natural lighting	0.655	0.272	0.890	0.809	2.336
Housing proximity to open green public area	0.545	0.342	0.888	0.802	2.426
Use solar energy systems	0.689	0.466	0.886	0.881	2.574
Design for the use of rainwater	0.625	0.309	0.889	0.746	2.598
Housing proximity to public transport service	0.607	0.202	0.890	0.650	2.607
Install rainwater treatment equipment	0.658	0.343	0.888	0.723	2.664
Install energy efficiency equipment in the house	0.693	0.459	0.886	0.763	2.680
Design to provide natural ventilation	0.642	0.389	0.888	0.739	2.730
Connection of pedestrian pathway to the outer pedestrian network	0.639	0.558	0.885	0.734	2.746
Install skylight for illumination	0.676	0.350	0.888	0.767	2.746
Design to reduce housing maintenance costs	0.578	0.466	0.886	0.731	2.812
Design plan for senior citizens	0.616	0.382	0.888	0.634	2.861
Design for reduction of energy consumption and embodied energy	0.605	0.380	0.888	0.666	2.893
Provides bicycle path and bicycle parking lot	0.678	0.510	0.886	0.714	2.893
Rational site environmental management plan	0.503	0.456	0.887	0.650	2.918
Low house price in relation to income	0.592	0.494	0.886	0.718	2.926
Design plan to ensure flexibility	0.613	0.441	0.887	0.719	2.943
Install thermal insulation in the house	0.707	0.306	0.889	0.867	2.992
Design plan to reduce waste disposal bills	0.461	0.368	0.888	0.698	2.992
Use eco-friendly methods of construction or new technology	0.686	0.428	0.887	0.838	2.992
Community center provided or planned	0.645	0.453	0.886	0.739	3.000
Use recycle materials construction &maintenance	0.792	0.288	0.889	0.760	3.016
Install storm water discharge system	0.687	0.416	0.887	0.643	3.016
Desirability of neighborhoods	0.710	0.382	0.888	0.616	3.016
Application of artificial sustainable space plan	0.657	0.240	0.890	0.686	3.025
Design plan for adaptability-for future changes	0.580	0.505	0.886	0.697	3.041
High green space area ratio	0.615	0.518	0.885	0.764	3.057
Install gray water use system for flushing, etc.	0.620	0.433	0.887	0.713	3.066
Maintain ecological value of the site	0.715	0.410	0.887	0.800	3.066
Design plan for reduction of CO_2 emissions	0.612	0.439	0.887	0.711	3.082
Housing proximity to workplace	0.700	0.245	0.890	0.645	3.115
Design plan for noise emission and control	0.536	0.493	0.886	0.782	3.131
Install sound absorption level of floor	0.586	0.519	0.886	0.668	3.156
Increase floor space index	0.614	0.324	0.889	0.635	3.164
Use minimal furniture	0.498	0.458	0.886	0.735	3.172
Densities providing social mix	0.592	0.391	0.888	0.700	3.197
Use of low-toxicity materials	0.618	0.386	0.888	0.633	3.213

saving. Households spent between 10 and 30% of the income for transportation to their workplaces, it is required that housing be planned near places of works in order to increase housing saving. In 2015 a World Bank's survey showed that Malaysians waste some 250 million hours in traffic holdup annually. Similarly, the number of private car will reduce if the affordable housings are located near workplaces in addition the large amount of reduction in carbon dioxide attributable to private cars ownerships. A major benefit of the mass transportation is reduction in the amounts of cars on the roads. Malaysia has the highest number of cars per household in the region. The just completed Phase 1 of the MRT line 1 from Sungai Buloh to Semantan is expected to take 160,000 cars off the roads. The numbers of cars used have impact on sustainability, because it also reduces pollution. Commuters will also use their times productively in the buses/trains instead of driving themselves. Malaysia is endowed with heavy rainfall and sunshine. The rainfall and sunshine can be 'designed' into the buildings. These should be parts of the design criteria. Making use of rainwater would reduce the payment of water bills and governments' subsidies on household waters. Surface runoff results in excess water that the soil cannot absorb and it is a major transporter of pollutant to the rivers, stream and lakes (Spiegel and Meadow, 2010). The government subsidies waters bills up to almost 50%. In terms of environmental impact, design to use rainwater would have impact on the incessant floods. It is not unexpected to discover that it is possible to install rainwater treatment equipment to affordable housing.

Factor Analysis

The factor analysis method was employed to identify the association among the 37 sustainability factors in order to investigate the pattern display among the factors. The data meet all the necessary requirements for a factor analysis. For instance, the factors are more than 10 and the respondents exceed twice the factors. Statistically, more than three respondents per factors are required for meaningful results. The Kaiser's Measure of Sampling Adequacy returned a high value of 0.712 and is significant. The data were subjected to the Principal Component Analysis. The Varimax rotation strategy was used. The resulting matrix was connected to the main and the division of variance was based on the degree of freedom. The Kaiser's normalization was used to normalize the row of the factor pattern. The results of the analyses suggest that the factors can be classified into nine groups (Table 14.7). The KMO is high and the relationship among the factors is significant (Table 14.8). The results are consistent with the Kaiser and Guttman rule. This is also evident in Fig. 14.3, as the function appears to level off with the 9th components. The R- matrix is 5.37E-006 and show lack of multicollinearity. The nine components explained 63.90% of the total variance. The eigenvalues for the nine components were more than one each. The first component weighed 21.43%, while the ninth component weighed 3.68%. The rotated component matrix is contained in Table 14.10. However three factors, namely "use recycle materials construction & maintenance" and "housing proximity to public transports", and "use minimal furniture" could not load into any component even after multiple trials. Thus the three were excluded in the subsequent factor analysis. However, in order to examine the

Table 14.7. Total variance explained.

Component	Initial Eigenvalues			Extraction Sums of Squared Loadings			Rotation Sums of Squared Loadings		
	Total	% of Variance	Cumulative %	Total	% of Variance	Cumulative %	Total	% of Variance	Cumulative %
1	6.642	21.427	21.427	6.642	21.427	21.427	3.042	9.814	9.814
2	2.772	8.943	30.370	2.772	8.943	30.370	2.511	8.102	17.916
3	2.012	6.491	36.861	2.012	6.491	36.861	2.392	7.716	25.631
4	1.732	5.587	42.448	1.732	5.587	42.448	2.351	7.583	33.214
5	1.543	4.976	47.424	1.543	4.976	47.424	2.128	6.863	40.077
6	1.461	4.713	52.136	1.461	4.713	52.136	1.981	6.389	46.467
7	1.348	4.350	56.486	1.348	4.350	56.486	1.920	6.195	52.662
8	1.158	3.735	60.222	1.158	3.735	60.222	1.795	5.790	58.452
9	1.141	3.681	63.902	1.141	3.681	63.902	1.690	5.450	63.902
10	0.978	3.155	67.057						
11	0.879	2.836	69.893						
12	0.862	2.781	72.674						
13	0.781	2.520	75.194						
14	0.739	2.383	77.577						
15	0.707	2.280	79.857						
16	0.670	2.160	82.017						
17	0.592	1.909	83.926						

18	0.569	1.837	85.763						
19	0.536	1.728	87.491						
20	0.507	1.635	89.127						
21	0.474	1.528	90.655						
22	0.436	1.406	92.061						
23	0.375	1.209	93.270						
24	0.371	1.197	94.467						
25	0.339	1.093	95.560						
26	0.307	0.989	96.550						
27	0.293	0.946	97.495						
28	0.248	0.799	98.294						
29	0.217	0.698	98.992						
30	0.177	0.570	99.563						
31	0.136	0.437	100.000						

Table 14.8. KMO and Bartlett's test.

Kaiser-Meyer-Olkin Measure of Sampling Adequacy.		0.712
Bartlett's Test of Sphericity	Approx. Chi-Square	1332.815
	df	465.000
	Sig.	0.000

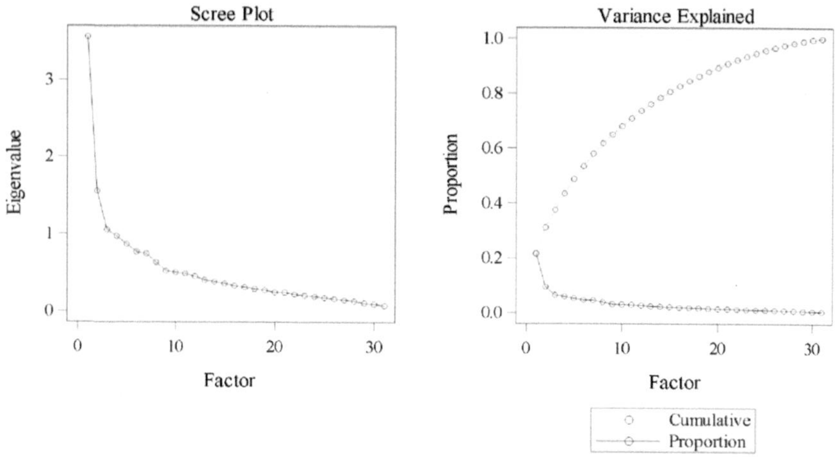

ScreePlot and Variance Explained

Figure 14.3. Scree plot.

relationship among the three a separate factor analysis was performed for the three factors. The results ($\chi 2$ (450) = 1332.815, p = 0.401) and KMO of 0.539 confirmed that they moderately related but the results are not significant. However, the three fall into one component and able to explain 39.455%. The reliability is high and the convergent validity is 0.627. The results of correlations among the three factors are contained in Table 14.9. Without question, the factors are related, for instance, furniture involves a lot of deforestation, which has significant impact on the amount of rainfall, erosion, floods, and disturbance of the ecosystem. Much of the timber is harvested from virgin forests. In between, lack of public transport in the vicinity will lead to the generation of carbon dioxide and in most cases also lead to noise pollution from private cars. Recycled materials and public transportation may be the better options for housing construction and maintenance and minimum furniture should be the first choice. The nine components included in the main factor analysis are explained next.

Operating cost explained 21.43% in the total variance and comprised of six factors. A second-order factor analysis combined these factors into a single factor of operating cost; Overall MSA = 0. 802, $\chi 2$ (15) = 166.605, $p < 0.001$) with Cronbach's alpha 0.770. The convergent validity is 0.693. The correlation ranges from 0.107 to 0.405 and all are statistically significant. The R-matrix is 0.244 more than 0.000001 thresholds. The six factors fall into two components. The first component comprises of use of low toxicity materials, low house price in relation to income and design plan for adaptability for future changes. The second component comprises of design to provide natural

Table 14.9. Correlations among the three factors.

Housing proximity to public transport service		Use eco-friendly methods of construction or new technology	Use minimal furniture
	Correlation Coefficient	–0.089	0.121
	Sig. (2-tailed)	0.328	0.183
	N	122	122

ventilation, design to reduce housing maintenance costs and design plan to ensure flexibility. The first component explained 32.4% and the second component explained 31.23%. The needs to reduce the operating costs of housing are on the increase because of the volatility in future expenditure, especially for modifications and utility costs. It is imperative that house price is considered alongside the cost of material used and future modifications because operating costs of a building are far outweighed its capital cost (Jewell and Flanagan, 2005). More so for sustainable housing, sustainable housing aims to lower housing operating cost and increase homebuyers' satisfaction. In the event of future modification and maintenance, natural and nontoxic materials should be selected. Sustainable housing is nontoxic, they are made from recycled materials/components and they can as well be recycled. Only if the capital cost and house price are related that sustainable affording housing could be achieved. Currently, housing affordability is still measured in terms of financial-ability. However, to the extent that the homebuyers will pay more on account of operating costs, even if the capital is low, the house is not sustainably affordable (Olanrewaju and Woon, 2017). Openings (i.e., windows) are used to increase natural ventilation of buildings, however, because of regulations and artificial restriction like adjoining properties, noise, pollution, and safety issues, the openings might not be used effectively. Correlations exist among flexibility adaptability and openings. For instance, openings on existing buildings are created for ventilation. If conditions outside the buildings perhaps due to pollution, noise or lighting are not conducive to the households, the windows might have to be permanently or temporarily shut down. Therefore, affordable housing should be flexible, adaptable, and maintainable and make use nontoxic materials both for constructions and future modifications. As Tam and Zeng, 2013, explained, housing should be flexible and adaptable at least up to 50 years.

The second factor was named **energy** and comprise of four factors with factor loading that ranges from 0.561 to 0.759. The factor explained 8.94% of the total variance. A second-order factor analysis combined these factors into a single factor of energy; Overall MSA = 0. 737, $\chi 2$ (6) = 88.708, $p < 0.001$) with Cronbach's alpha 0.709. The convergent validity is 0.734. The extracted communalities range from 0.445 to 0.575. The correlation ranges from 0.282 to 0.495 and all are statistically significant. The R-matrix is 0.474 more than 0.000001 thresholds. The four factors collectively explained 54.07% of the model. As may be seen, these factors are related. The operating costs, including for water, electricity, telephone are associated among themselves yet collectively relate to a temperature in the housing as well as on it environments. Cost of energy is very high especially during the heat wave occasioned by climate changes. The amount of energy that housing consumes is huge during their lifespan. Housing orientation and technologies can help save a lot of energy and water

Table 14.10. Rotated component matrix.

Factor	Operating	Energy	Pollution	Water	Green	Location	Ecology	Social	Place
Design plan for ensure flexibility	0.737								
Design to reduce housing maintenance costs	0.714								
Low house price in relation to income	0.641								
Design to provide natural ventilation	0.581								
Use of low-toxicity materials	0.577								
Design plan for adaptability-for future changes	0.523								
Install energy efficiency equipment in the house		0.759							
Install thermal insulation in the house		0.737							
Install gray water use system for flushing, etc.		0.593							
Use solar energy systems		0.561							
Increase floor space index			0.679						
Rational site environmental management plan			0.576						
Design plan for reduction of CO_2 emissions			0.523						
Design to provide natural lighting			−0.428						
Design for the use of rainwater				0.698					
Design for reduction of energy consumption and embodied energy				0.673					
Install rainwater treatment equipment				0.638					
Install storm water discharge system				0.603					

Use eco-friendly methods of construction or new technology					0.843				
High green space area ratio					0.476				
Install skylight for illumination						0.777			
Provides bicycle path and bicycle parking lot						0.614			
Housing proximity to workplace						–0.600			
Housing proximity to open green public area							0.672		
Maintain ecological value of the site							0.618		
Connection of pedestrian pathway to the outer pedestrian network							0.509		
Application of artificial sustainable space plan								0.807	
Community center provided or planned								0.487	
Design plan to reduce waste disposal bills								0.445	
Design plan for senior citizens									0.712
Desirability of neighborhoods									0.712

use of buildings. The air in Malaysian buildings is 'stuffy' because of lack of air velocity in the buildings as winds in Malaysia are not strong and coupled with high humidity. As more people spent time indoor because of technology (e.g., internet, smartphone, and online business), more energy will be required for air conditioning, fans, cooking, lighting, and computers. Renewable energy source and passive design are beneficial. Lack of soft landscaping leads to overheating is common in Malaysia but at the same time, flash floods are also common because of surface run-off from heavy rainfall. The impact of heating is leading to a high increase in energy and water usage. Many households have installed thermal comfort in their homes. However such additional expenses can increase the capital cost of the buildings. In order to reduce expenditure on thermal comfort, passive heating and passive design, may be suggested. The Urban Heat Island (UHI) effect is identified as a major cause leading to increase in demand for thermal comfort regulation because of the high increase in temperature in the cities. Exhaust heat that is discharged into the atmosphere from air conditioning/chiller plant outlets are some of the causes of heat waves. The design teams can specify ranges of equipment to safe and conserved energy. There is a proliferation of equipment that can be installed in the building to generate and store energy in the buildings.

Pollution comprises of four factors with factor loading that ranges from –0.423 to 0.679. The factor explained 5.59% of the total variance. A second-order factor analysis combined these factors into a single factor of pollution; Overall MSA = 0. 572, $\chi2$ (6) = 45.63 $p < 0.001$). Because one of the factors was negative, it is not appropriate to calculate reliability because it violates reliability model assumptions. The R-matrix is 0.681 more than 0.000001 thresholds. The factors load into two separate components. The first component comprises of design plan to reduce CO_2 emission and ration site environment plan and the second component comprises of increase floor space index and design to provide natural lighting. The convergent validities are 0.800 and 0.18 respectively. The extracted communalities range from 0.610 to 0.853. The correlation ranges from –0.189 to 0.408 and all are statistically significant. The four factors collectively explained 53.39% of the model. In general the factors, however, are related, rational site environmental management involves construction on brownfield, this often impacts on the spatial index. A lot of pollution emissions can be generated from vehicles and the housing itself, housing is carefully designed and constructed in brownfield and an irregular site because of planning. However, the practical implication of the results is that to increase floor space index, rational site environmental management plan and design plan for reduction of CO_2 emissions, the stakeholders must forgo design to provide natural lighting. In other words, there is a trade off. Forgoing one sustainability measure in order to achieve another, is a common barrier to achieve full sustainability in buildings in the UK (Williams and Dair, 2007).

Water comprises of four factors with factor loading that ranges from 0.603 to 0.698. The factor explained 6.491% of the total variance. A second-order factor analysis combined these factors into a single factor of energy; Overall MSA = 0. 734, $\chi2$ (6) = 73.633, $p < 0.001$) with Cronbach's alpha 0.690. The convergent validity is 0.720. The extracted communalities range from 0.443 to 0.595. The correlation ranges from 0.275 to 0.422 and all are statistically significant. The R-matrix is 0.538 more than 0.000001 thresholds. The entire factor load into one component loading ranges from 0.665 to 0.771. The four factors collectively explained 51.83% of the model. Freshwater is a scarce resource in most countries, water audit should be conducted to obtain information on water use and improve quality of water. While rainfall in Malaysia is heavy, this resource not is used effectively, and households still depend on process water even for washing cars, flushing, and gardening. Impact of climate changes (i.e., floods) can be reduced through making use of various water harvest techniques. The impact of floods is great, not only that the building may be totally destroyed but water source may be contaminated, insurance companies may lose their revenues and homeowners might have to live in deplorable conditions before their homes are repaired or rebuilt in some cases. With the tropical nature of Malaysia, it is obvious that the temperature will be on the rise, so there is a need for means to improve thermal comfort. However, because the water content in the warm air is high, it will be accompanied by heavy rainfall. Regulatory frameworks are available from the government to encourage both the developers and homeowners to use rainwater retention system to reduce the surface runoff. Subsidies on water consumptions might need to be taken away to serve as conservation incentives to homeowners. Alternatively, policies regarding preserving natural green areas or man-made parks need to be enforced for the ground to be recharged as a groundwater reservoir.

Green comprises of two factors with factor loading that ranges from 0.476 to 0.843. The factor explained 4.976% of the total variance. A second-order factor analysis combined these factors into a single factor of green; Overall MSA = 0. 500, $\chi2$ (1) = 31.305 $p < 0.001$) with Cronbach's alpha 0.645. The convergent validity is 0. 860. The extracted communality was 0.739. The correlation is 0.478 and all are statistically significant. The R-matrix is 0.771 more than 0.000001 thresholds. The two factors collectively explained 73.920% in the model. As previously explained, some construction management strategies like lean construction and value management can be initiated to provide sustainable housings. The design team should keep abreast with the latest technologies and identify construction procurement strategies that can protect the environment, reduces cost construction. Sustainable materials require sustainable methods or natural methods for manufacturing, construction. Cement (a major proportion of concrete) production consumes a lot of energy for its production and can store large energy as embodied energy. The amount of concrete use should be lessened to thermal comfort because concrete has a peculiar feature of absorbing heat during the daytime and the released heat into the building during the night.

Location comprises of three factors with factor loading that ranges from –0.59 to 0.777. The factor explained 4.713% of the total variance. A second-order factor analysis combined these factors into a single factor of location; Overall MSA = 0. 531, $\chi2$ (3) = 40.472 $p < 0.001$). Because one of the factors has a negative loading, it is not appropriate to calculate reliability because it violates reliability model assumptions. The convergent validity is 0.383. The extracted communalities range from 0.246 to 0.724. The correlation ranges from –0.117 to 0.494 and all are statistically significant. The R-matrix is 0.712 more than 0.000001 thresholds. The four factors collectively explained 53.39% of the model. One of the best ways to improve sustainability is through passive planning, whereby the distance from home to a place of place work should be close in order to save time and transportation. Sustainable homeowners use bicycles in order reduce pollution and increase the quality of life because the distance of the housing to major places like a workplace, hospitals and markets are carefully planned, so homeowners can use a bicycle for transportation. Interpreting Tam and Zeng's (2013) suggestion, affordable housing is within a radius of 15 km from most frequency places. Though 15 km is reachable by bike, the weather (i.e., heavy and unpredictable rain, high sunshine, and high humidity) in Malaysia might place some restrictions on this and because of this, up to 10 km would be suitable for Malaysia. In addition, renting bicycles to intending users could be a strategic measure.

Ecology comprises of three factors with factor loading that ranges from 0.509 to 0.672. The factor explained 4.350% of the total variance. A second-order factor analysis combined these factors into a single factor of ecology; Overall MSA = 0.649, $\chi2$ (3) = 46.831 $p < 0.001$) with Cronbach's alpha 0.644. The convergent validity is 0. 765. The extracted communalities are ranges from 0.538 to 622. The correlation ranges from 0.342 to 423 and all are statistically significant. The R-matrix is 0.675 more than 0.000001 thresholds. The factors collectively explained 58.55% of the model. There should be a relationship between access to green areas and natural ecosystem or Greenfield. The design team and urban planners should use affordable housing and their landscaping to deliver, protect, secure and enhance the natural ecosystem. The ecological values of the place should be designed and constructed to a place of workplaces in which the household walk or use bicycles to places they often visit. This is consistent with most of the construction-related indicators and assessments tools like BREEAM, LEEDS, PBRS and QSAS, Green star system, green mark assessment and HK-BEAM. With respect to indicators, the ecological value of a site, protection of ecological features, reuse contaminated lands, site developments, open space, heat island effect landscape management vegetation protection are considered (Al-Jebouri et al., 2017). Ecological footprint analysis ensures that housing projects are carefully sited in order not to destroy the ecosystem in the vicinity (Blair et al., 2004).

Social comprises of three factors with factor loading that ranged from 0.445 to 0.807. The factor explained 3.735% of the total variance. A second-order factor analysis combined these factors into a single factor of place; Overall MSA = 0.610, $\chi2$ (3) = 26.577 $p < 0.001$) with Cronbach's alpha 0.539. The convergent validity is 0. 809. The extracted communalities range from .489 to 589. The correlation ranges from 0.224 to 0.311 and all are statistically significant. The R-matrix is 0.800 more than 0.000001 thresholds. The three factors collectively explained 52.180% of the model. The interpretations of the statistics are that the factors are related and can be organized for a particular objective.

The **place** comprises of two factors with the factor loading of 0.712. The factor explained 3.681% of the total variance. A second-order factor analysis combined these factors into a single factor of social; Overall MSA = 0.500, $\chi2$ (1) = 19.345 $p < 0.001$) with Cronbach's alpha 0.557. The convergent validity is 0. 693. The extracted communalities range 0.693. The correlation is 0.387 and is statistically significant. The R-matrix is 0.851 more than 0.000001 thresholds. The three factors collectively explained 69.33% of the model. While designers and developers do not always consider the need of seniors, the elderly and those with mobility difficulties in affordable housing, thought should be given to all those who would use the house immediately and in future and those in the neighborhoods. Both the internal configurations of the housing and environment need to be appealing to those with restricted motilities

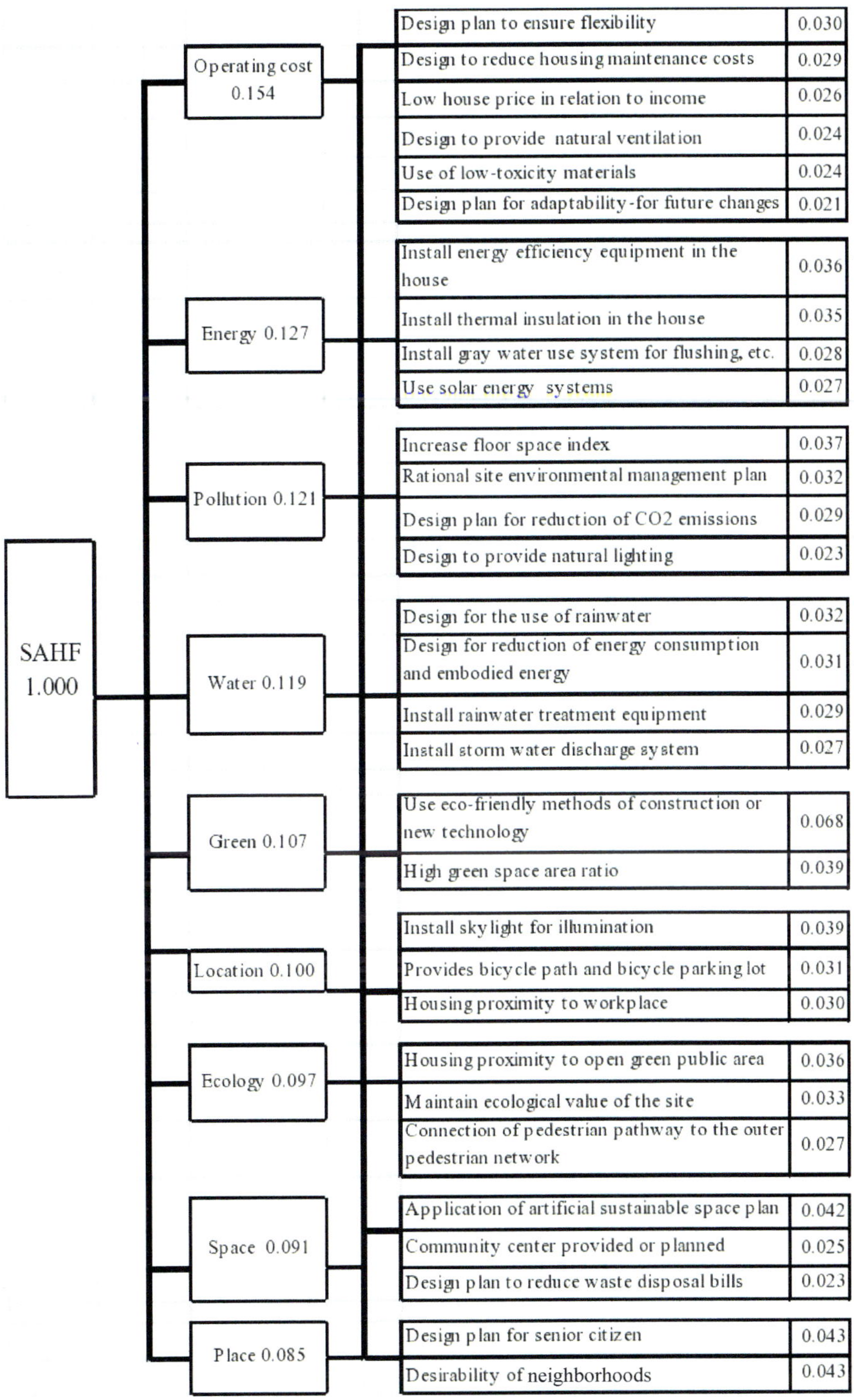

Figure 14.4. Sustainable affordable housing framework.

together. Otherwise, total sustainability is not attainable. At the strategic level, the first use of the framework is to establish that homebuyers demand sustainable affordable housing and secondly for the developers to focus on the nine components.

At the tactical level, it is important for housing developers, is to ensure that all the factors attributable to each of the factors are adequately addressed or implemented in the housing delivery. However, this does mean that the other three factors of housing proximity to public transport service use eco-friendly methods of construction or new technology and use minimal furniture should be discarded because they could not be integrated into the factor analysis. An attempt should be made to integrate these factors at the component level. Deductively, the main purpose of the framework is to allow the housing organization to focus on homebuyers requirements. This framework provides a systemic component for sustainable affordable housing and offers valuable insight on how developers can supply sustainable housing especially at the operation level.

Trade Off or Opportunity Cost

Earlier the various CSFs for sustainable affordable housing were explained. Achieving sustainable affordable housing is however, conflicting, challenging and sometimes conflicting. To supply sustainable building is not without cost. That this will be problematic in projects in the built environment has been noted by William and Dair (2007), as the need to have one indicator could to a trade off for another indicator. Put simply, as previously outlined, the capital cost of sustainable affordable housing is more than the cost of conventional buildings. It is also a common knowledge that it is difficult to offer buildings that meet all the criteria for the client without a compromise. To deliver projects at the lower cost would imply that either time or quality or both suffer. A high quality building could imply high cost or more time or both. Similarly, to achieve a benefit of sustainable housing one or more benefits might be traded. Therefore, the opportunity cost of the benefits should be carefully considered. It is not surprising that this research found that it is not possible to install energy efficiency equipment in the house and have sustainable space plan at the sametime. Also it is important to ensure either thermal insulation in the house is installed or natural lighting, requirements of the senior citizen, flexibility and increase in floor space index are provided, see Table 14.11.

Summary and Suggestions to the Housing Industry

This research aims to investigate the possibility of delivery sustainable affordable housing and with the primary purpose of developing a framework for their implementation. Although the benefits of a sustainable affordable housing may not be immediately apparent , it will add values to homebuyers and occupants' investments over a long time. It will also benefit developers because developers use ecological features to market their buildings. The research has produced a workable framework for affordable delivery. This framework will guide architects and engineers in the design, production, and operation stages of affordable housing. It will be useful to homebuyers and occupants of the existing housing towards achieving sustainability.

Table 14.11. Trade off of critical success factors sustainable affordable housing.

	Install energy efficiency equipment in the house	**Install thermal insulation in the house**	**Install gray water use system for flushing, etc.**	**Install rainwater treatment equipment**	**Design to provide natural lighting**	**Design for the use of rainwater**	**Community center provided or planned**	**Housing proximity to workplace**	**Housing proximity to public transport service**	**Design for reduction of energy consumption and embodied energy**	**Housing proximity to open green public area**	**Maintain ecological value of the site**
Design to provide natural lighting		–0.008										
Housing proximity to workplace					–0.011			1				
Housing proximity to public transport service							–0.052		1			
Design plan for senior citizen		–0.032										
Design plan for adaptability-for future changes					–0.054							
Use minimal furniture						–0.054						
Densities providing social mix										–0.079		
Design plan to ensure flexibility		–0.067		–0.006								
Design plan to reduce waste disposal bills					–0.01				–0.007			
Design to provide natural ventilation			–0.029	–0.032						–0.093		
Design to reduce housing maintenance costs				–0.043								
Provides bicycle path and bicycle parking lot								–0.117				

Table 14.11 contd. ...

...Table 14.11 contd.

	Install energy efficiency equipment in the house	**Install thermal insulation in the house**	**Install gray water use system for flushing, etc.**	**Install rainwater treatment equipment**	**Design to provide natural lighting**	**Design for the use of rainwater**	**Community center provided or planned**	**Housing proximity to workplace**	**Housing proximity to public transport service**	**Design for reduction of energy consumption and embodied energy**	**Housing proximity to open green public area**	**Maintain ecological value of the site**
Increase floor space index		–0.043			–0.189	–0.034					–0.009	
Use eco-friendly methods of construction or new technology						–0.085			–0.067			
Install skylight for illumination								–0.241				–0.027
Design plan for reduction of CO2 emissions									–0.037			
High green space area ratio									–0.087			
Design plan for noise emission and control									–0.044			
Application of artificial sustainable space plan	–0.016		–0.003	–0.017						–0.049		

Although there is previous research on how to implement sustainability in housing nascent though on affordable housing, they are complex, therefore, this research offers a relatively simple yet empirical framework to implement sustainability in affordable housing. However, its main defect is that the indicators are 'small' and do not include process and focus more on technical aspects in sustainable affordable housing, however, the indicators can be increased without losing its value. Some of the aspects of sustainable affordable housing require attitudinal changes. At a practical level, sustainable affordable housing entails a reduction in energy and water consumption and reduction in pollution arising from housing construction and operation, reduction in the cost of ownership and operation through transportation, future modifications. The framework adds virtually nothing to housing construction costs yet will increase the value of the housing significantly. The government would need to modify the existing policy to incorporate such techniques in planning, designing, and construction of affordable housing. It is extensively recognized that the sustainability requirements need to be addressed and government and other stakeholders including developers, contractors, and third-party agencies have introduced measures towards the objectives. Incorporating sustainability during design, construction and operation into affordable housing have strategic consequences on the performance of the delivered housing. This study develops a sustainable affordable housing project framework for the housing industry. The establishment SAHF would allow housing developers to approximately determine if their affordable housing is sustainable or not especially during the design, construction and operation stages. However, the framework is not meant to be definitive and prescriptive but indicative because it is not easy to accurately measure sustainability in housing partly due to behavior of the home occupants. The framework is adaptable to building types. A sustainable housing audit is a review for existing housings to determine if the materials, process, and method could reduce energy and water consumption and to increase occupants' comforts and satisfaction. However, various sustainable techniques may be theoretically and practically possible to incorporate into affordable housing, it appears that their incorporation is naïvely hindered by a sequence of obstacles. The embryonic nature of sustainable affordable housing, resulting in lack of information or consensus on what specifically, a sustainable affordable housing looks or behaves like is a major problem in advancing knowledge on this topic.

The theoretical contribution of this research is threefold: (a) creating awareness on knowledge of the housing industry on sustainability, (b) on examining the possibility of implementing the sustainability criteria in affordable housing using the critical success factor approach; (c) developing knowledge on the implementation of sustainability in affordable housing at the operational level all of which are currently nascent. However, the practical implication requires further scrutiny, for example, it is not clear if the method of normalization of weightage of the components and factors are theoretically grounded. There exists a possibility that the factors need expansion. Although a model does not require to include 'all' factors, in as much as the major factors are included, this is deemed sufficient. However, future research could examine the possibility of expanding or even reducing the current factors. Future research should examine the experience of the homebuyers on the same factors. This will among others, allow the examination of the 'measurement gap' because of it very likely that homebuyers attach different weightage to these factors.

Acknowledgement

The research presented in this paper was supported in full by a grant from the "FRGS"; project: Analytical Investigation of Problems in Housing Supply in Malaysia. Project number: FRGS/1/2015/TK06/UTAR/02/2.

References

Abdul, Y. and R. Quartermaine. 2014. Delivering Sustainable Buildings: Savings and Payback. IHS Bre Press.

Akadiri, P. O., E. A. Chinyio and P. O. Olomolaiye. 2012. Design of a sustainable building: A conceptual framework for implementing sustainability in the building sector. Buildings 2(2): 126–152.

Al–Jebouri, M. F., M. S. Saleh, S. N. Raman, R. A. A. B. O. Rahmat and A. K. Shaaban. 2017. Toward a national sustainable building assessment system in Oman: Assessment categories and their performance indicators. Sustainable Cities and Society 31: 122–135.

Ang, S. A. Olanrewaju, F. C. Chia and S. Y. Tan. 2016. Awareness on Sustainable Affordable Housing among Homebuyers in Malaysia. In Proc. International Conference on Sustainable Construction & Structures, Melaka, Malaysia, 6–7 December 2016.

Assaf, S. A., A. A. Bubshaitr and F. AL-Muwasheer. 2010. Factors affecting affordable housing cost in Saudi Arabia. International Journal of Housing Markets and Analysis 3: 290–307.

Atombo, C., K. C. J. Dzantor and A. A. Agbo. 2015. Integration of sustainable construction in project management: A case study in Ghana. International Journal of Construction Engineering and Management 4(1): 13–25.

Blair, J., D. Prasad, B. Judd, R. Zehner, V. Soebarto and R. Hyde. 2004. Affordability and sustainability outcomes: a triple bottom line assessment of traditional development and master planned communities - Volume 1. Australian Housing and Urban Research Institute. UNSW-UWS Research Centre in collaboration with the Southern and Queensland Research Centre. June 2004 AHURI Final Report No. 63 ISSN: 1834-7223, ISBN: 1 920941 28 2.

BNM (Bank Negara Malaysia. 2017. Demystifying the Affordable Housing Issue in Malaysia. Box Article in Annual Report. Kuala Lumpur.

Bragança, L., R. Mateus and H. Koukkari. 2010. Building sustainability assessment. Sustainability 2(7): 2010–2023.

CIOB (The Chartered of Institute of Building). 2017. Basics of Sustainability, 2 Climate Change, Carbon Action, 2050.

Cohen, L., L. Manion and K. Morrison. 2011. Surveys, longitudinal, cross-sectional and trend studies. Research Methods in Education, 7th edition. Abingdon: Routledge, 261–4.

Cole, J. R. 2011. Motivating stakeholders to deliver environmental change. Building Research & Information 39(5): 431–435.

Dempsey, N., C. Brown and G. Bramley. 2012. The key to sustainable urban development in UK cities? The influence of density on social sustainability. Progress in Planning 77: 89–141.

Energy Commission. 2013. Peninsular Malaysia Electricity Supply Industry Outlook 2013. ISBN: 978-967-12023-0-2, ST(P)06/07/2013. www.st.gov.my.

Energy Commission. 2015. Malaysia Energy Statistic Handbook 2015, Suruhanjaya Tenaga (Energy Commission), Putrajaya, Malaysia, www.st.gov.my, ISSN No.: 2289-6953.

Ganiyu, B. O., J. A. Fapohunda and R. Haldenwang. 2017. Sustainable housing financing model to reduce South Africa housing deficit. International Journal of Housing Markets and Analysis 10(3): 410–430.

Ganiyu, B. O., J. A. Fapohunda and R. Haldenwang. 2015. Construction approaches to enhance sustainability in affordable housing in developing countries. pp. 101–107. *In*: Sustainable Technologies (WCST), 2015 World Congress on. IEEE.

Glaeser, E. L. and J. Gyourko. 2003. The impact of building restrictions on housing affordability. FRB New York—Economic Policy Review, V9(2, Jun): 21–39.

Government of Malaysia. 2010. Tenth Malaysian Plan 2011–2015. Putrajaya: Economic Planning Unit Prime Minister's Department.

Hamid, Z. A., A. F. Roslan, M. C. Ali, F. C. Hung, M. S. M. Noor and N. M. Kilau. 2014. Towards a national green building rating system for Malaysia. Malaysian Constr. Res. J. 14: 0–16.

Joslin, R. and R. Müller. 2016. The relationship between project governance and project success. International Journal of Project Management 34: 613–626.

Kamali, M. and K. N. Hewage. 2015, June. Performance indicators for sustainability assessment of buildings. pp. 8–10. *In*: Proceedings of the International Construction Specialty Conference of the Canadian Society for Civil Engineering (ICSC), Vancouver, BC, Canada.

Keeble, B. R. 1988. The Brundtland report: Our common future. Medicine and War 4(1): 17–25.

Killip, G. 2006. The housing maintenance and sustainability debate. *In*: Proceedings of the Annual Research Conference of The Royal Institution of Chartered Surveyors, London.

Ibrahim, F. A., M. W. M. Shafiei, I. Said and R. Ismail. 2013. Malaysian housing developers' readiness in green homes development. World Applied Sciences Journal 30: 221–225.

Miller, W. and L. Buys. 2013. Factors influencing sustainability outcomes of housing in subtropical Australia. Smart and Sustainable Built Environment 2(1): 60–83.

Mora, E. P. 2007. Life cycle, sustainability and the transcendent quality of building materials. Building and Environment 42: 1329–1334.

Nguyen, T. M. 2005. Does affordable housing detrimentally affect property values? A review of the literature. Journal of Planning Literature, Vol. 20, No. 1

Olanrewaju, A., S. Y. Tan, L. T. Lee, F. Ayob and A. Ang. 2016. Investigating the compatibility of affordable housing with sustainability criteria: A conceptual framework. pp. 228–240. *In*: Proceeding—Putrajaya International Built Environment, Technology and Engineering Conference (PIBEC2016), 24 – 25 September, 2016. Bangi, Malaysia. ISBN 978-967-13952-8-8.

Patterson, W. 2007. Keeping the Lights On—Towards Sustainable Electricity. Earthscan. London.

Pitt, V. 2013. Top sustainable house builders: Next Generation Initiative. [Accessed on 00 00 00 and Available at http://www.building.co.uk/top-sustainable-housebuilders-nextgenerationinitiative/5064570.]

Priemus, H. 2005. How to make housing sustainable? The Dutch experience. Environment and Planning B: Planning and Design 32: 5–19.

Sanchez, F. and K. Sobolev. 2010. Nanotechnology in concrete—a review. Construction and Building Materials 24: 2060–2071.

Sayce, S., A. Sundberg and B. Clements. 2010. Is Sustainability Reflected in Commercial Property Prices: An Analysis of the Evidence Base.

SEDA (Sustainable Energy Development Authority Malaysia, 2012) Renewable Energy Status in Malaysia. Available at: http://www.mida.gov.my/env3/uploads/events/Sabah04122012/SEDA.pdf. [Accessed 22 March 2015].

Shad, R., M. Khorrami and M. Ghaemi. 2017. Developing an Iranian green building assessment tool using decision making methods and geographical information system: Case study in Mashhad city. Renewable and Sustainable Energy Reviews, pp. 324–340.

Shen, L., L. J., V. W Hao and H. Yao. 2007. A checklist for assessing sustainability performance of construction projects. Journal of Civil Engineering and Management, Vol XIII, No 4: 273–281.

Sherwin, D. 2000. A review of overall models for maintenance management. Journal of Quality Maintenance Engineering 6(3): 138–164.

Spiegel, R. and D. Meadows. 2010. Green Building Materials: A Guide to Product Selection and Specification. John Wiley & Sons.

Susilawati, C. and W. F. Miller. 2013. Sustainable and affordable housing: a myth or reality. In Proceedings of the 19th CIB World Building Congress (pp. 1–14). Queensland University of Technology.

Tam, W. Y. V. and X. S. Zeng. 2013. Sustainable performance indicators for Australian residential buildings. Journal of Legal Affairs and Dispute Resolution in Engineering and Construction (5): 168–179.

Tankard, C. 2016. Smart buildings need joined-up security. Network Security 10(2016): 20.

UN (United National Environment Programme Report, 2014) Climate change costs to escalate. NST (News Sunday Times) 07 December, 2014 pp.38.

UNEP (United Nations Environment Programme. 2014. Sustainability Metrics: Translation and Impact on Property Investment and Management. A report by the Property Working Group of the United Nations Environment Programme Finance Initiative, May 2014.

Williams, K. and C. Dair. 2007. What is stopping sustainable building in England? Barriers experienced by stakeholders in delivering sustainable developments. Sustainable Development 15(3): 135–147.

Wood, B. R. 2009. Building maintenance. Oxford: Blackwell Publishing.

World Economic Forum (World Economic Forum Industry Agenda Council on the Future of Real Estate & Urbanization). 2016. Environmental Sustainability Principles for the Real Estate Industry REF120116 www.weforum.org.

Zavr, S. M., R. Žarnić and J. Šelih. 2009. Multicriterial sustainability assessment of Residential buildings. Technological and economic development of economy. Baltic Journal on Sustainability 15(4): 612–630.

CHAPTER 15

Cost Prediction for Green Affordable Housing in Nigeria

*Olubunmi Comfort Ade-Ojo** and *Deji Rufus Ogunsemi*

Introduction

Housing is one of the basic needs of an individual and the quality of life of a person is directly/indirectly affected by the quality of housing. If housing is believed to be a basic component in determining the quality of life of a person, without quality housing, a person might not be able to meet other basic needs and participate effectively in the society according to Adetokunbo (2012) and Akinyode and Tareef (2014). Despite these assertions, meeting the housing needs of people has been a problem due to the geometrical rate of urbanization (Adejumo, 2008; Olotuah and Bobadoye, 2009). Housing is one of the major outputs of the construction industry. The quantity and quality of this product is a reflection of the construction industry in any country. Apart from civil engineering projects like roads and the likes, housing constitutes a large chunk of infrastructure for the development of any nation. A greater percentage of both individual and public resources are committed to the provision or acquisition of buildings and infrastructures including housing. Therefore, consideration for cost takes a prime position in the decision to build or not to build. The owner, either a private individual or government is usually concerned and takes particular interest in the likely cost of the project. In lieu of this, the owner is weary of anything that may introduce additional cost or that can increase the marginal cost of the project. Hence, green building construction has been viewed with suspicion when it comes to the cost of construction. Thus, the availability of cost predictive tool can be a lee way for the adoption of green building development in Nigeria.

The construction industry all around the world is at the center of every economic development. Construction activities are geared towards improving the social and economic life of human beings through the provision of diverse infrastructures. Thus,

Federal University of Technology Akure, Nigeria. Email: dejifeyi@gmail.com
* Corresponding author: oluwabunmiade@gmail.com

as human population increases, there is a corresponding increase in the volume of social and economic activities to meet the need of the people. It therefore implies that the increase in the social and economic activity automatically creates an imbalance in the sustainability equation. This imbalance can only be addressed by implementing green building and this is what discussions on sustainable building development seek to mitigate (Emas, 2015; Windapo and Rotimi, 2012). In its bid to provide required facilities needed to compliment the ever increasing human growth, the construction industry uncontrollably leaves in its wake untold environmental degradation. The industry consumes 45% of energy generated across the world and generates 40% of all manmade waste (Bashir et al., 2010; Hussin et al., 2013). The built environment, which is the sole product of construction activities, also affects the sustainability equation with waste and environmental pollution (water, air and land). The built environment according to Shen et al. (2007) contributes about 30% of ecological problems the world over, ranging from the destruction of flora and fauna in the quest for infrastructural and building development to the generation of greenhouse gases and carbon emissions. The challenge therefore is in the provision of quality housing with consideration for sustainability while maintaining an affordable cost of construction.

The outcry against the huge deficit in housing provision in the country has been on for decades. Despite the efforts made by various past administrations in Nigeria, the housing deficit is estimated at about 60% while 75% of urban dwellers are said to live in slums according to Olotuah and Bobadoye (2009). The dual problems of environmental degradation and housing shortage and rather poor quality dwellings is bringing to the fore the necessity for green building development in Nigeria. However, since every construction activity results into cost to the owner, the cost impact of having a green building comes to the front burner. As a relatively new innovation and yet to be fully integrated it is viewed like any other new technology which introduces additional cost to the project.

Affordability refers to having the financial means for, or to be able to bear the cost of something with little or no inconvenience. However, literature has shown that lack of long term mortgage finance and high cost are in the forefront of housing provision in Nigeria (Adejumo, 2008; Olotuah & Bobadoye, 2009; Adetokunbo, 2012; Ibimilua and Oyewole, 2015). Considering the economic capability of an average Nigerian with not too good economic indices (Adedeji and Olotuah, 2012), adopting green building development with implied cost increase presents an additional burden financially. However, this cost mirage can be resolved with the use of cost predictive models. Hence, the study is aimed at considering the development of cost predictive model for the provision of green affordable housing in Nigeria. The methodology involves a desk review of literature on Nigeria housing needs, green building principles, provision of green affordable housing and the development of cost predictive models in enhancing green building development for low/medium income households in Nigeria.

Green Building Construction

The California's Sustainable Buildings Task Force report (Sustainable Building Task Force and the State and Consumer Services Agency, 2003) notes that Green buildings are the same as Sustainable or High performance buildings. The United States Green Building Council (USGBC) also describes green buildings in line with the SBTF report. Hence the term sustainable and Green are synonymous. The building construction industry in Nigeria is confronted with two hydra headed problems of housing shortage and environmental degradation. This is due to the increased use and consumption of environmental and natural resources in the provision of buildings and infrastructural facilities. The worldwide threatening global warming is the result of constant increase in the economic and social activities of human beings with its inversely proportional effect on the environment which forms the tripod of sustainable development. The need to mitigate the negative impact on the environment gave rise to the consideration for environmentally friendly buildings otherwise referred to as green buildings, sustainable buildings, ecofriendly buildings and the likes.

Green building is one of the developmental models arising from the Brundtland commission report (1987) on the need to adopt sustainable development to meet human needs and not increase environmental problems. Others models include Eco-dwellings and Sustainable buildings which are intended to minimize the deteriorating effect of building construction on the environment especially the destruction of natural resources like land and loss of viable ecosystem needed to support human existence. The resultant effect of massive building construction to meet the need of the present generation is environmental degradation being witnessed in deforestation in the northern part of the country, gully erosion in the southeast and flooding in the southwest. Green building development will not only reduce the negative impacts on the environment but also guarantees the optimum satisfaction and comfort of the present user. It will not only ensure that environmental resources are used responsibly but that they are used to efficiently satisfy the present need while guaranteeing the satisfaction of the future need as well.

Briefly, looking at the principle of sustainable development and green building construction, sustainable development is geared towards harmonizing the three aspects of environmental, social and economic concerns (Boyd and Kimmet, 2005; Shen et al., 2007; Hussin et al., 2013). A development is sustainable when it is able to meet the need of the present and not jeopardize the ability of the coming generations to meet their needs. Hence, green building is achievable when its development does not unnecessarily consume natural resources, uses renewable energy sources, records minimal impact on the environment, promotes water conservation, pollution reduction, waste minimization during construction and effective waste disposal system (Morelli, 2011; Bal et al., 2013). Unfortunately enough, there is no significant proof to show that the construction industry in Nigeria has embraced the principle of sustainable development (Windapo and Rotimi, 2012; Nduka and Ogunsanmi, 2015).

Nigeria Housing Need

The Nigerian National Housing Policy according to Waziri and Roosli (2013) and Olofinji (2015) defines Housing as the process of providing functional shelter properly set in a neighborhood which is supported by sustainable maintenance of the built environment for the day to day living and activities of an individual and families within the communities. This laudable dream of providing a functional shelter has remained in the pipeline as reported in literature. The Nigerian housing deficit increased from 12 million in 2007 to 18 million in 2014 (Okpoechi, 2014).

The 2013 demographic data shows that about 60% of the population is less than 45 years (Nigerian Bureau of Statistics, 2014). Nigeria has an estimated population of 180 million people (National Bureau of Statistics, 2015). This implies that a larger percentage of the population is in need of good quality housing both in the now and in the future. Literature reveals that efforts to meet the ever increasing housing shortage in Nigeria dates back to the pre-colonial era (Olotuah and Aiyetan, 2006; Adejumo, 2008; Olotuah and Bobadoye, 2009; Akinyode and Tareef, 2014). The provision of adequate housing to meet the present housing deficit in the country has been and will continue to be a herculean task. The implication is that in the next few years the deficiency in the provision of housing would have multiplied and all the effort made by government in the past till the present will amount to scratching the surface. The National Population Commission (NPC) projects that a sum of US$ 60billion at 2014 market rate will be needed in the first five years to drive the economic development of the country (Tsokar, 2015). Among other sectors, housing will need a sum of US$ 5 billion. According to Hussin et al. (2013), about 48% of the world's resources and energy used is linked to the construction and maintenance of buildings. This high capital outlay needed in the provision of quality infrastructure and housing development underscores the fact that available financial resources be used judiciously in the now and sustainably for the future.

Green Affordable Housing for Nigeria

Affordability is commonly defined as not spending more than 30% of household income on housing (Aribigbola, 2011). Affordable housing refers to dwellings whose total housing costs are deemed "affordable" to a group of people within a specified income range (Adejumo, 2008). The total housing costs include energy costs which could be up to 20% of the household income and 25% of the total housing cost in the United States. Adejumo posits that there is nothing like affordable housing in Nigeria as of today as has been presented. This he explained by drawing a distinction between public housing which are owned by the government and Social housing (rental housing) which may be owned and managed by the state, non-profit organizations, or by a combination of the two with the aim of providing affordable housing. The argument is based on the premise that provision of low income housing Nigeria since the National Housing Policy in 1980s has been through both public and private institutions. However, the prices of the completed structure are always beyond the reach of intended users. Hence the buildings are sold to the highest bidder (Akinyode and Tareef, 2014). According to Emiedafe (2015), Nigeria housing affordability analysis shows that a low income earner can only afford a two million naira building while the cost of an average 3-beroom flat costs a sum of eight million naira. This is definitely beyond the reach of the common man. Green building practices apart from being environmentally sustainable, can also improve quality of life, particularly for people with low- and moderate incomes. The most salient features of green affordable housing are energy use, material use/

durability and a healthy indoor environment. An affordable housing is described as a reasonably priced housing that incorporates sustainable features in the United States. The United State acquires green building practices for publicly owned buildings, lower energy cost burden and improved health (Marable, 2012).

Constructing environmentally friendly buildings has been greatly criticized because of the perceived increase in price (Kats, 2003; Langdon, 2004; Lowe et al., 2006; Alshamrani, 2016). However, in the report to assess the cost and financial benefit of green buildings, Kats (2003), Kats et al. (2003) and Langdon (2007a) observed that most green buildings cost a premium of about 2%, and yields up to 10 times as much over the entire life of the building. The report showed that there is a lack of adequate knowledge of up-front cost vs. life-cycle cost, noting that the savings in money come from more efficient use of utilities which result in decreased energy bills. On the other hand, Langdon (2007) found that some green buildings, over a 20 year life period yielded US$53 to US$71 (7,155–9,585in naira) per square foot back on investment (10,706–14,342 in 2016). According to Langdon, LEED and Energy Star certified buildings achieve significantly higher rents for commercial real estate market, sale prices and occupancy rates within the period. This perception has been a major challenge in the adoption of green buildings generally (Kats, 2003; Langdon, 2004; Lowe et al., 2006). The suspicion that green building implies increased cost of development is also due to lack of understanding of the principles of green buildings, lack of integrative practices to promote mutual understanding and implementation of the building process towards sustainable development (Alshamrani, 2016).

Challenges to Green Affordable Housing in Nigeria

The purpose of mass housing (low cost housing) is to provide decent housing at reasonably reduced costs, to households unable to afford the heavy investment of acquiring land and building houses at prevailing market rates (Okpoechi, 2014). However researches have shown that there are some basic factors impeding low/medium housing provision in Nigeria. Such factors include rapid urbanization arising from rapid population growth, rural-urban drift, and high cost of building materials, dearth of indigenous technology and skilled personnel, inadequate financial structure as well as poor managerial skill of our mortgage institutions (Obi and Ubani, 2014). Others are identified as poverty/affordability gap, ineffective housing finance, shortage of infrastructural facilities, bureaucracies in land acquisition, and high cost of land registration and titling (Emiedafe, 2015). Low per capita income after inflation and high interest rates are other factors militating against housing provision in Nigeria. The lack of willingness to accept new construction practices which include the green building development and lack of financial provision especially for low income households are part of the constraints (Olotuah and Bobadoye, 2009).

The aforementioned coupled with the impression that constructing green building leads to additional cost has affected its level of acceptance and adaptability in general. This is quite important going by the 2010 estimate that about 70% of the Nigerian population live below the poverty line (National Bureau of Statistics, 2012). While the housing stock in the country is far from being adequate, the existing ones are built without quality standard to stand the test of time and also be able to provide minimum comfort in the face of global warming. Consequently, Abolore (2012) noted that sustainable construction has not received the necessary attention it requires to positively impact on the way construction projects are executed in Nigeria just like other developing countries. Therefore the need to improve the quantity and quality of the housing stock in Nigeria cannot be met without implementing the principles of green building development.

Green Principles for Housing Development

A Green Building refers to a high-performance building designed, built, operated and disposed of in a resource-efficient manner. Green buildings are designed to minimize the overall (negative) impact of the building on the built environment, human health and the natural environment (Kats, 2003; Langdon, 2004; Langdon, 2007) . Green building practices refer to design and construction techniques meant to reduce waste, promote the efficient use of resources and lessen the ecological impact of the built environment. Green building features include the choice of site and orientation, efficient use of materials and resources, indoor environmental quality and innovation. According to the National Association of Realtors (NAR) green buildings are healthier, more economical, energy efficient and have much smaller environmental footprint than conventional buildings (National Association of Realtors, 2006). The success of green building development depends on its ability to leave a lighter foot print on the environment through conservation of resources and creating a balance between energy – efficient, cost effective and low maintenance products to meet the construction needs. It is thus finding a balance between home building and the sustainable environment.

The Green Building practice encompasses the building design concerns of economy, utility, durability, and comfort (Intergovernmental Panel on Climate Change-IPCC, 2007). When the decision to have a green building is made early in the design process, it is possible to maximize the green potential, minimize redesign, and assure the overall success and economic viability of the green elements of the building project. The decision to build green should be made before the site is selected, as many of the green criteria are affected by site characteristics and some sites are inappropriate for certain green projects. In addition to being environmentally sustainable, such practices can also improve quality of life, particularly for people with low- and moderate incomes. Well-sited green buildings can lower utility costs, decrease exposure to harmful pollutants, and reduce transportation costs by providing access to a greater range of community amenities. Reductions in energy and water consumption and storm water runoff can also lessen the strain on local utility infrastructure, providing benefits to the larger community.

Developing a green building requires a process to optimize every element of the design and then the impact and interrelationship of various different elements and systems within the building and site are re-evaluated, integrated, and optimized as part of a whole building solution (Lee and Guerin, 2009). The design will not be fully optimized if the interrelationships between the building site, site features, the path of the sun, and the location and orientation of the building and elements such as windows and external shading devices are not considered early in the design process. This is likely to result in a very inefficient building. These parameters to a great extent determine the quality and effectiveness of natural day lighting. It is an integrated design process which requires that all of the design professionals work cooperatively towards a common goal from the inception of the project.

Although, the practices, or technologies, employed in green building are continuously evolving, they differ from region to region. The fundamental principles of green building development are the same which are: Siting and structure design efficiency, Energy efficiency, Water efficiency, Material efficiency, Indoor environmental quality enhancement, Operations and maintenance optimization, and Waste and toxics reduction (US EPA, 2011; World Green Building Council, 2013). The essence of green building depends on the optimization of one or more of these principles. In designing environmentally optimal buildings, the objective is to minimize the total environmental impact associated with all life-cycle stages of the building project.

Site Design

The design stage forms the foundation for every building project. The stage is very important to the project life cycle and has the largest impact on cost and performance (Hegazy, 2002). According to Pushkar et al. (2005) sustainable site design is expected to minimize urban sprawl and needless destruction of valuable land, habitat and green space, which results from inefficient low-density development. Encourage higher density urban development, urban re-development and urban renewal, and brownfield development as a means to preserve valuable green space. Preserve key environmental assets through careful examination of each site. It should engage in a design and construction process that minimizes site disturbance and which values, preserves and actually restores or regenerates valuable habitat, green space and associated eco-systems that are vital to sustaining life (Pushkar et al., 2005). The land use Act currently in force in Nigeria does not give room for the attainment of this requirement in the provision of affordable housing. The provision of the land use Act has been greatly criticized in literature as hampering access to land. The difficulty in the processing of land titles, certificate of occupancies and related documents are contributory factors to the development of urban sprawls (Akeju, 2007; Adejumo, 2008; Adedeji and Olotuah, 2012; Ibimilua and Oyewole, 2015).

Water Quality and Conservation

The key principle is to preserve the existing natural water cycle and design site and building improvements such that they closely emulate the site's natural "pre-development" hydrological systems. It places emphasis on the retention of storm water and on-site infiltration and ground water recharge. It aims to minimize the unnecessary and inefficient use of potable water on the site while maximizing the recycling and reuse of water, including harvested rainwater, storm water, and gray water. Green building facilities makes more use of water that is collected, used, purified and re-used on –site. Reducing water consumption and protecting water quality are key objectives in sustainable buildings (Shiklomanov, 1998).

The construction industry is said to be responsible for more than half of carbon emission, water consumption and land fill wastes in the UK with 13% of the raw materials used (Bashir et al., 2010).

According to the report, about 35% of human water use is unsustainable noting that the percentage was likely to increase if climate change worsens, populations increase, aquifers become progressively depleted and supplies become polluted and unsanitary. Humans currently use 40–50% of the globally available freshwater in the approximate proportion of 70% for agriculture, 22% for industry, and 8% for domestic purposes and the total volume is progressively increasing. The low cost houses are poorly serviced and existing water systems are in a deplorable state (Adetokunbo, 2012).

Energy and Environment

Green building development produces a high performance building which uses less energy thereby minimizing its adverse impacts on the environment (air, water, land, natural resources) through optimized building siting, optimized building design, material selection, and aggressive use of energy conservation measures. The resulting building performance should exceed minimum International Energy Code (IEC) compliance level by 30 to 40% or more. It maximizes the use of renewable energy and other low impact energy sources. Embodied energy make up to 30% of the overall life cycle energy consumption of buildings. The operating energy use is reduced by minimizing energy leakage through the building envelope, high performance windows and extra insulations in walls, ceilings and floors. Others include passive solar building designs in which windows and walls are oriented towards tree shades. Effective window placement for day-lighting is also employed to provide more of natural light against the use of artificial lighting during the day (Simpson et al., 1993). Power generation is generally the most expensive feature to add to a building. In the study carried out to assess the environmental impacts of building construction projects in Nigeria, Ijigah et al. (2013) observed that the amount of energy required for Heating, Ventilation and Air Conditioning (HVAC) in Nigeria is 50% compared to only 15% required for lighting. Apart from its contribution to high operational cost, this also increases the greenhouse gas emission especially in urban centers.

Indoor Environmental Quality

The objective of this principle is to provide a healthy, comfortable and productive indoor environment for building occupants and visitors. It provides a building design, which affords the users the best possible conditions in terms of indoor air quality, ventilation, and thermal comfort, access to natural ventilation and day-lighting and effective control of the acoustical environment. In a desk study to review literature on liveability of selected housing estates in Nigeria, the study reports the low quality of houses and its services in the estates (Mohit and Iyanda, 2016). The sick building syndrome is one of the problems associated with low income housing since buildings rely on properly designed ventilation system, either naturally or mechanically powered for adequate ventilation of cleaner air. It has been reported that human beings spend more than 90% of their time indoors while 87% is spent at home (Jenkins et al., 1992; TSI Incorporated, 2011; Al horr et al., 2016), Fig. 15.1. A high performance luminous environment through the careful integration of daylight and electrical light sources will improve the lighting quality and energy performance of a structure. Personal temperature and airflow control over the HVAC system backed with a properly designed building envelope will increase a building's thermal quality (The U.S. Green Building Council, 2013). The Indoor Environmental Quality (IEQ) category in LEED standards as one of the five environmental categories was created to provide comfort, well-being, and productivity of occupants. The LEED IEQ category addresses design and construction guidelines especially: Indoor Air Quality (IAQ), thermal quality, and lighting quality (Lee and Guerin, 2009). The amount of moisture accumulation in the building has strong implication to the indoor air quality. A well-insulated and tightly sealed building envelope will reduce moisture problems. Nonetheless, adequate ventilation is necessary to eliminate moisture from sources indoors, such as human metabolic processes, cooking, bathing, cleaning, and other activities.

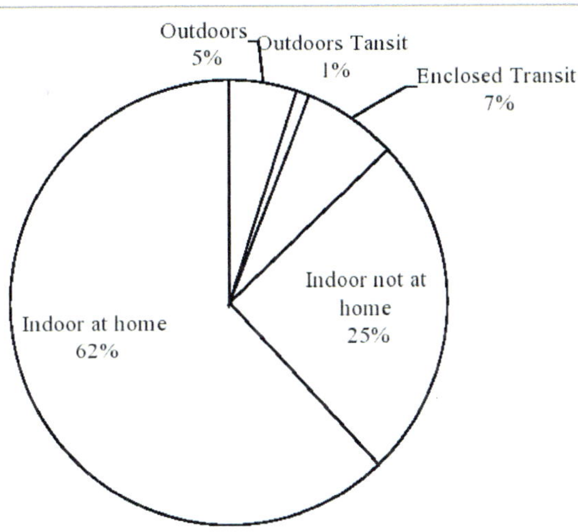

Figure 15.1. Percentage time spent indoors by humans. Source: Jenkins et al. (1992)

Materials and Resources

Unnecessarily high cost of building materials is one of the major factors attributed for inadequate housing provision, affecting both quantity and quality (Obi and Ubani, 2014; Emiedafe, 2015). Construction projects require huge capital outlay. It is posited that to combat the challenge of housing deficit estimated at 17 million, Nigeria will need to provide 700,000 units of housing annually (Emiedafe, 2015). A sum of N59.5 trillion will be required to meet this need. The assessment of cost and time performance of construction projects under the due process policy in Nigeria showed that a cost overrun of 19%. The due process policy was put in place to checkmate the high cost of construction project in Nigeria (Ade-Ojo and Babalola, 2013). Going by these figures, the country and the construction industry in Nigeria stands to lose a large chunk of its financial resources in the provision of affordable.

Implementing the green building principle will minimize the use of non-renewable construction materials and other resources such as energy and water through efficient engineering, design, planning and construction and effective recycling of construction debris. Maximize the use of recycled content materials, modern resource efficient engineered materials, and resource efficient composite type structural systems wherever possible. Maximize the use of re-usable, renewable, sustainably managed, bio-based materials (Intergovernmental Panel on Climate Change-IPCC, 2007). Because of the high volume of wastes associated with the conventional construction process, up to 35% of the investment goes as waste (Hussin et al., 2013). A huge amount of financial resources will continue to go down the drain thereby hampering the ability of the construction industry to provide the housing much needed to meet the need of the populace. Apart from the wastefulness and inefficiency in the provision of housing, the construction industry in Nigeria faces the peculiar challenges due to inefficient energy, transportation systems and waste management (Africa Business Insight, 2010; Abolore, 2012). The problem of housing shortage, poor quality building and its facilities and environmental degradation due to the increased use and consumption of environmental and natural resources in the provision of affordable housing will be minimized if the principles of Green building development is embraced. However, to encourage the implementation of green building principles, the fear of incurring additional cost burden must be addressed by establishing the likely cost of the low/medium income housing right from the design stage through the use of cost predictive models.

Cost Prediction for Green Affordable Housing

The perception that green buildings generally are more expensive than conventional buildings is gradually dissipating with increasing awareness and availability of cost data on green buildings. Also of importance is the improved knowledge and skill of designers and engineers in implementing sustainable features and green technologies in their designs. While some green prerequisites have little or no financial implication on the project others have significant cost implication on the project (Langdon, 2007a; Tatari and Kucukvar, 2011). Examining the cost of green building development (Langdon, 2004) suggested the need to assess the cost of sustainable building designs through models before financial decisions are made.

Green building cost can be assessed as an addition from three different perspectives, i.e., add-on to the actual project. In the first stance, sustainable building development is seen as an additional task or building type rather than a decision process. This perception does not give room for the consideration of sustainable building right from the design stage hence it is wrong. Another approach to assessing the cost is by checking against the budget. This is the most prevalently used because the cost data for the project are readily available. However this approach presumes that the original budget is accurate. Since budgets are not usually accurate at the end of the day, using this method also has its limitations. The third approach of comparing cost with similar projects presents the best alternative. The shortcoming is that not many similar green building projects are available for comparison (Langdon, 2004; 2007). Whatever approach used, there is the need for a professional to estimate the cost of the green project.

Construction Cost Estimation

The life cycle costs estimates are based on estimates and projections from historical evidence for reality not on verification of quantitative data. Determining the cost implication of green building development has been an uphill task due to the fact that developers place premium on initial costs while sustainable building development is more of a life time experience (what comes in the future). According to Langdon (2007a), a 2% increase in the initial cost of a project will yield 10% more on the return from savings in future costs. Hence, earlier studies have proposed the use of life cycle costing for the assessment of green building developments (Ade-Ojo and Fasuyi, 2013). Therefore cost assessment of green building developments have greatly depended on using the life cycle-cost and life-cycle assessment to establish basis for cost comparison (Rahim, et al., 2014). The challenge however is that life cycle-cost analysis depends on long term cost data for calculations while decisions to build green are made at the earlier stage of the projects. Thus the collection of real life cost data over a long period of time for life- cycle cost assessment has not been feasible and cost data are relatively more available in the developed world than in Nigeria. In lieu of this, resolutions carried out to assess cost in relation to green building development has adopted the use of cost projections over a period of life span to arrive at the estimated future costs of the project (Liu, 2015). Previous studies have identified that energy and water costs constitute a larger percentage of future costs, i.e., operational and maintenance costs (Langdon, 2007b).

Cost Prediction Models for Building Construction

Cost estimation is a fundamental aspect of construction project development. This usually involves the estimation of the quantity of labor, utilities, floor span, materials, sales, overhead, time and costs for a set of series of time periods. It is frequently done without perfect sample of còst data or adequate sample size (Smith and Mason, 1997). Since estimation is usually done for new projects with no quality historical data, the models make use of scanty and approximate information. The success of any construction project depends largely on the level of accuracy of its cost estimates. Estimating construction cost requires some level of expertize and highly dependent on the nature, size and complexity of the project. Due to the limited nature of information at the design stage and the need to predict the cost of a proposed project with little margin of error, cost estimation methods have given way to the use of cost prediction models.

Prediction models use computer applications and therefore are able to give more accurate and reliable cost estimates with limited information. Thus it is suitable for estimating the cost of a project right before the final decisions are made. Prediction models are based on mathematical models which are defined with the help of parameters estimated from available data. Such prediction models include the Regression Analysis (RA), Fuzzy logic (FL), Case Based reasoning (CBR) and Artificial Neural Network (ANN). The choice of the prediction model is determined by the ease of operation, familiarity, speed, degree of accuracy and available data (Lowe et al., 2006; Adegbile, 2013; Alshamrani, 2016).

The Regression Analysis (RA) model has various applications for cost estimation. According to (Smith and Mason, 1997), it is used more by cost estimators. Regression analysis uses a series of mathematical equations to estimate fixed and variable costs using computer software. It uses a series of mathematical equations to find the best possible fit of the line to the data points and thus tends to be accurate. The Fuzzy logic is limited in use due to the need for quantitative information. It is does not depend primarily on historical data like the Regression Analysis (RA) and the Artificial Neural Network (ANN). The ANN is data driven. It allows iteration from random figures to the final model. It is not dependent on assumptions. It is useful in predicting time series, financial phenomenon etc. The ANN architecture follows a network relationship. It represents relationship between variables and their connection. Neural networks have been reported to give substantial improvement to cost estimation over linear and multiple regressions (Smith and Mason, 1997)

In assessing the cost of construction projects, different cost prediction models have been deployed by researchers using computer aided methods such as the artificial intelligence. Günaydin and Doğan (2004) developed and tested a model of cost estimation for structural systems of reinforced concrete using ANN. Another cost prediction model was developed by Lowe et al. (2006) for predicting construction cost of buildings using multiple regression. The study was set to provide a benchmarking for an earlier developed ANN using the same data set. The ANN was used to develop a model for estimating building construction projects cost by El-sawalhi and Shehatto (2014). Other cost estimating models included the comparison of Regression Analysis, Neural Networks and Case Based Reasoning (CBR) as construction cost estimating models. The study noted that the use of Regression Analysis also known as multiple regressions dated back to the 1970s. According to Kim et al. (2004) and Kim et al. (2013), RA is a powerful tool for both analytical and predictive techniques, it is however not appropriate for non-linear relationships. The CBR has also been explored as a cost estimating model. The use of artificial intelligence models have the difficulty in obtaining required set of rules that can elicit the knowledge and lack the capability to learn by themselves, the CBR is considered as a better alternative. The CBR attempts to solve new problems by recognizing their similarity to a known problem. It thus adapts the solutions used in solving the previous problems (An et al., 2007; Koo et al., 2011). Despite the mirage of cost prediction models, Tatari and Kucukvar (2011) on cost premium prediction for certified green buildings using the neural network approach opined that there were very few decision models targeted specifically at green buildings. Nonetheless, researches have shown that the ANN is better suited for cost prediction models. Since the decisions to build green are better made at the design stage, the use of ANN is being proposed for the cost prediction for affordable green housing in Nigeria. This will afford the developer either public or private to determine early enough the likely cost of the proposed housing scheme if it will fall within the affordable price range or not.

Artificial Neural Network for Cost Prediction Model

The fear of incurring additional cost has been generally agreed to inhibit the tendency towards green building development (Issa et al., 2010; Samari et al., 2013). However, researches have shown over time that green buildings do not cost more than conventional buildings. It was observed that the variation in cost is directly related to the complexity of the building and the level of LEED certification desired (Langdon, 2007a). This underscores the fact that building project costs are dependent on the choices made between various design variables. Hence the decision to build or not is taken at the design stage (Mapp et al., 2011). Bragança et al. (2014) noted that the design stage is the best time to incorporate green building requirements into the building process. Estimating construction cost at the design stage is quite difficult owing to the fact that project information is scarce and not fully defined. Therefore the use of traditional cost estimating techniques have given way to the use of modern techniques such as the artificial intelligence (Günaydın and Doğan, 2004).

Artificial intelligence involves the use of computer simulation methods to predict the cost of construction projects. Artificial intelligence techniques such as the artificial neural network among many others have been used in cost estimating at different stages for different purposes. It is commonly used for cost estimating due to its ability to simulate the human brain using little data. Kim et al. (2004) in estimating construction cost of building elements used the data on the actual construction cost of 350 residential buildings projects in Seoul, Korea. However, the purpose of the study was on improving the accuracy of cost estimating techniques using ANN. In 20013, Kim et al. (2013) compared costs estimating methods for school buildings using regression analysis, neural network and support vector machine. The study collected construction costs of 217 school building projects in Kyeongi province, Korea. The result showed that the ANN was the most accurate and reliable of the three predictive models.

Günaydın and Doğan (2004) noted that the traditional cost estimating procedures follow a quantity procedure. But the emergence of artificial intelligence tools such as the NN made it possible to assess multi and non-linear relationships. Cost estimates using the ANN are adjudged to be more realistic and accurate than other modern techniques such as the RA and CBR (Kim et al., 2004; El-sawalhi and Shehatto, 2014; Shafiee et al., 2015). This is because the ANN has the ability to understand and simulate complex functions and can accommodate drivers' variables at a time. The disadvantage is that it depends mainly on historical data. Waziri (2010), used the ANN model for predicting construction cost of institutional buildings in Nigeria. The study used details of 510 institutional projects. The input variables for the network model included the building height, construction duration, external wall area, gross external floor area, number of floors and proportion of openings on external walls. These five variables were identified as the key predictor variables. The study reported that the ANN model developed was satisfactory and the training time was less than five minutes

Tatari and Kucukvar (2011) used the ANN approach to develop a model for cost premium of 74 certified green buildings. The study used seven input variables while the cost premium was entered as the output variable. Eighty percent of the cases were used for training while the remaining 20% were used to test the ANN. The study showed an R^2 value of 0.983 which was higher than the generally accepted value of 0.7. Hence, the prediction rate was said to be significantly high for the developed ANN model. Out of the numerous application of ANN to cost prediction model, a cost premium prediction model for certified green buildings was developed using the ANN (Tatari and Kucukvar, 2011).

Conclusion

The review of literature indicates that efforts made at providing low/medium income housing in Nigeria have been unfruitful. One prominent factor is the high cost of the so called low/medium income houses after completion. Secondly, the fear of incurring excessive cost by adopting green building principles is a challenge to its adoption. And lastly, the use of cost predictive models will enhance the development of green affordable housing in Nigeria by giving the stakeholders a tool for making informed design decisions. To enhance the provision of green affordable housing in Nigeria, the study recommends that stakeholders should ensure that the likely cost implication of the housing projects is established against the ability of the intended owners to acquire the buildings. Having reviewed different cost predictive models, the study therefore recommends the use of Artificial Neural Network for the development of cost predictive model for green affordable housing in Nigeria.

References

Abolore, A. A. 2012. Comparative study of environmental sustainability in building construction in nigeria and malaysia. Journal of Emerging Trends in Economics and Management Sciences 3(6): 951–961.

Ade-Ojo, C. O. and O. A. Fasuyi. 2013. Cost-In-Use: A panacea for sustainable building development in Nigeria. International Journal of Business and Management Invention 2(3): 1–5.

Ade-Ojo, C. O. and A. A. Babalola. 2013. Cost and Time Performance of Construction Projects under The Due Process Reform In Nigeria 3(6): 1–6.

Adedeji, Y. M. D. and A. O. Olotuah. 2012. An Evaluation of Accessibility of Low-Income Earners to Housing Finance in Nigeria 7(1): 23–31. https://doi.org/10.5829/idosi.aejsr.2012.7.1.1101.

Adegbile, B. O. 2013. Assessment and adaptation of an appropriate green building rating system for Nigeria. Journal of Environment and Earth Science 3(1): 1–11.

Adejumo, A. A. 2008. Some Thoughts On Affordable and Social Housing in Nigeria. Retrieved from http://www.nigeriavillagesquare.com/articles/akintokunbo-a-adejumo/some-thoughts-on-affordable-and-social-housing-in-nigeria.html, cited online 10 December 2008.

Adetokunbo. O. I. 2012. Housing, Neighbourhood Quality and Quality Of Life in Public Housing in Lagos, Nigeria. International Journal for Housing Science 36(4): 231–240.

Africa Business Insight. 2010. Challenges and Benefits of "Going Green" in Nigeria.

Akeju, A. A. 2007. Challenges to Providing Affordable Housing in Nigeria. In 2nd Emerging Urban Africa International Conference on Housing Finance in Nigeria (p. 5). Abuja.

Akinyode, B. F. and H. K.Tareef. 2014. Bridging the gap between housing demand and housing supply in nigerian urban centres: A review of government intervention so far. British Journal of Arts and Social Sciences 18(2): 94–107.

Al horr, Y., M. Arif, M. Katafygiotou, A. Mazroei, A. Kaushik and E. Elsarrag. 2016. Impact of indoor environmental quality on occupant well-being and comfort: A review of the literature. International Journal of Sustainable Built Environment 5(1): 1–11. https://doi.org/10.1016/j.ijsbe.2016.03.006.

Alshamrani, O. S. 2016. Construction cost prediction model for conventional and sustainable college buildings in North America Othman. Journal of Taibah University for Science 1–9. https://doi.org/10.1016/j.jtusci.2016.01.004.

An, S., G. Kim and K. Kang. 2007. A case-based reasoning cost estimating model using experience by analytic hierarchy process. Building and Environment 42: 2573–2579. https://doi.org/10.1016/j.buildenv.2006.06.007.

Aribigbola, A. 2011. Housing Affordability as a Factor in the creation of sustainable environment in developing world: The example of akure. Nigeria. Journal of Human Economics 35(2): 121–131.

Bal, M., D. Bryde, D. Fearon and E. Ochieng. 2013. Stakeholder Engagement: Achieving Sustainability in the Construction Sector (February). https://doi.org/10.3390/su5020695.

Bashir, A., S. Suresh, D. G. Proverbs and R. Gameson. 2010. Barriers towards the sustainable implementation of lean construction in the United Kingdom construction organisations. pp. 1–8. *In*: Arcom Doctoral Workshop.

Boyd, T. and P. Kimmet. 2005. The triple bottom line approach to property performance evaluation. Australian Cooperative Research Centre for Construction Innovation.ation. Retrieved from http://www.prres.net/Proceedings/..%5CPapers%5CBoyd_The_Triple_Bottom_Line_Approach.Pdf.

Bragança, L., S. M. Vieira and J. B. Andrade. 2014. Early Stage Design Decisions: The Way to Achieve Sustainable Buildings at Lower Costs. The ScientificWorld Journal, na, 8. https://doi.org/10.1155/2014/365364.

Brundtland, G. (ed.). 1987. Our Common Future: Report of the World Commission on Environment and Development. World Commission on Environment and Development (Vol. 4). https://doi.org/10.1080/07488008808408783.

El-sawalhi, N. I. and O. Shehatto. 2014. A neural network model for building construction projects cost estimating. Journal of Construction Engineering and Project Management 4(4): 9–16.

Emas, R. 2015. The concept of sustainable development: definition and defining principles. Brief for GSDR, 1–3.

Emiedafe, W. 2015. Housing in Nigeria: Why We Desperately Need 17 Million More Homes. Sapient Vendors LTD.

Günaydın, H. M. and S. Z. Doğan. 2004. A neural network approach for early cost estimation of structural systems of buildings. International Journal of Project Management 22(7): 595–602. https://doi.org/10.1016/j.ijproman.2004.04.002.

Hegazy, T. 2002. Computer-Based Construction Project Management. Prentice Hall. Retrieved from www.civil.uwaterloo/tarek.

Hussin, J. M., I. A. Rahman and A. H. Memon. 2013. The way forward in sustainable construction: Issues and challenges. International Journal of Advances in Applied Sciences 2(1): 15–24. https://doi.org/dx.doi.org/10.11591/ijaas.v2i1.1321.

Ibimilua, A. F. and I. A. Oyewole. 2015. Housing Policy in Nigeria : An Overview. American International Journal of Contemporary Research 5(2): 53–59. https://doi.org/ISSN 2162-139X (Print), 2162-142X (Online.

Ijigah, E. A., R. A. Jimoh, B. O. Aruleba and A. B. Ade. 2013. An assessment of environmental impacts of building construction projects. Civil and Environmental Research.

Intergovernmental Panel on Climate Change-IPCC. 2007. Summary for Policymakers. *In*: Climate Change 2007: The Physical Science Basis. Contribution of Working Group I to the Fourth Assessment Report of the Intergovernmental Panel on Climate Change [Solomon, S. D. Qin, M. Manning, Z. Chen, M. Marquis, K.B. Aver. htts://doi.org/10.1038/446727a.

Issa, M. H., J. H. Rankin and A. J. Christian. 2010. Canadian practitioners ' perception of research work investigating the cost premiums , long-term costs and health and productivity benefits of green buildings. Building and Environment 45(7): 1698–1711. https://doi.org/10.1016/j.buildenv.2010.01.020.

Jenkins, P. L., T. J. Phillips, E. J. Mulberg and S. P. Hui. 1992. Activity patterns of Californians_ Use of and proximity to indoor pollutant sources - ScienceDirect. Atmospheric Environment 26A(12): 2141–2148. https://doi.org/https://doi.org/10.1016/0960-1686(92)90402-7.

Kats, G., E. Capital, L. Alevantis, A. Berman, E. Mills and J. Perlman. 2003. The costs and financial benefits of green buildings. A Report to California's Sustainable Building Task Force. Retrieved from http://scholar.google.com/scholar?hl=en&btnG=Search&q=intitle:The+Costs+and+Financial+Benefits+of

+Green+Buildings+A+Report+to+California+?+s+Sustainable+Building+Task+Force#7%5Cnhttp://www.calrecycle.ca.gov/greenbuilding/Design/CostBenefit/Report.pdf%5Cnh.

Kats, G. H. 2003. Green Building Costs and Financial Benefits. USA.

Kim, G. H., J. -E Yoon, S. H. An, H. H. Cho and K. I. Kang. 2004. Neural network model incorporating a genetic algorithm in estimating construction costs. Building and Environment 39(11): 1333–1340. https://doi.org/10.1016/j.buildenv.2004.03.009.

Kim, G., S. H. An and K. I. Kang. 2004. Comparison of construction cost estimating models based on regression analysis, neural networks, and case-based reasoning. Building and Environment 39(10): 1235–1242. https://doi.org/10.1016/j.buildenv.2004.02.013.

Kim, G., J. Shin, S. Kim and Y. Shin. 2013. Comparison of school building construction costs estimation methods using Regression Analysis, Neural Network, and Support Vector Machine. Journal of Building Construction and Planning Research 1: 1–7.

Koo, C., T. Hong and C. Hyun. 2011. Expert systems with applications the development of a construction cost prediction model with improved prediction capacity using the advanced CBR approach. Expert Systems With Applications, 38(7): 8597–8606. https://doi.org/10.1016/j.eswa.2011.01.063.

Langdon, D. 2004. Examining the Cost of Green. Santa Monica. Califonia.

Langdon, D. 2007a. Cost of Green Revisited: Reexamining the Feasibility and Cost Impact of Sustainable Design in the Light of Increased Market Adoption. Carlifonia. https://doi.org/10.1093/jhmas/jrr055.

Langdon, D. 2007b. Life Cycle Costing (LCC) as a contribution to sustainable construction: a common methodology. Retrieved from http://scholar.google.com/scholar?hl=en&btnG=Search&q=intitle:Life+cycle+costing+(+LCC+)+as+a+contribution+to+sustainable+construction+:+a+common+methodology#0%5Cnhttp://scholar.google.com/scholar?hl=en&btnG=Search&q=intitle:Life+Cycle+Costing+(+LCC+)+as+.

Langdon, D. 2007c. The cost & benefit of achieving Green buildings. Australia. Retrieved from www.davislangdon.com.

Lee, Y. S. and D. A. Guerin. 2009. Indoor environmental quality related to occupant satisfaction and performance in LEED-certified buildings. Indoor and Built Environment 18(4): 293–300. https://doi.org/10.1177/1420326X09105455.

Lowe, D. J., M. W. Emsley and A. Harding. 2006. Predicting construction cost using multiple regression techniques. ASCE Journal of Construction Engineering and Management 132(7): 750–758. https://doi.org/10.1061/(ASCE)0733.

Mapp, C., M. C. Nobe and B. Dunbar. 2011. The cost of LEED — an analysis of the construction costs of LEED and NON-LEED banks. Journal of Sustainable Real Estate, 3(1), 254–273. https://doi.org/10.5555/jsre.3.1.m702v24r70455440.

Marable, K. 2012. Green Building, Energy Efficiency & Healthy Housing. Enterprise Community Partners, Inc.

Mohit, M. A. and S. A. Iyanda. 2016. Liveability and Low-income Housing in Nigeria. Procedia - Social and Behavioral Sciences 222: 863–871. https://doi.org/10.1016/j.sbspro.2016.05.198.

Morelli, J. 2011. Environmental sustainability: A Definition for environmental professionals. Journal of Environmental Sustainability 1(1): 1–10. https://doi.org/10.14448/jes.01.0002.

National Association of Realtors. 2006. Green Building. Retrieved from http://www.realtor.com/.

National Bureau of Statistics. 2012. Nigeria Poverty Profile 2010.

National Bureau of Statistics. 2015. Demographic Statistics. Demographic Statistics Division. https://doi.org/10.1038/1381007d0.

Nduka, D. O. and O. E. Ogunsanmi. 2015. Construction professionals' perception on green building awareness and accruable benefits in construction projects in Nigeria. Covenant Journal of Research in Built Environment 3(2): 30–52.

Nigerian Bureau of Statistics. 2014. Statistical Report on Women and Men in Nigeria.

Obi, A. N. I., and O. Ubani. 2014. Dynamics of Housing Affordability In Nigeria. Civil and Environmental Research 6(3): 79–84.

Okpoechi, C. U. 2014. Middle-income Housing in Nigeria. Architecture Research 4(1A): 9–14. https://doi.org/doi: 10.5923/s.arch.201401.02.

Olofinji, L. 2015. An Overview of The National Housing Policy. Real Estate.

Olotuah, A. O. and A. O. Aiyetan. 2006. Sustainable Low-Cost Housing Provision in Nigeria : A Bottom-Up Participatory. *In*: D. Boyd (ed.). 22nd Annual ARCOM Conference (Vol. 1, pp. 633–639). Birmingham, UK: Association of Researchers in Construction Management.

Olotuah, A. O. and S. A. Bobadoye. 2009. Sustainable housing provision for the urban poor: A review of public sector intervention in Nigeria. The Built & Human Environment Review 2(May): 51–63.

Pushkar, S., R. Becker and A. Katz. 2005. A methodology for design of environmentally optimal buildings by variable grouping. Building and Environment. https://doi.org/10.1016/j.buildenv.2004.09.004.

Rahim, F. A., S. A. Muzaffar, N. S. Mohd Yusoff, N. Zainon and C. Wang. 2014. Sustainable construction through life cycle costing. Journal of Building Performance 5(1): 2180–2106. Retrieved from http://spaj.ukm.my/jsb/index.php/jbp/index.

Samari, M., N. Godrati, R. Esmaeilifar, P. Olfat and M. Shafiei. 2013. The investigation of the barriers in developing green building in Malaysia. Modern Applied Science 2(7).

Shafiee, A., A. Alvanchi and S. Biglary. 2015. A neural network based model for cost estimation of industrial building at the project's definition phase. pp. 1–9. *In*: 5th International/11th Construction Specialty Conference.

Shen, L., J. L. Hao, V. W. Tam and H. Yao. 2007. A checklist for assessing sustainability performance of construction projects. Journal of Civil Engineering and Management 13(4): 273–281. https://doi.org/10.1080/13923730.2007.9636447.

Shiklomanov, I. A. 1998. World Water Resources. A new appraisal and assessment for the 21st century. United Nations Educational, Scientific and Cultural Organization, 40.

Simpson, J. R., E. G. McPherson, J. M. Lichterand and R. A. Rowntree. 1993. Potential of tree shade for reducing building energy use in the Sacramento valley. Journal of Arboriculture 22(1): 10–18.

Smith, A. E. and A. K. Mason.1997. Cost estimation predictive modeling: regression versus neural network. Engineering Economist 42(2): 137–161. https://doi.org/10.1080/00137919708903174.

Sustainable Building Task Force and the State and Consumer Services Agency. 2003. Building Better Buildings: An Update on State Sustainable Building Initiatives.

Tatari, O. and M. Kucukvar. 2011. Cost premium prediction of certi fi ed green buildings : A neural network approach. Building and Environment 46(5): 1081–1086. https://doi.org/10.1016/j.buildenv.2010.11.009.

The U.S. Green Building Council. 2013. the Business Case for Green Building. Retrieved from http://www.worldgbc.org/activities/business-case/.

TSI Incorporated. 2011. Indoor Air Quality Handbook.

Tsokar, K. 2015. Bridging infrastructure development gaps in Nigeria, p. on line.

US EPA. 2011. Green Building. Retrieved from http://www.epa.gov/greenbuilding/.

Waziri, A. G. and R. Roosli. 2013. Housing Policies and Programmes in Nigeria : A Review of the Concept and Implementation © Society for Business and Management Dynamics 3(2): 60–68.

Waziri, S. B. 2010. An artificial neural network model for predicting construction costs of institutional building projects in Nigeria. pp. 63–72. *In*: Laryea, S., R. Leiringer and W. Hughes (eds.). West Africa Built Environment Research (WABER) Conference. Accra, Ghana: School of Construction Management and Engineering.

Windapo, A. O. and J. O. Rotimi. 2012. Contemporary issues in building collapse and its implications for sustainable development. Buildings 2, 283299. https://doi.org/10.3390/buildings2030283.

World Green Building Council. 2013. The Business Case for Green Building 50–55.

CHAPTER 16

Structural Settlement for the Pile Foundation Design using Eurocode and its Implications on Green Affordable Housing in Penang, Malaysia

Zafarullah Nizamani, Lee Zhi En* and *Akihiko Nakayama*

Introduction

Penang, Malaysia is located on the northwest coast of Peninsular Malaysia on the coast of Malacca Straits. It is considered as a state with limited land compared to the population in that area. Due to the rapid development as well as the growth of human population, the concept of maximum land utilization is increasing significantly in cities where available land is a scarce commodity. Coastal area consists of marine clay or soft clay lying under the ground. Therefore, high-rise affordable housing projects are being developed on coastal reclaimed land. Due to the soil condition, the robust foundation is the most important consideration for the structures. Durability and stability require that sustainable but affordable and good performance building foundations should be provided. Design codes provide the guideline for safe, economical and affordable housing to the public. These structures generally have two basic components, i.e., superstructure and substructure. The substructure is the part, which lies below the ground level and hidden from public eye throughout the design life of the structure. Before deciding the type of foundations, some factors such as affordable, technical, operational, environmental and safety aspects which play a significant role in its selection, should be evaluated. This part of the structure is the most important part

Universiti Tunku Abdul Rahman, Malaysia.
* Corresponding author: zafarullah@utar.edu.my

from the design point of view. (1) it will remain invisible unless cracks appear in the superstructure, (2) if placed on the soft soil the foundation may have a differential settlement, (3) if it is placed where sewerage lines are passing underneath then there is a danger of rupture of the pile and mixing with the underground sweat water aquifer. Therefore, it is very important to select a suitable foundation, which is not only safe but also economical and sustainable for affordable housing. When it comes to the foundation for high-rise buildings, the type of foundations is either raft or pile. The raft foundation is not only expensive due to the amount of excavation, amount of reinforced concrete, producing a hazard to the public such as the settlement of slopes and damaging underwater pipelines/cables. The pile foundation provides the best solution for the design of the green building by reducing the amount of risk and natural hazard.

The main aim of any foundation is to avoid the differential settlement, which can cause serious damage to the superstructure and may induce cracks in the structural elements. There are various advantages attributed to the pile foundations such as saving of materials, savings due to the reduced amount of excavations of the foundations, economical and lesser time required for the construction. The avoidance of excavation provides reduced health and safety risk, by avoiding the slope sliding which can cause a serious hazard. The design should have enough safety as well as stability against failure. Pile foundation can help to avoid environmental and health hazard risk caused by shallow foundations such as raft foundations as they are considered green and environmentally friendly. The hazards, which include the raft, are due to the reason of mixing of contaminated groundwater to the permeable groundwater during excavations. Generally, the amount of environmental impact due to the use of equipment for excavation of raft is high as compared to the pile foundation. This is in due to carbon content, excavation, and disposal of the material.

The life cycle of a building structure requires that it should be designed for all phases of its life such as planning, design, construction, operation, extension and finally its demolition. Piled foundations are suitable for buildings due to many reasons such as weak soil, reclaimed land, earthquake-prone areas and structures near coasts or sea such as coastal area. Eurocode 7 provides a guideline, for the allowable pile settlements at working loads, should be within the acceptable limits. Pile foundations, as compared to pile raft or raft foundations can use less amount of reinforced concrete and thus provide an economical, green, fast and sustainable foundation.

The affordable and economical housing mainly requires efficient building design, less material usage, quality of the work and thus they should be not only affordable but also environmentally friendly and sustainable structural systems in the long-term. The structural engineer needs to adopt an improved design-based strategy for improving the affordability of the building. The affordable apartment structures should have some basic functional requirements with regard to their purpose. Increased demand for affordable green housing, due to the environmental awareness and rising costs for the land has forced citizens to look for vertical buildings. The affordable high-rise buildings have forced many design engineers, contractors and builders to look into pile foundations, which are not only robust but economical in the long time period and sustainable. The affordable housing design should also emphasize on sustainable design principles. The structures must not only meet code requirements of today but also foresee and must be able to withstand the rigorous future assessments at the end of its design life for any future extension. Pile foundation provides a sustainable solution such as its carbon footprint is less as compared to the shallow raft foundation. Due to the uncertainty of change in resistance or load pile foundation provides long-standing benefits considering future requirements of the design code.

The reaction from the superstructure, i.e., the loads acting on the support produce pile displacement at the base. Determination of pile displacement is a part of the foundation analysis where the settlement of the pile is calculated. One of the methods to determine this settlement is by using the finite element method. This work required two-separate simulation software, for the design of superstructure and substructure of the building. The simulation software is able to perform design and analysis of structures. The design of residential multi-story reinforced concrete buildings in Malaysia is generally based on Eurocode, EC2 or BS 8110. The Eurocode has defined two limit states out of which ultimate limit state requires the strength of the structure to be evaluated against failure, which may cause injuries and economic losses to the occupants. Piles for affordable housing are analyzed using numerical simulation through finite element software. The environmental/green effects and sustainability of this type of foundation are also determined by using PLAXIS 3D software.

Background of the Research

Highland Towers, a 12-story apartment building in Malaysia constructed in 1977 and collapsed in 1993 with loss of 48 lives. Soil erosion due to rainwater from the hill and extreme monsoon water caused a landslide, which affected the retaining wall at the perimeter. The landslide containing 100000 square metres of mud. This force acted on one of the tower blocks, the foundation gave way and whole structure collapsed, while almost the whole structure was intact except its foundation. The project engineer was accused of carelessness for not taking into consideration the lateral loads on the foundations. When it comes to the design of structures or buildings, the foundation design acts as one of the most important criteria, as the foundation serves as a medium to transmit the loads or weight from the superstructures to the underground or hard strata layer. The failure such as shrinkage, settlement, cracking can occur on the structures within the design life of the structure due to the improper design of foundation (Leow, 2005). Therefore, to ensure that the structures are able to perform their function in the best condition, the design of the foundation system must be sufficient and safe to withstand the loads and weight of the superstructures.

Due to the rapid increase in the population, people have a greater dependence on high-rise buildings. The needs and the requirement on the construction of high-rise building become a major concern of the engineers. Due to the limitation of land used in most of the cases, engineers are required to build the structures on loose sand. A high-rise building comes along with heavy load, which can fail if the foundation is not designed with adequate analysis and safety factors. Appropriate design of the foundation is necessary to transmit the load of the superstructures to the underground layer (Mohammadizadeh and Mohammadizadeh, 2016).

For a high-rise building, normally the foundation system is designed by using a deep foundation system. A deep foundation system is more costly than a shallow foundation, in terms of economics, a deep foundation is only adopted when a shallow foundation is not capable to support the structures. When the soil at the surface is weak or unsuitable to design for shallow foundation system, or the firm and hard strata is very deep that it cannot be reached by the shallow foundation, the deep foundation systems are required. Commonly found deep foundation systems are piles, piers, and caissons (Kulkarni, 2009).

In the design and assessment of new high-rise building or existing structures, the prediction of the performance on the supported system such as pile foundation serves as an important factor in the foundation design system. When structures such as pile are built in the soil, the process of the response of the soil could affect the motion of the structure, or vice versa, where one of the examples is the displacement, can be explained in the phenomenon of dynamic soil-structure interaction (Ravishankar and Satyam 2013). A numerical modelling method using a model test in geotechnical engineering leads to the advantage on the simulation of the complex foundation systems with controlled conditions and situations, as well as gives the opportunity to acquire the information of the fundamental mechanisms operating in these systems (Lok et al. 2000).

Land reclamation refers to the process of creating new land from coastal areas such as oceans, river beds or lake beds. Sand from the ocean or other material such as heavy rock or cement are deposited and filled in the selected area, then the soil is filled until the desired height is reached. In Malaysia, some of the states such as Penang, Melaka, and Johor undergo this type of land reclamation method for construction purposes. In Penang Island, due to land limitation, buildings and structures are required to be built on the coastal area where pile foundations are mostly used for the building on the reclaimed land. Over Penang, the sea around Permatang Damar Laut near Bayan Lepas has been used for land reclamation and recently the land reclamation is applied in front of Queensbay Mall, south from the first Penang Bridge.

In terms of the method of the installation, piling construction can be divided into two methods, driven or bored, or in other words displacement or replacement pile. A displacement pile, is where the pile is directly forced into the ground by using machinery and displaces the ground. For a replacement pile, or a bored pile, the soil is removed by using excavation and then the pile is inserted and replaced with cast-*in situ* concrete. Replacement pile or the bored pile is more favourable to sustainable development, as the process does not involve the vibration; hence produce lesser noise to the environment.

Foundation for High-Rise Building

Humans depend on high-rise building due to several factors. The number of the affordable high-rise buildings is significantly increasing in many countries. In most of the country, their high-rise building is mostly supported by either pile foundation, or the combination of raft and pile (Kayvani, 2014). Table 16.1 shows a few case studies for the foundation system used for high-rise buildings.

Table 16.1. Foundation used for different high-rise buildings.

Country	No. of story/ building height	Type of foundation	Author
Hong Kong, IFC Tower II	88	Raft foundation, end bearing bored pile on bedrock	John et al., 2004
Hong Kong, Union Square	108	Shaft grouted barrette pile	John et al., 2004
Poland, Supermarket hall	-	Pile foundation	Rybak et al., 2008
Germany, CITY-TOWER	121 m	Combined pile raft foundation	Katzenbach and Schmitt, 2004
Doha, Qatar	74, 400 m	Pile-supported raft	Poulos, 2011
Germany, Messe-Torhaus	30	Combined pile raft foundation	Katzenbach et al., 2005
Germany, Messeturm	60, 256 m	Pile-supported raft	Katzenbach et al., 2005

Type of Piles and Simulation Software

According to (Kulkarni, 2009), the type of pile can be classified based on the following criteria; (a) Construction material such as concrete, timber, steel or composite materials, (b) Shape of the pile such as cylindrical, tapered or under ream, (c) Load transfer method such as skin friction, end-bearing or tension, (d) Method of installation such as bored, driven or vibrated. Driven piles are used for the offshore jacket platform because they are stable in soft ground and bored piles are considered fit where the ground has minimum vibration. To simulate the model of the pile by using software, one of the key to be considered is the material properties of the pile. Some of the material properties can be found where they are needed to define in most of the simulation software, such as Young's modulus, Poisson's ratio, unit weight. The material properties of the pile, which had been used in literature and software used are summarized in Table 16.2.

The application of Finite Element Method (FEM) for modelling test in geotechnical engineering field provides a variety of advantages of simulating the complex structure and the knowledge of the fundamental mechanism operating behind these systems. The results of the model test are used as a benchmark for the analysis to make the prediction behaviour of the model (Rayhani and Naggar, 2008). The developed

Table 16.2. Engineering properties of the pile used in the finite element method.

Location	Young's Modulus, E (kN/m^2)	Poisson's ratio (v)	Unit weight (kN/m^3)	Software used	Author
Bangkok, Thailand	26.0 x 10^6	0.2	24	Plaxis 3D	Amornfa et al., 2012
Pavaratty and Thrissur, India	25.0 x 10^6	0.38	24	Plaxis 3D	Sreechithra and Niranjana, 2017
Brooklyn, New York	25.0 x 10^6	0.2	24	Plaxis 3D	Mohammadizadeh and Mohammadizadeh, 2016
Coimbature, India	15.0 x 10^3	0.25	32	ANSYS	Amutha and Satheeskumar, 2017
Torino and Sistiana, Italy	32.0 x 10^6	0.2	25	Plaxis 2D	Moniz, 2014

simulation tool can be divided into two categories, which is the structural design and analysis software such as ETABS, Scia Engineer and ESTEEM, and foundation design and analysis software such as Plaxis, STAAD, ABAQUS, and FLAC3D. ETABS and Scia Engineer software have been developed more than 30 years ago and are known to the industry for Building Analysis and Design Software (Zheng and Chin, 2011). Structures such as bridge or highway roads can be modelled and analyze under certain situations by using ETABS and Scia Engineer. Different types of foundation system can be modelled by using Plaxis 3D as well as ABAQUS. Engineering properties of the subsoil and also the pile can be defined in this software depending on the user.

Subsoil Condition and Water Table

Investigating the subsoil condition of the building is essential in the geotechnical engineering field as shown in Table 16.3. Therefore, in the numerical simulation of the foundation, subsoil profile has to be known to define the ground condition inside the software. As similar to the pile, engineering properties such as saturated unit weight, unsaturated unit weight, Young's modulus, Poisson's ratio, with extra cohesion and angle of friction value of each layer of the soil are the necessary parameters to be defined in the software.

Other than the material properties of the soil, the groundwater level has to be taken into the consideration during the modelling process. The foundation needs to consider subsoil water pressure during its construction. The lesser the depth, it is more difficult to build a strong foundation as the soil may have a low bearing capacity (Das and Biswas, 2014). As the water table will cause the unit weight of the soil to change from unsaturated to become saturated, therefore affect the bearing capacity of the soil. The foundation system acts as an interface element to transmit the load from the superstructures to the underlying soil with a wide area to reduce the pressure. The foundation can fail based on several factors, such as drag down and heave, unequal support of an earthquake (Srivastava et al., 2012).

Table 16.3. Material properties of the subsoil (Amornfa et al., 2012).

Material	Depth	Young's modulus (kN/m^2)	Poisson's ratio	Unit weight (kN/m^3)	Friction angle (degree)	Cohesion (kN/m^2)
Soft clay	0–15	3000	0.495	15.2	-	20
Stiff clay	15–25	40500	0.495	18.4	-	90
1st sand	25–35	80000	0.3	19.4	35.8	-
Hard clay	35–45	150000	0.495	19.8	-	300
2nd sand	45–55	200000	0.25	20.1	36.2	-

Drag Down and Heave

When the footing of the foundation is located at a compressible soil especially plastic soil, the foundation may fail by drag down or heave. In plastic soil, upward movement, with heave some distance away, often accompanies the new settlement drag down. This kind of failure leads to the differential settlement and heaving of the soil lying under the foundation. The change in volume in the plastic or expansive soil due to the moisture variation can cause the damage. One of the reasons, which can lead to the variation of the moisture content in the soil, is due to the improper design of the design system that does not consider runoff the water properly (Nor et al., 2014).

Unequal Support

It is well known that not every soil is homogenous, where the soil on every spot is varying and having a different bearing capacity. When the footing of the foundation is lying on heterogeneous soil, the unequal load distribution can lead to the differential settlement of the foundation. One of the unequal support failures can be seen in the leaning tower of Pisa. The main factor that causes the tilting of the tower is because the soil underneath at the south side of the building is more compressible comparing to the north side; therefore, the tower is tilting towards the south side (Bajaj and Choudhary, 2014). This unequal distribution of load hence leads to the unequal support of the foundation, causing the leaning of the Pisa tower.

Earthquake

When there is an earthquake, the violent shaking causes the foundation and superstructure of the building to move along with the ground. The structure vibrates and shakes in an irregular way because of the inertia of its own weight (Srivastava et al., 2012). In the case of the Kocaeli earthquake 1999, Turkey, a lot of buildings in the area failed and collapsed after the earthquake. Several failure factors can be found such as vertical displacement, ground heave, a tilt of building, differential settlement and most importantly liquefaction (Sancio et al., 2004). Soil liquefaction can be observed in most of the large-scale earthquakes, where a saturated soil loses its strength due to an applied stress, i.e., earthquake. The soil starts to behave like a liquid and loses its strength and ability to resist any shear and they look like quicksand (Srivastava et al., 2012).

Eurocode 7

The design of bridges, buildings and including pile foundation designs are dependent on the design codes. The basis of structural design (EN, 1990) and Eurocode 1 is for the action design, Eurocode 8 is for the earthquake resistance design. Eurocode 7 consists of two parts, which are the EN 1997-1 Geotechnical design part 1: General rules and the other part is EN 1997-2 Geotechnical design part 2: Ground investigation and testing. Part 1 general rules show the principles for geotechnical design in terms of limit state design, where these principles are relevant to the calculation of geotechnical action on the structural elements which are contacting with the ground such as footings, piles. For part 2, generally covers the ground investigation elements including the field and laboratory test for soil and rocks such as cone penetration test, standard penetration test and so on (Frank, 2006). In the Eurocode Section 7.5.2.1 loading procedure, the code states that displacement or settlement of the pile should not exceed 10% of the pile diameter.

Methodology

In United States Department of Housing and Urban Development had launched an online tool in 2001 and upgraded it in 2013 for Affordable Housing Design Advisor (Design Advisor). This tool helps to find properties, which are made of good quality design that needs to consider well-planned and designed buildings. Good design is a critical difference between affordable housing which succeeds and one that does not (Evans, 2014).

Superstructure

The model is designed as a high-rise building, a with 12-story height. For the superstructure of the building, the model is generated using ETABS. ETABS is a simulation tool, which can perform the various functions including modelling, analysis and design, and also output and display (Kalny, 2013). The section properties such as the concrete grade, dimension or thickness of each type of member are summarized in Table 16.4.

The building floor will experience the live load, which is the variable load, where in this case it will be 2.0 kN/m^2 according to Eurocode 1. For the permanent load, the software calculates the self-weight of each structural member; however, the floor will be at the value of 2.0 kN/m^2 according to Eurocode 1 as it is designed as a waterproofed floor with standard finishing. For the Ultimate limit State, load combination is used with safety factors of 1.5 and 1.35 for live and permanent loads respectively.

Table 16.4. Section properties of structural members.

Structural Element	Column	Beam	Slab	Shear wall
Concrete Grades	C30/37	C25/30	C25/30	C30/37
Dimension/thickness	600 mm (d) x 600 mm (w)	600 mm (d) x 300 mm (w)	250 mm	300 mm

Substructure

Pile displacement is a part of the foundation analysis where the settlement of the pile is calculated along with the load applied. The foundation to be used for the building is pile foundation, where the building will be simulated on the soil condition of coastal area due to its location. The pile foundation design of this model is done by using plaxid 3D version 1.1 simulation software. Plaxis is a software that related more to geotechnical engineering. Similar to ETABS, plaxis allows the user to perform a variety of functions such as modelling, design and analysis, and also output and display. The main function is that the analysis part of this software is able to predict the displacement of the pile or foundation as a result.

Length of the Pile, its Selection and Pile Cap

The length of the pile that applied to this pile foundation design includes 10 m, 13 m, 20 m, 30 m, and 55 m. The length of the pile can be defined in the section of "workplane" in Plaxis 3D. The pile can be defined in the section "pile" in Plaxis 3D.

The type and the diameter of the pile is a massive circular pile with a 0.4 m diameter in this chapter. The pile is designed as a concrete pile with concrete grade C25/30, with the unit weight of 25 kN/m^2, Young's modulus of 30e6 kN/m^2 and the Poisson's ratio of 0.2. The pile cap can be defined in the section of "floor" in Plaxis 3D. The material properties of pile cap are designed as the same of the material properties of the pile, which is the concrete grade C25/30 with 25 kN/m^2 unit weight, Young's modulus of 30E6 kN/m^2 and the Poisson's ratio of 0.2.

Site Characteristic and Geotechnical Reports

Borehole description and geotechnical description of parameters need to be included in the pile design. The borehole is the section where the subsoil condition and water level are defined. The water level is set to be 14.5 m below ground level at the Batu Ferrenghi Penang island (Azmi, 2014).

The subsoil condition is applied by referring to one of the case study of reclaimed land in Jelutong, Georgetown Penang where the land is in the coastal area (Chen and Tan, 2002). The subsoil condition is only obtained in the form of the type and depth of the soil, along with the Standard Penetration Test (SPT) for each layer. By using the information such as the type of the soil and SPT value obtained from the literature, and to refer to (Bowles, 1997) the other information such as unit weight, Young's modulus, Poisson's ratio, cohesion value, the angle of friction can be obtained. The table shows the subsoil condition.

2D and 3D Mesh

Plaxis 3D is able to help the user to generate the 2D and 3D mesh by just clicking on the "generate 2D mesh" and "generate 3D mesh" keys in the software. Before the user proceeds to the analysis and calculation stage, the 2D and 3D mesh of the model have to be generated as this is required by the software to proceed with next step.

Calculation Phase

Just before the calculation stage, the phases of calculation have to be defined first, where it can be done by clicking the section "phases" in plaxis 3D. There are total five phases, which are initial, phase, pile, excavation, pile cap and lastly loading. The important part is in the loading phase, where the user has to insert the load in unit kilo newton to the pile. The reaction and load at the base of the building are divided into five groups to apply in the loading phase. Table 16.6 shows the summary reaction at the base of the structure that will be applied in the loading phase.

Calculation

After defining the calculation phase, the last stage is the calculation stage where the software will calculate the displacement of the pile according to the load insert by the user.

Table 16.5. Subsoil condition of case study (Bowles, 1997).

Layer			**1**	**2**	**3**	**4**
Soil	Notation	Unit	Heterogeneous landfill	Very soft to soft silty clay	Medium stiff or medium dense silty clay or silty sand	Very dense or hard soil layer
Soil Depth		m	0–4.5	4.5–9.5	9.5–54.5	54.5–65
SPT value	N	n	15	2	18	50
Material Model			Mohr-Coulomb	Mohr-Coulomb	Mohr-Coulomb	Mohr-Coulomb
Material Type			Drained	Drained	Drained	Drained
Unsaturated Unit Weight	γ	kN/m^3	18.065	13.352	18.536	21.992
Saturated Unit Weight	γsat	kN/m^3	19.767	17.280	20.160	23.563
Young's Modulus	E	kN/m^2	8000	9576.052	47880.259	95760.518
Poisson's Ratio	v		0.3	0.4	0.4	0.35
Cohesion	c	kN/m^2	20	23.940	210.673	383.042
Angle of Friction	φ	°	32.5	28.5	34.0	40.5

Table 16.6. The summarize reaction at the base through ETABS.

Pile Group	Reaction (kN)	Average reaction (kN)	No. of column
1	2000–3999	3273	7
2	4000–5999	4793	6
3	6000–7999	6581	9
4	8000–9999	8978	5
5	10000–12000	11292	5

Result and Discussion

The results in terms of displacement of the pile are generated by using a different load, and also with a different length of pile and the type of pile with either single or group pile. The allowable displacement or settlement of the pile will be ensured to fall within 10% of the pile diameter as the maximum allowable displacement of the pile stated in Eurocode 7 if not pile is considered to have failed (Simpson, 2011).

Superstructure

Figure 16.1 shows the 3D view of the completed model of the 12-story high building. The ETABS software can obtain the reaction at the column. In order to proceed to the foundation design of the building, the reaction which is the load applied to the column is necessary. Table 16.7 shows the result generated by ETABS in term of the reaction of the column at the base.

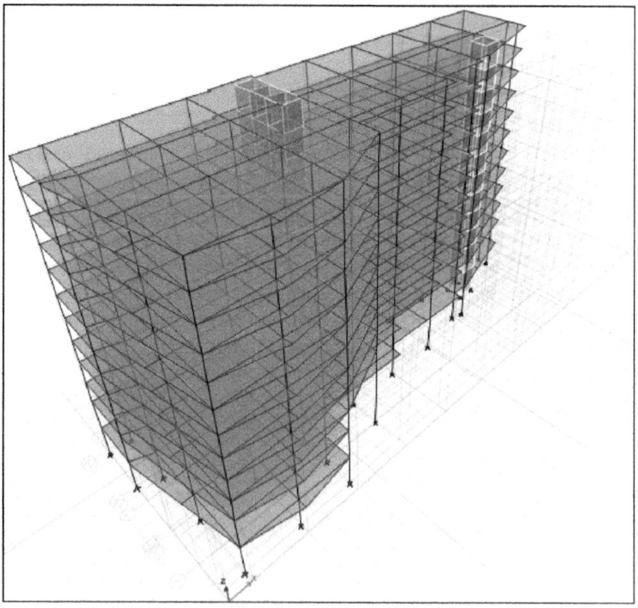

Figure 16.1. Three D view of the 12-story height hotel model.

Table 16.7. Summarization of the reaction Fz in table form.

Grid	Fz (kN)	Grid	Fz (kN)	Grid	Fz (kN)
A10	2511	J8	3940	D3	5372
B10	3815	K8	3550	E3	4097
C10	6388	B4	9041	B2	5413
D10	6266	C4	11717	C2	8238
F10	6221	D4	9344	D2	5188
G10	6525	E4	7622	G8	11444
H10	6484	F4	8998	H8	9271
J10	6148	G4	6436	C8	11271
K10	3710	H6	4226	D8	10673
A9	4463	I5	2464	F8	11356
B8	7137	K7	2918		

Table 16.7 shows that the columns at the C4, C8, D8, F8, and G8 of the structure have the highest reaction, and the column at A10, K7, K8, and K10 have the lowest reaction are located at the edge of the structure. Similar to every matter, this is due to the centre of mass theory, whereby the average position of all the weight by the members will be at the centroid of the whole structure.

Foundation

The foundation of this 12-story building is designed as a pile foundation, where it is done by using Plaxis 3D simulation software. This will cover the load against displacement result as well as the different result by using the varying length of the pile.

Load and Displacement

Single Pile

The Figures 16.2–16.6 show five different types of piles for displacement of the single pile with five different lengths under five different loads according to the reaction. The figures show that with an increase of height displacement decreases. When comparing the Figs. 16.2–16.6, by providing the pile with increasing length, the displacement is decreased. This is due to the reason that the capacity provided by the pile is increased by increasing length. The pile diameter fixed for first two cases, i.e., for single and double piles is 0.4 m. For this section, the limit for the displacement or settlement is 40 mm, based on the allowable displacement of 10% of pile diameter. For load group 1, as shown in Fig. 16.1, the 10 m pile failed. For load group 2, the pile length with 10m and 13m failed to achieve the 10% target displacement. For the load groups 3, 4, 5, none of the five provided length including 10 m, 13 m, 20 m, 30 m, and 55 m is able to reach the limit. This is probably due to the water level depth of the coastal area is near to the ground level, and leads to the changing in moisture in the soil, and hence the soil loses its strength.

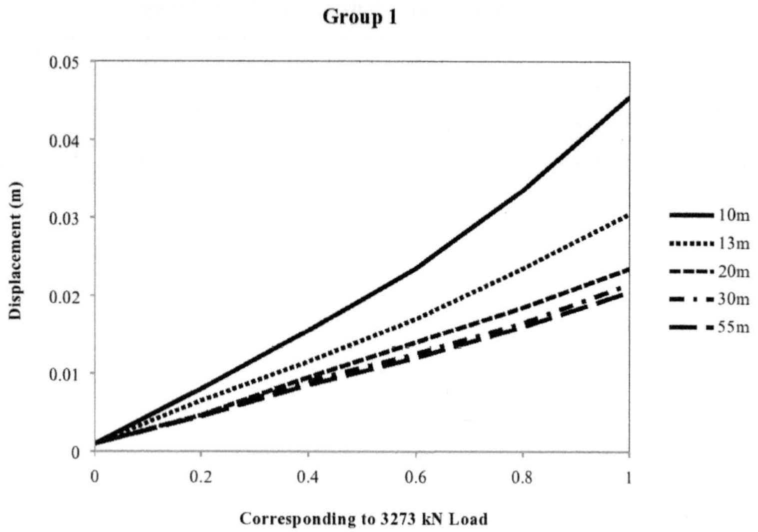

Figure 16.2. Graph of displacement vs. load of 3273 kN for single pile.

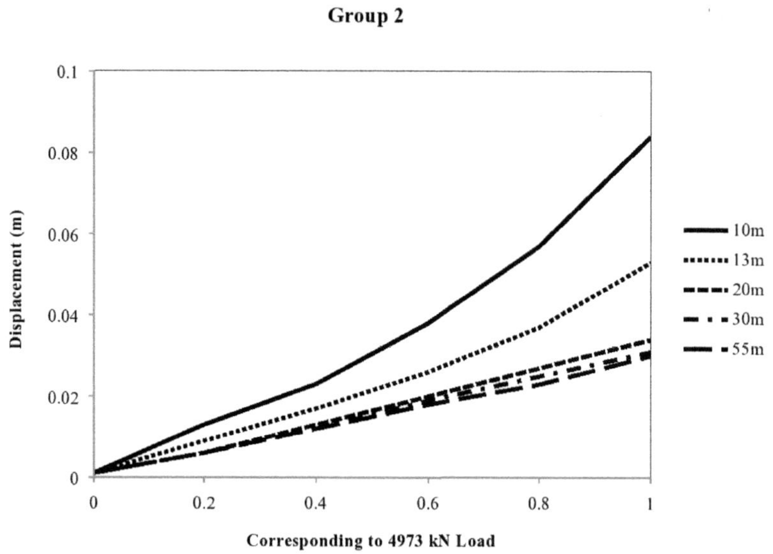

Figure 16.3. Graph of displacement vs. load of 4793 kN for single pile.

The effect of displacement for piles in Fig. 16.2 between 20–55 m is very little. It can be seen that due to this reason 20 m pile would be sufficient to be recommended for this load group. This would not only save the cost of the building but also the work can be finished in less time. The environmental effect will also be significant on the holistic structure. With less material and time consumed the buildings cost will decrease significantly which will make it as affordable for the local people. Note: The value at Y-axis is adjusted for the better view.

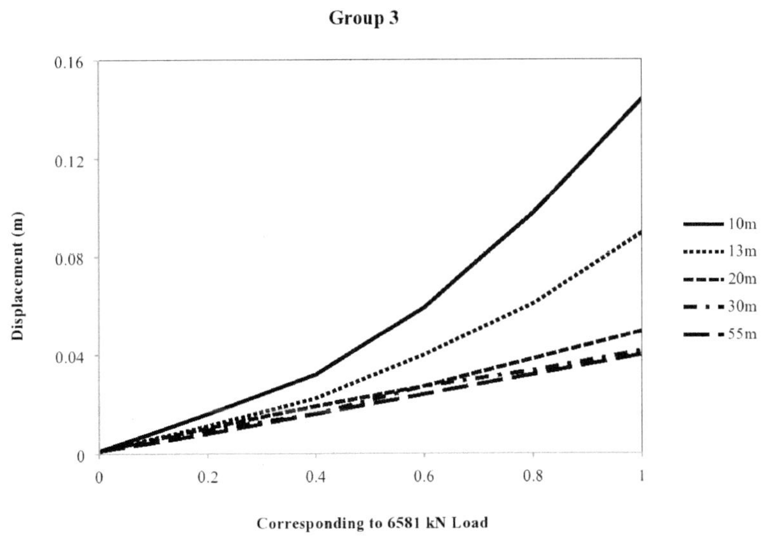

Figure 16.4. Graph of displacement vs. load of 6581 kN for single pile.

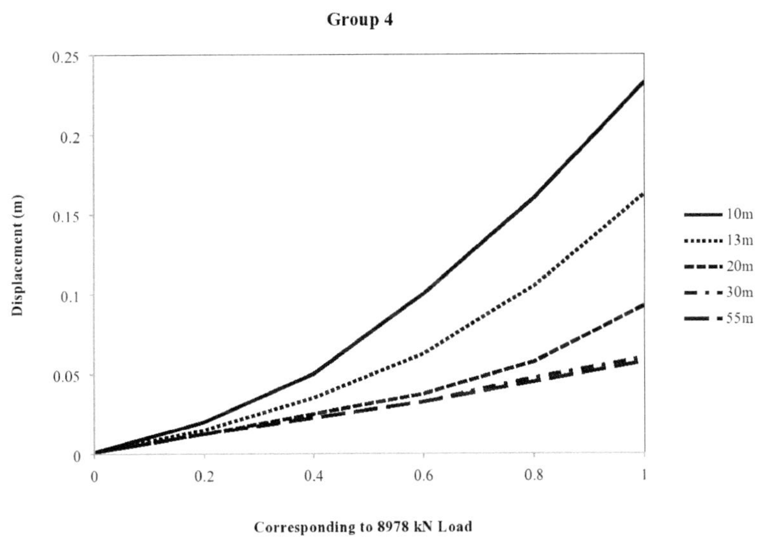

Figure 16.5. Graph of displacement vs. load of 8978 kN for single pile.

Double Pile

This section is for those column reactions where single pile failed to produce target displacement. Thus for this case group pile considered is the double pile. The result of displacement vs. graph is shown in Figs. 16.7–16.11. Similar to the case with the single pile, the displacement is increasing along with the increasing load. By providing an increased length of the piles, however, the difference is that with the same length

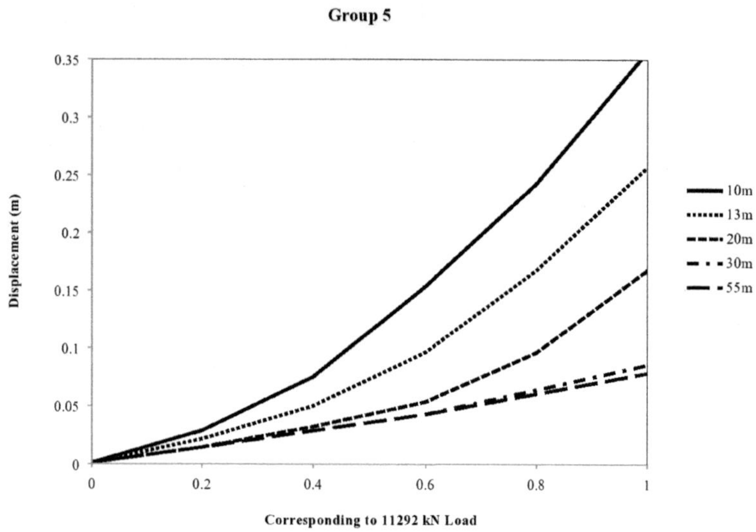

Figure 16.6. Graph of displacement vs. load of 11292 kN for single pile.

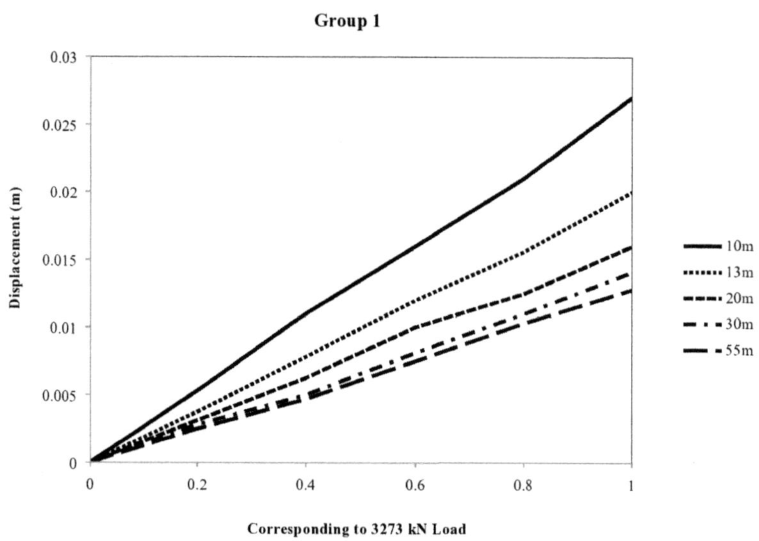

Figure 16.7. Graph of displacement vs. load of 3273 kN for double pile.

of the pile and under same reaction load from the structure, a double pile yields a lower result of displacement compared to the single pile. This result can be seen by referring to the graph, wherein the load group 1, even the displacement of 10 m pile length falls within 40 mm of displacement range. By comparing to the highest load with the longest diameter provided, the displacement is 44 mm which is more than 40 m as shown in the EC7. Therefore, in this case, an alternative method has to be applied to solve the problem, which is by providing the larger diameter of the pile.

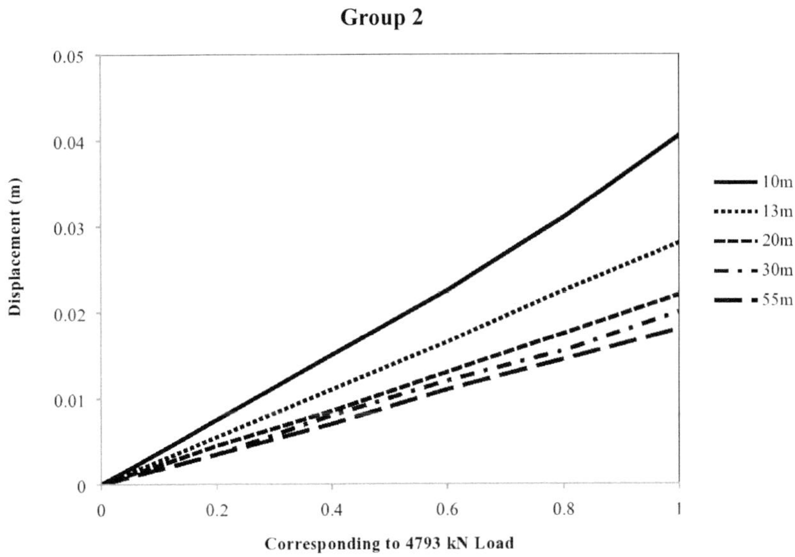

Figure 16.8. Graph of displacement vs. load of 4793 kN for double pile.

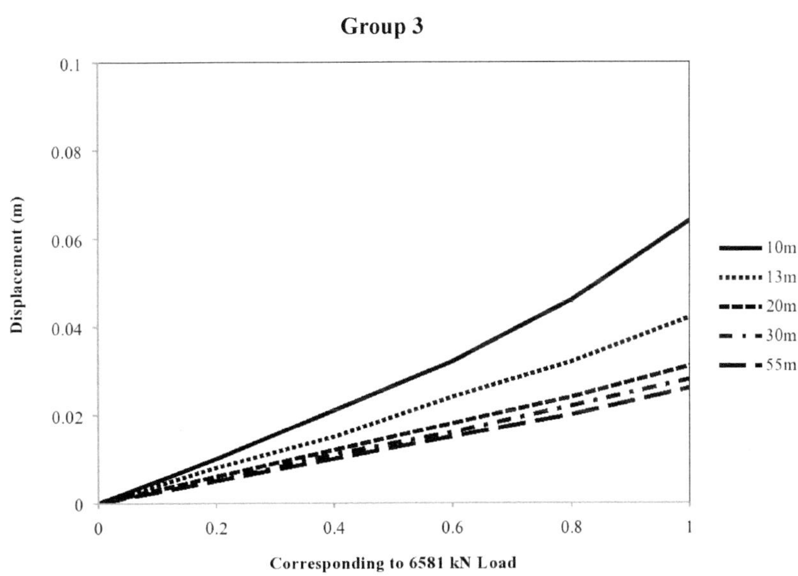

Figure 16.9. Graph of displacement vs. load of 6581 kN for double pile.

The maximum displacement (Umax) of the pile in a single and double pile is summarized in Table 16.8 and Table 16.9. The result shows that the length and displacement of the piles for 0.4 m diameter pile.

The optimum pile length is listed in Table 16.10. The list show that piles with displacement of 55 mm has higher displacement then allowed by the code. Therefore, it has to be rechecked by using more than one pile or increase the length of the pile.

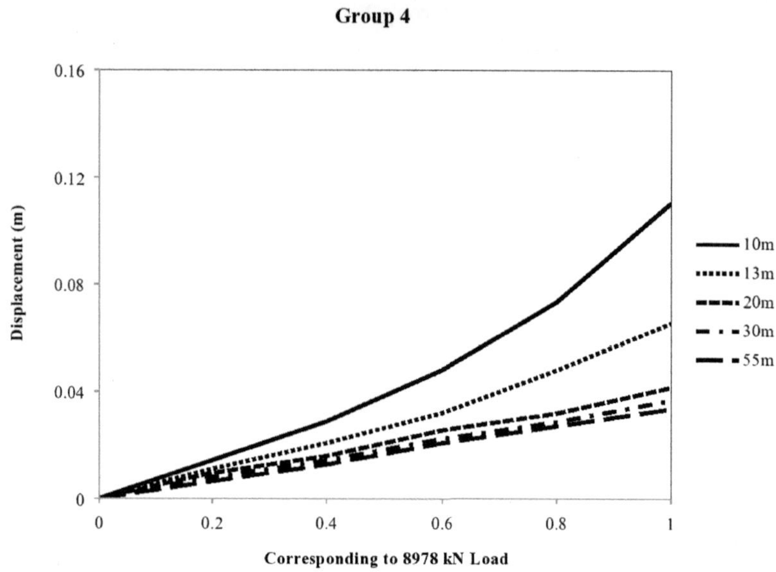

Figure 16.10. Graph of displacement vs. load of 8978 kN for double pile.

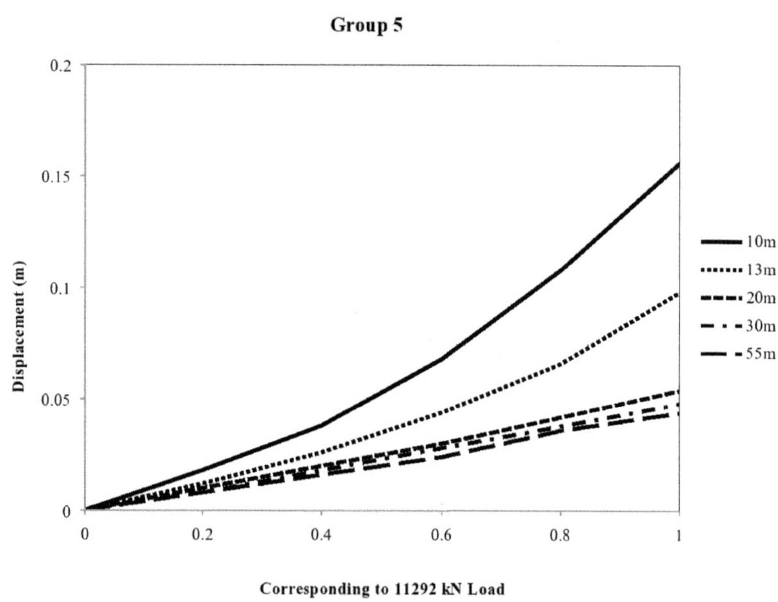

Figure 16.11. Graph of displacement vs. load of 11292 kN for double pile.

Alternative Method with 0.5 m Diameter

According to Eurocode part 1, where it states that the maximum limit of the displacement should fall within 10% of the pile diameter, wherein this chapter it is 40 mm. The numbers in asterisks in Table 16.10 indicates that with the maximum length

Table 16.8. Maximum displacement of single pile.

Column Reactions									
3273 kN		**4793 kN**		**6581 kN**		**8978 kN**		**11292 kN**	
Length (m)	**Umax (mm)**	**Length (m)**	**Umax (mm)**	**Length (m)**	**Umax (mm)**	**Length (m)**	**Umax (mm)**	**Length (m)**	**Umax (mm)**
10	44	10	83	10	142	10	237	10	348
13	30	13	52	13	90	13	165	13	253
20	23	20	33	20	49	20	95	20	162
30	21	30	31	30	42	30	61	30	84
55	20	55	29	55	41	55	58	55	77

Table 16.9. Maximum displacement of double pile.

3273 kN		**4793 kN**		**6581 kN**		**8978 kN**		**11292 kN**	
Length (m)	**Umax (mm)**	**Length (m)**	**Umax (mm)**	**Length (m)**	**Umax (mm)**	**Length (m)**	**Umax (mm)**	**Length (m)**	**Umax (mm)**
10	26	10	41	10	64	10	107	10	157
13	19	13	29	13	41	13	66	13	98
20	15	20	22	20	30	20	42	20	54
30	13	30	20	30	27	30	37	30	47
55	13	55	18	55	25	55	35	55	44

Table 16.10. Optimum pile length of single pile and double pile under different load group.

Load group	**Single pile displacement (mm)**	**Double pile displacement (mm)**
Group 1	13	13
Group 2	20	13
Group 3	55	20
Group 4	55	30
Group 5	55	55

of pile provided, the maximum displacement of the pile is exceeded 40 mm. For load groups 3 and 4, although it is more than 40 mm displacement, the double pile can be applied for the support as the result shows the maximum displacement falls within 40 mm. For load group 5, the maximum displacement is exceeding 40 mm in both single and double pile with 55 m pile length. Therefore, an alternative method is applied which is by changing the diameter of the pile from to 0.5 m and fixing them as double pile with the length of 30 m. Figures 16.12–16.16 show the comparison between 0.4 m and 0.5 m pile diameter with both having a fixed length of 30 m and designed as a double pile. The load vs displacement graph is plotted under the same five groups of loads. The results show that with the increasing of diameter of the pile, the pile with 0.5 m diameter has lower displacement comparing to the 0.4 m pile.

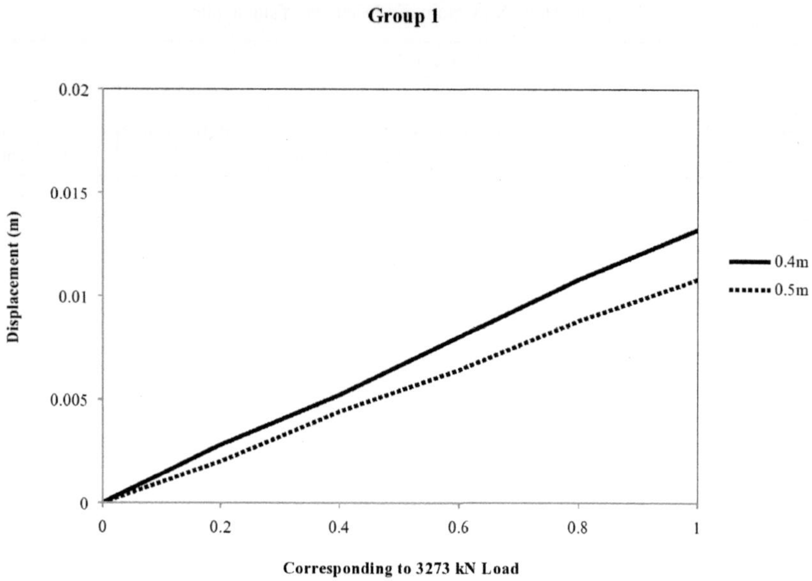

Figure 16.12. Graph of displacement vs. load of 3273 kN for 0.4 m and 0.5 m diameter pile.

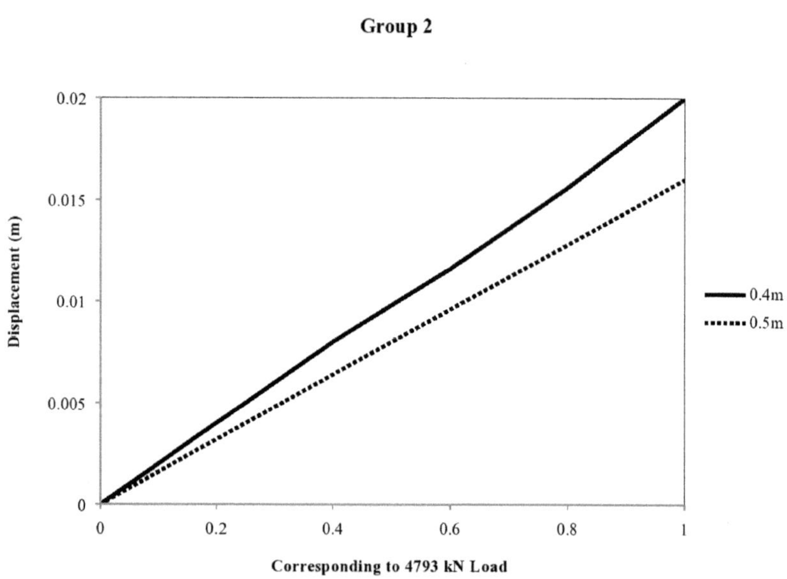

Figure 16.13. Graph of displacement vs. load of 4793 kN for 0.4 m and 0.5 m diameter pile.

The maximum displacement between two different diameters of the pile is plotted in Table 16.11, and is plotted as a graph form in Fig. 16.17.

Based on the results, a pile designed with larger diameter can provide a decrease in the displacement and thus a pile with 0.5 m diameter produces only 38 mm of

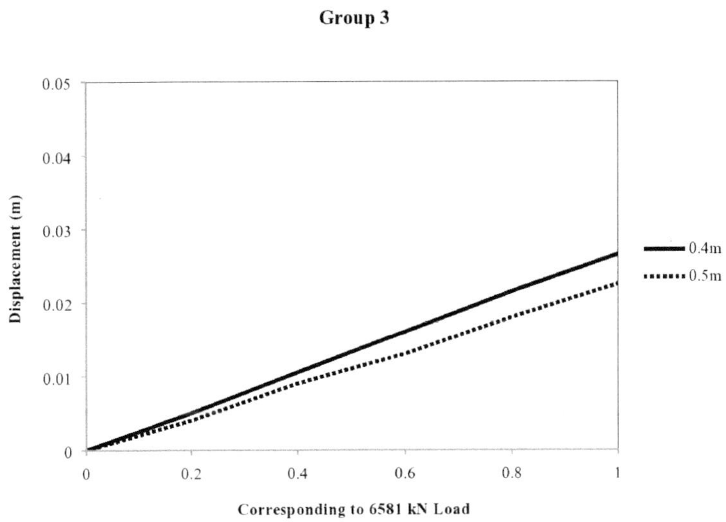

Figure 16.14. Graph of displacement vs. load of 6581 kN for 0.4 m and 0.5 m diameter pile.

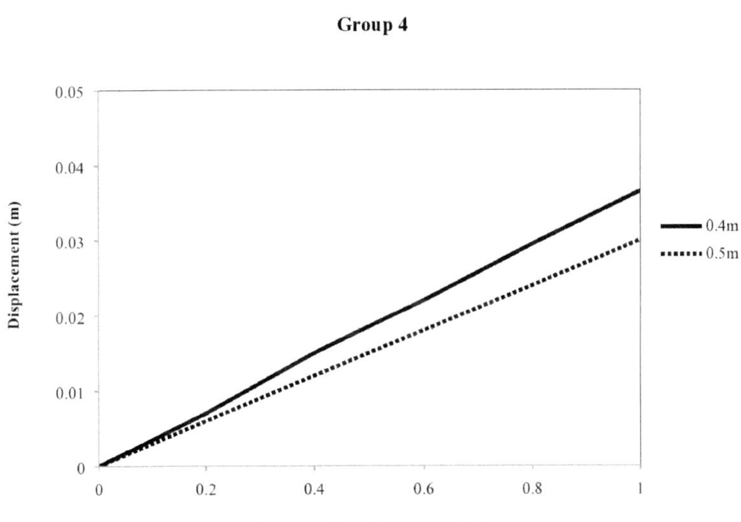

Figure 16.15. Graph of displacement vs. load of 8978 kN for 0.4 m and 0.5 m diameter pile.

Table 16.11. Maximum displacement of 0.4m and 0.5m diameter pile.

Column reaction (kN)	0.4 m	0.5 m
	Displacement (mm)	Displacement (mm)
3273	13	11
4793	20	16
6581	27	23
8978	37	31
11292	47	38

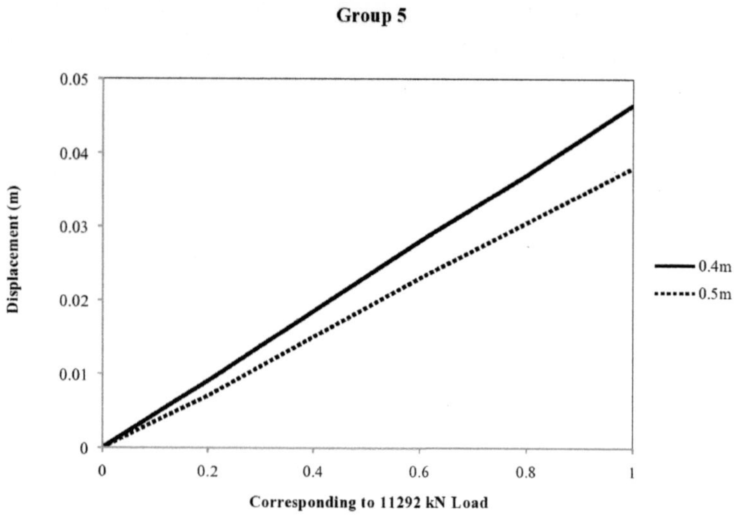

Figure 16.16. Graph of displacement vs. load of 11292 kN for 0.4 m and 0.5 m diameter pile.

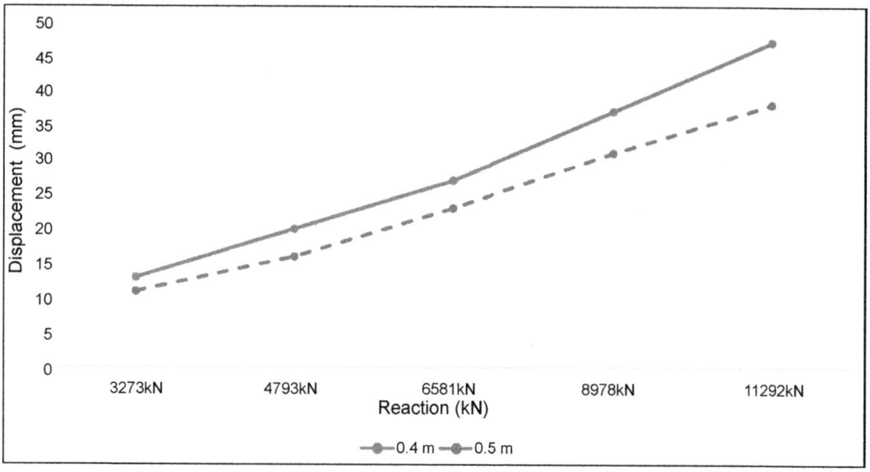

Figure 16.17. Maximum displacement of 0.4 m and 0.5 m pile diameter.

displacement which falls within 10% of the diameter, in fact, 50 mm. However, the pile with 0.4 m diameter shows a displacement of 47 mm where it exceeds the 10% of its diameter, i.e., 40 mm. The result shows that the alternative method by changing the diameter of the pile can be applied to the economic structures if the same diameter is to be selected. Otherwise, if the change of diameter is allowed then for economical and affordable housing a different diameter should be used.

Cost Estimation

The cost of the pile foundation is estimated based on the material of concrete and reinforcement bar. A summary of the price of total price of concrete and reinforcement bar is listed in Tables 16.12–16.14. Figure 16.18 shows the distribution of the cost among five groups of piles and pile cap. Based on (Quantity Surveyor Online n.d.), material cost of concrete grade C25 is RM200 per cubic metre in Malaysia, therefore the material cost to be used is listed in Tables 16.12–16.13, it covers the total volume of the concrete as well as the price of the concrete. Table 16.14 shows the calculation of the material price of reinforcement bar. Figure 16.18 shows how much s every group of the pile costs. Apart from the cost for pile cap, significantly the pile designed under with the highest load which is group 5 that requires higher length and diameter of the pile, hence it occupies the second largest portion among the material cost of concrete.

The weight of reinforcement in kg/m^3 for the pile cap falls between 110–150 kg/m^3, therefore if it is taken as 130 kg/m^3. The total volume of concrete needed for

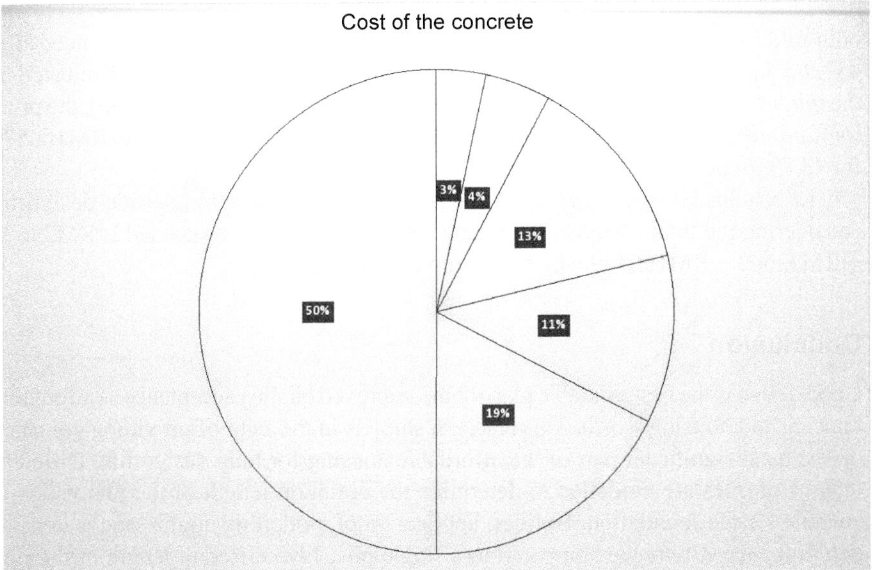

Group 1 = 3%, Group 2 = 5%, Group 3 = 13%, Group 4 = 11%, Group 5 = 18% and Pile cap = 50%

Figure 16.18. Distribution of the cost among five groups of piles and pile cap.

Table 16.12. Estimated volume and cost of the concrete of pile.

Pile Group	Reaction (kN)	Columns	Pile Type	Pile Length (m)	Pile diameter (m)	Concrete Volume (m^3)	Material cost (RM)
1	3273	7	single	13	0.4	11.44	2288
2	4793	6	single	20	0.4	15.08	3016
3	6581	9	double	20	0.4	45.24	9048
4	8978	5	double	30	0.4	37.70	7540
5	11292	5	double	30	0.5	58.90	11780
Total						168.36	33672

Table 16.13. Estimated volume and cost of the concrete of the pile cap.

Dimension	Single pile	Double pile	Total:
	2 m x 2 m x 1 m = 4 m^3	**3 m x 2 m x 1 m = 6 m^3**	
Total no. of pile cap	7 + 6 = 13	9 + 5 + 5 = 19	32
Total volume	54 m^3	114 m^3	168 m^3
Material cost	RM10800	RM22800	RM33600

Table 16.14. Estimated volume and cost of the reinforcement bar.

130 kg/m3 x 336.36 m^3 =	43726.8 kg/43.73 tonne
43.73 tonne x RM2300/tonne =	RM100579

both pile and pile cap is 336.36 m^3, and the weight of reinforcement bar needed is 43726.8 kg or equals to 43.73 tonne. Mild steel bar R25 mm diameter is selected as the reinforcement bar. Based on the price of (Quantity Surveyor Online n.d.), the price for mild steel R25 mm diameter bar is RM2300 per tonne, which equals to RM100579 for 43.73 tonne of reinforcement bar.

In conclusion, the estimated total material cost of this foundation design by considering the total volume of concrete and reinforcement bar needed is RM33672 + RM33600 + RM100579 which sums up into RM167851.

Conclusion

Good design is the best available plan for the improved quality, acceptance as affordable housing and housing worth. This research study is in the light of providing guidance on the most significant part of the affordable housing for long design life. Different lengths of piles are modelled to determine the optimum length of the pile which is suitable for the foundation. Besides, application of modelling double pile is done to determine the difference compared to a single pile. Five different length of the pile are modelled under five groups of loads generated by the superstructure model. The length of the pile applied in this study includes 10 m, 13 m, 20 m, 30 m and lastly 55 m. Similar to a single pile, the pile is modelled as a double pile, with the same length and also under same load. This is to determine the effect of displacement by providing a pile in a group form, and it shows positive result where a double pile yields lower displacement.

The optimum parameter of the pile is determined by using Euro code 7 that states that the maximum displacement should be less than 10% of the pile diameter, which is 40 mm in this chapter. For the column under load 3273 kN, is designed with 13 m single pile with 0.4 diameters. For 4793 kN load, the pile should be designed with 20 m single pile with 0.4 diameters. For the third load with 6581 kN, the pile should be designed as a double pile, along with 20 m length and 0.4 m diameter. For the column that has the second highest load, which is 8978 kN, the pile should be designed as a double pile with 30 m length and 0.4 m diameter. For the last case, the highest load, which is 11292 kN, the design of the pile, is based on alternative method, which is a double pile with 30 m length and 0.5 m diameter, then the displacement is within the maximum allowable limit.

Acknowledgement

This material is based upon work supported by the Universiti Abdul Rahman (UTAR) under Grant No. IPSR/RMC/UTARRF/2016-C1/Z1.

References

Amornfa, K., N. Phienwej and P. Kitpayuck. 2012. Current practice on foundation design of high-rise buildings in Bangkok, Thailand. Lowland Technology International 14(2): 70–83.

Amutha, K. and V. Satheeskumar. 2017. Experimental and numerical modelling of composite piled raft with cushion. *In*: International Conference on Geotechniques for Infrastructure Projects. Thiruvananthapuram, 27–28 February 2017.

Azmi, M. 2014. Study on slope stability of Penang island considering earthquake and rainfall effects. Available at https://dx.doi.org/10.14989/doctor.k18226.

Bajaj, R. and S. Choudhary. 2014. Outstanding structure: the leaning tower of Pisa. Journal of Civil Engineering and Environmental Technology 1(5): 80–83.

Bowles, J. E. 1997. Foundation analysis and design. 5th ed. Singapore: McGraw-Hill Companies.

Chen, C. S. and S. M. Tan. 2002. A case history of a coastal land reclamation project. *In*: GSM-IEM Forum on Engineering Geology & Geotechnics in Coastal Development. Petaling Jaya.

Das, A. and S. Biswas. 2014. The effect of water table on bearing capacity. International Journal of Innovative Research in Science, Engineering and Technology 3(6): 82–86.

Davies, J., J. Lui, J. Pappin, K.K. Yin and C.W. Law. 2004. The foundation design for two super high-rise buildings in Hong Kong. *In*: CTBUH 2004 Seoul Conference, 10–13 October 2004.

Evans, D. 2014. Bringing the Power of Design to Affordable Housing: The History and Evolution of the Affordable Housing Design Advisor. Cityscape, A Journal of Policy Development and Research. 16.

Frank, R. 2006. Design of pile foundations following Eurocode 7. *In*: Slovenian Geotechnical Society, Proceedings XIII Danube-European Conference on Geotechnical Engineering. Ljubljana 29–31.

Kalny, O. 2013. CSI knowledge base. https://wiki.csiamerica.com/display/etabs/Home.

Katzenbach, R. and A. Schmitt. 2004. High-rise buildings in Germany soil-structure interaction of deep foundations. *In*: Missouri University of Science and Technology, International Conference on Case Histories in Geotechnical Engineering. New York. 13–17.

Katzenbach, R., G. Bachmann, G. Boled-Mekasha and H. Ramm. 2005. Combined pile raft foundations (cprf): an appropriate solution for the foundations of high-rise buildings. Slovak Journal of Civil Engineering. 19–29.

Kayvani, K. 2014. Design of high-rise buildings: past, present and future. *In*: 23rd Australian Conference on the Mechanics of Structures and Materials (ACMSM23). Byron Bay, Australia 9–12.

Kulkarni, S. R. 2009. Foundations of high rise buildings. *In*: Amrutvahini College of Engineering, International Conference (ITECH-09). Sangamner.

Leow, J. H. 2005. A study of building foundations in Malaysia. Degree. University of Southern Queensland.

Lok, T. M. H., J. M. Pestana and R. B. Seed. 2000. Numerical modelling of seismic soil-pile-superstructure interaction. In: 12th World Conference on Earthquake Engineering. New Zealand.

Moniz, L. C. M. 2014. Foundations and soil treatment using the full displacement piles technique.

Mohammadizadeh, M. and M. Mohammadizadeh. 2016. Numerical modelling of a dynamic pile-soil interaction in layered soil media. *In*: 4th Geo-China International Conference. Shandong, China.

Nor, A. H. M., M. I. M. Masirin and M. E. Sanik. 2014. Site investigation of road drains for rural road on Batu Pahat soft clay (BPSC). Journal of Mechanical and Civil Engineering 11(2): 12–19.

Poulos, H. G. 2011. The design of high-rise building foundations. *In*: International conference on Geotechnics for Sustainable Development Geotec. Hanoi.

Quantity Surveyor Online, n.d. Prices of building materials.: <http://www.quantitysurveyoronline.com.my/materials-prices.html.

Ravishankar, P. and D. N. Satyam. 2013. Numerical modelling to study soil structure interaction for tall asymmetrical building. *In*: Centre for Earthquake Engineering, International Institute of Information Technology, International Conference on Earthquake Geotechnical Engineering. Istanbul, Turkey.

Rayhani, M. H. T. and El M. H. Naggar. 2008. Numerical modelling of seismic response of rigid foundation on soft soil. International Journal of Geomechanics.

Rybak, J., D. Sobala and G. Tkaczynski. 2008. A case study of piling project and testing in Poland. Science, Technology and Practice, Jaime Alberto dos Santos (ed). 175–178.

Sancio, R. B., J. D., Bray, T. Durgunoglu and A. Onalp. 2004. Performance of buildings over liquefiable ground in Adapazari, Turkey. *In*: 13th World Conference on Earthquake Engineering. Vancouver, B.C., Canada.

Saravanan, V. K. 2011. Cost effective and sustainable practices for piling construction in the Uae. Degree. Heriot-Watt University.

Simpson, B. 2011. Concise Eurocodes: geotechnical design. London: British Standards Institution.

Sreechithra, P. and K. Niranjana. 2017. Numerical analysis of pile raft system. *In*: International Conference on Geotechniques for Infrastructure Projects. Thiruvananthapuram.

Srivastava, A., C. R. Goyal and A. Jain. 2012. Review of Causes of Foundation Failures and their Possible Preventive and Remedial Measures.

Zheng, J. J. and S. P. Chin. 2011. Application of numerical modelling method on the dynamic response of long span structure to vehicular load. Procedia Engineering 14: 997–1004.

CHAPTER 17

Sustainable Construction and Waste Management in Malaysian Residential Projects

Hui Chen Chu[1]

Introduction

Sustainability in construction has now become inevitable with growing conscience and knowledge on how much damage development has done to the environment with greenhouse effect, climate change, soil erosion, flash floods, pollution and nuisance in general. Social impact from construction activities has also been disturbing with high influx of foreign workers coupled with health and safety issues at construction sites. From the economic perspective, even stronger reasons to ensure sustainability in construction when it is an instigator of other GDP factors, such as manufacturing, mining and quarrying. Construction has to shift towards sustainability because it cannot be done away with; it provides shelters for residence, buildings for commerce and infrastructures for commuting and facilities. All in all, construction is essential.

With increasing population every year (Department of Statistics Malaysia, 2016), residential projects are a necessity which the government are expected to ensure the nation of its sufficiency in affordable housing. In Malaysia, commitment by the government had been evidently reflected in the national Five Yearly Development Plans. Up until now, massive amounts of low-cost housing under various affordable housing schemes such as Housing for the hardcore poor (PPRT), People Housing Project (PPR), 1Malaysia Civil Servants Housing (PPA1M) and Public Listed National Housing Corporation (SPNB) had been delivered. Credit has to be given to the private sector in their joint participation with the local government in achieving the above mentioned projects (IBCC, 2016). Unfortunately, with the drop in Malaysia's annual GDP growth rate, it also affects the residential market. There had been a severe drop

[1] Universiti Tunku Abdul Rahman, Malaysia.

in the percentage of change in volume and in value of transaction for residential properties after 2014. This can be observed in Figs. 17.1 and 17.2.

With reference to Zainul Abidin back in 2009 on Sustainable Construction in Malaysia, there was only one sustainable housing project in Malaysia. Namely, project Tanarimba at Janda Baik, Pahang from 2006. Currently, there are many more sustainable housing projects which can be observed from the Green Building Index (GBI) listing. There are 29 housing projects which had been completed and verified as green housing projects, a further 135 housing projects had submitted for design assessment (http://new.greenbuildingindex.org, July 2017). GBI listing also contains two completed and verified green townships and another 10 more townships in the design assessment list (http://new.greenbuildingindex.org, July 2017).

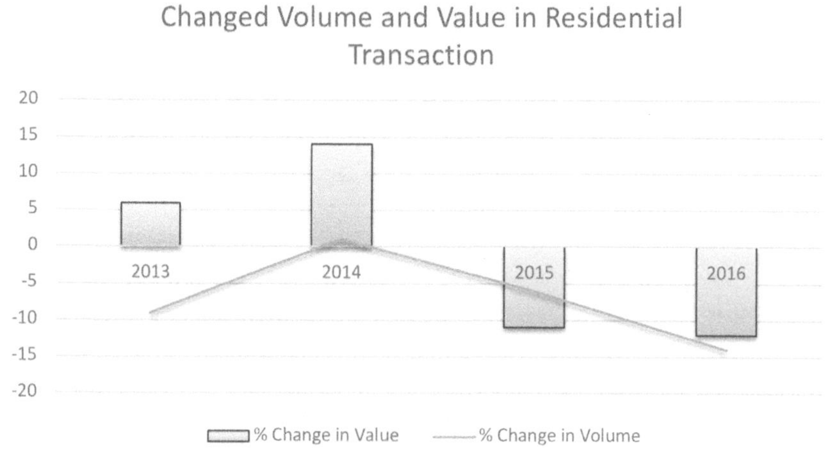

Figure 17.1. Changed volume and value in residential Transaction. Source: National Property Information Centre, 2018.

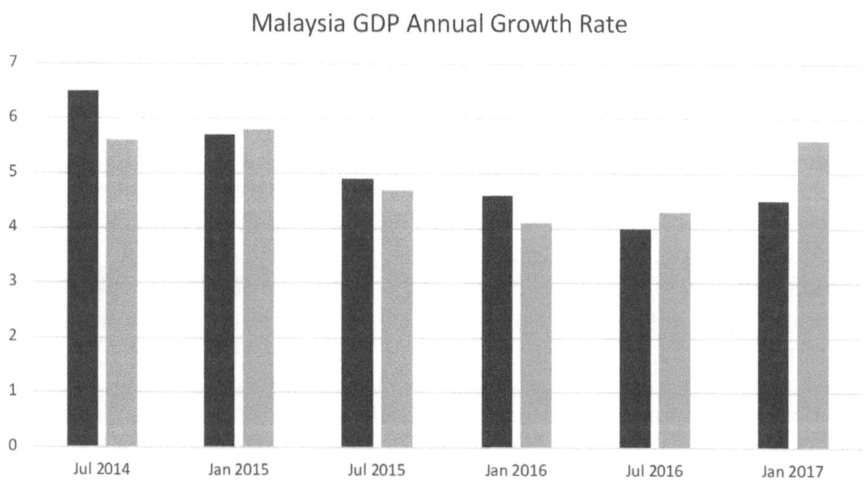

Figure 17.2. Malaysia GDP annual growth rate. Source: Tradingeconomics.com Department of Statistics Malaysia.

Waste Management Practices

In general, sustainable construction comes in various forms and practices; it could be by way of using sustainable materials, sustainable construction methods, technology and practices or even the application of a sustainable design for the building; which is to be applied at different stages in a construction process and to be implemented by different parties in the project. Therefore, for construction to successfully achieve sustainability, every party involved would need to contribute their part, be it an idea, a practice or a better choice of material. Ideally, sustainability is to be practiced as a complete package from design to construction and then its maintenance but strategies adopted by various countries are influenced by a number of factors, such as geographical area, population density, transportation infrastructure, and environmental regulations (Bai and Sutanto, 2002; Tan and Khoo, 2006). Developed and developing countries would set different goals and frameworks. Developed countries would focus on energy conservation and restoration works and its implementation through policies or mandatory laws and regulations whereas for developing countries, the focus would be on new buildings, its whole life cycle perspective and its implementation are merely voluntary policies (Adshead, 2011). Voluntary policies are being practised in most developing countries because they do not have a policy which they could be certain or prove appropriate to be implemented mandatorily.

At a glance from the National Land Use Report 2012, shown in Chart 3, Malaysia still has abundant forestry area. In fact, there has been a growth in forestry area. Preservation of such areas would be crucial for sustainability of the ecosystem as well as the tourism industry. Development is inevitable, thus the importance to ensure that all area used for development would be put to good use, that waste would take up only a negligible area. With the increase in developed area, there had been an obvious decrease in agricultural land use. However, this does not interpret directly that agricultural activity has been dropped. The department of statistics had in fact reported doubling value for crops and livestock in Fig. 17.4. The reason for the contradicting

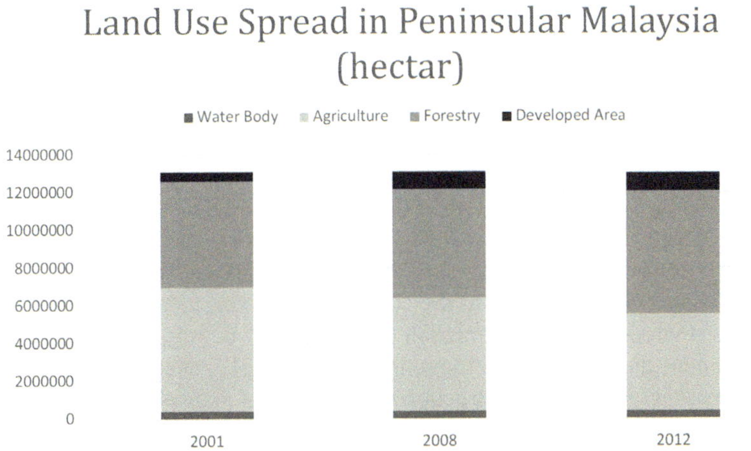

Figure 17.3. Land area and land use spread in peninsular malaysia. Source: National Land Use Information Centre, 2012.

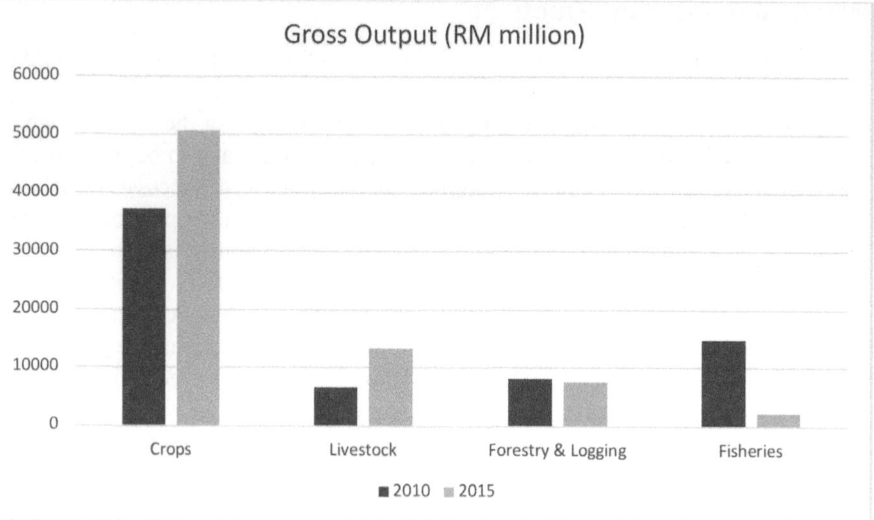

Figure 17.4. Economic Census 2016 (Agriculture Gross Output). Source: Department of Statistics Malaysia, 2016.

indication between these two figures, Figs. 17.3 and 17.4 is due to modernization in agriculture. Livestock has been reared in enclosed buildings such as in chicken rearing and vegetables planted indoor by means of aquaculture. Therefore, agriculture has also become a major boost for the construction industry.

The National Land Use Report 2012 also indicated a considerable drop in land area for 2012 from 2008, about 0.25%. This involves in issues of rising sea level, an effect of global warming where further reference could be made to the National Geographic report from 2015 at its webpage (http://www.nationalgeographic.com, July 2017). Perhaps in future, with the advancement in technology and a change of lifestyle, to achieve zero waste by 2020 as such reported in Kamikatsu, Katsuura District, Tokushima, Japan could be possible for the rest of the world. They currently have 45 categories of waste to be recycled, reused and composted. The village is recycling and reusing more than 80% of their non-organic waste, about a 20% higher rate of the national Japanese average (South China Morning Post, 21 February, 2017).

In Malaysia, construction waste management had gained its emphasis as it appears to be one of the themes under the R&D Themes and Titles of the Construction Research Institute of Malaysia (CREAM) 2008/2009 Research Fund List. CREAM encourages the development of comprehensive assessment of waste management technology, management towards Life Cycle Assessment (LCA) approach, emission inventory for Green House Gas (GHG) from the construction sector, mechanism to establish supply chain management of construction waste material and waste database portal for creation of new cluster industries for construction. In 2012, one of the potential key research areas on sustainable construction and green technology published in CREAM research focus webpage is the development of standards, guidelines, policies and best practices. This could be elaborated and linked with the waste hierarchy. What about the size of construction waste.

A "waste management hierarchy" would generally prioritize efforts in minimizing wastes production, which can be classified as preventive actions. This is followed by actions to recover and transform wastes to be usable, and the last resort, disposal. This waste management hierarchy had been adopted by most industrialized nations as the bedrock of strategy development for achieving sustainable waste management systems (Barton et al., 1996; Tan and Khoo, 2006). Be it preventive or curative, both actions are relatively dependent. Where preventive actions are not entirely successful, curative actions follow suit. None the less, curative actions could be an enhancement to better preventive measures. If a systematic curative action is in place, that system can affect the process of construction, and even the planning stage. Take for example the effect of proper documentation practices. Where an organization has no practice of proper documentation, flow of work could be affected due to the difficulty in finding reference documents and accuracy of information contents in such documents are often doubted. This will result in reduced efficiency at work. Not only efficiency is reduced, but the situation allows greater room for the effect of human factor, such as communication problems, conflicts, missing information and erroneous information. Putting a systematic documentation practice in place provides proper reference archive. Subsequent documents generated can have more accurate contents or lesser errors. When there is a proper reference archive, there is less need for human interaction, thus, lesser conflicts and lesser communication problems. A change can give a chain of effects. Therefore, even though an action is curative in nature, it could contribute to further development of preventive actions.

In the past, waste minimization practices were heavily reliant on clients and designers to use low waste construction technologies or to address opportunities in the advancement of construction operations and technology (Poon et al., 2000; Majdalani et al., 2005). It is not exactly a controllable element as there are so many variants affecting its outcome. Initial studies on construction waste management were basically to convince people of its economic impact in the construction industry of the United Kingdom (UK) (Formoso et al., 2002). At that point, the objective was to review the cost of construction waste against project cost, to show the significance of the cost of construction waste. At present, there are other more alarming concerns for the need of implementing construction waste management especially from the environmental aspects. Issues which had become noticeably alarming over the years are the direct impact of construction and demolition waste, the amount of waste generated and the area required for disposal of such physical material waste from the construction industry (Skoyles, 1976; Bossink and Brouwers, 1998; Forsythe and Marsden, 1999; Poon et al., 2000; Enhassi, 2000; Tam et al., 2007; Wan et al., 2009). These issues addressed the concept of sustainable development. Sustainable development concerns the achievement of economic growth without eroding natural resources, upsetting the ecosystem or causing pollution and such for development to remain as a continuous activity (Enhassi, 1999). Thus, the objective is to keep construction a continuous activity.

A waste management system includes waste collection and sorting, recovery and recycling of secondary materials, biological treatment of organic materials, thermal treatment or landfill (McDougall and Hruska, 2000; Tan and Khoo, 2006). "Waste management hierarchy" mostly prioritize minimization, which is followed by recovery and transformation, and lastly, the unavoidable option of disposal. This hierarchy is generally adopted by most industrialized nations in their sustainable waste management strategies (Barton et al., 1996; Tan and Khoo, 2006). The rationale of prioritizing minimum waste is when the amount of waste produced becomes insignificant, disposing waste will not be an issue provided that the content is not hazardous. Previous research concluded that wet trades and traditional construction methods, which are methods using *in situ* formwork and concrete, cutting and bending of materials, are the main causes of waste generated on sites (Gavilan and Bernold, 1994; Poon et al., 2000). Modern building technologies introducing precast parts, such as large panel metal formwork, precast concrete façade elements and internal drywall partition would be an ideal solution but its feasibility will only apply to buildings of standard design, of considerable quantity but the construction site to be of considerate distance from the precast manufacturing plants. This is the geographical challenge faced by the industry in its effort of minimizing waste. New developments are generally in new areas. A precast manufacturing plant cannot afford to move itself to remain near to new developments, thus transportation costs will have to be incurred. Unfortunately, transportation costs are not cheap, therefore, offsetting any savings obtained from reduced waste generated from traditional construction methods. For some instances where the contractors have limited resource management skills, or the manufacturing plant is located quite a distance away, it would be faster and cheaper for the contractor to do cast *in situ* than to wait for the parts to be transported over for assembly, especially if it is a low rise project such as a housing development.

Documenting Construction Waste

Waste minimization is seldom considered because the costs of wasted materials and waste disposal charges are insignificant compared to the cost of the entire project (Poon et al., 2000). The contractor is usually satisfied from the outcome whereby he had generated considerable profit at the end of construction. He will not be bothered by the cost of wastage because he had made enough. In other words, such waste had been paid for by the client. Therefore, physical waste will still be prominent. The general conclusion obtained from these studies is that waste minimization practice are heavily reliant on clients and designers in encouraging the use of low waste construction technologies or to address opportunities in the advancement of construction operations and technology (Poon et al., 2000; Majdalani et al., 2005). Where a client is willing to pay a price in lieu of the contractor's profit, the cost of waste, there will be little initiative expected from the contractor to practice waste minimization. Where designers do not provide specifications for low waste construction technologies, the contractor will proceed with his regular methods of which he is familiar with and in his opinion, which is the cheapest or most convenient method.

At present, there are complications in determining the magnitude of waste generated from the construction industry. Companies are mostly unaware of the magnitude of waste generated on site. Very few of the sites had organized records on the actual delivery, storage and consumption of materials (Formosa et al., 2002). This indicates a lack of transparency in the performance of the production system (Greif,

1991). Consequently, many studies had been conducted to determine such magnitude of waste generated by the industry, such as to obtain in percentage, wasted materials in weight in relation to the amount of materials defined by design (Skoyles, 1976), packaging waste in relation to the volume of materials purchased (HK Polytechnic & HK Construction Association, 1993) and direct waste in relation to the weight of purchased amount (Bossink and Brouwers, 1998). The purpose of these studies is to identify major waste components so as to propose solutions either to prevent or minimize such waste. Therefore, it is important, a pre-requisite in fact, to identify the type and magnitude of such waste before one can come up with a practical solution, whether to prevent or minimize such waste or to replace such material altogether.

Construction waste is most often appreciated as the physical material waste generated from the industry because of its ease for direct quantification in the form of weight, volume and cost. It is also because of its physical form that complications and nuisance arises. Due to the bulk of waste generated, there are complications in transporting such waste. Proper transport and perhaps a designated route and dump site would be required for the prevention of nuisance to the public. However, most contractors will just succumb to the ease of illegal dumping and illegal dumping of construction waste is still of concerning rates in Malaysia (Pereira et al., 2007; Nagapan et al., 2013, The Star Online, 11 May 2015). Hong Kong for example, is facing problems with the lack of landfill space. Such shortage jeopardizes the use of landfills for other wastes such as domestic and hazardous waste (Poon et al., 2000; Tam et al., 2007; Wan et al., 2009). To curb the matter, Hong Kong has an Environmental Protection Department (EPD) to keep track of construction and demolition waste generated and disposed (Poon et al., 2000).

A local study conducted by Nur Hasnida Mohd Ramli (2010) on "A computerized construction waste exchange system" provides an online platform of information on construction waste for trading purposes. Its targeted users are the contractors, developers, local council and the Construction Industry Development Board of Malaysia (CIDB). Previous local researches of relevant area studied the development of framework for waste management policy and strategy (Wan-A-Kadir, 1997), waste minimization benefit and obstacles for solid industrial waste in Malaysia (Mallak et al., 2014) and a study on the potential of construction waste minimization and recycling (Begum, 2005). All these studies generally highlighted the need to manage Malaysian construction waste for sustainable development because there are apparent environmental issues. Nevertheless, to identify areas of potential improvement, it would be necessary to obtain a measure in the amount of waste generated (Formosa et al., 2002). There had been a wide range of measures used for monitoring construction wastes, such as excess consumption of materials (Skoyles, 1976; Bossink and Brouwers, 1996), quality failure costs (Cnudde, 1991), maintenance, repair costs, accidents and non-productive time (Oglesby, 1989; Formosa et al., 2002). Malaysian construction scenario has yet to resolve issues arising from the physical material waste generated from the industry. Local researchers found no comprehensive data on the types of waste generated by the construction industry but reported on apparent illegal dumping, Zainun et al., 2016. The lack of information on construction waste records hinders evaluation on the feasibility of construction waste recycling industry. Without a proper system in place, it is almost impossible to track final destination of construction waste upon leaving the site. Setbacks such as dump site facilities assigned refusing to take in construction waste only contribute to more frequent illegal dumping which causes public nuisance thus, being reported in the newspapers column (The Star Online, 2010).

Thus, more recent studies had shown efforts in quantifying wastes generated and to the need to develop database for documenting such information. For example, *Quantification of Construction Waste Generated in Residential Housing Projects via Heap Survey Sampling with the Method of Visual Estimation: A Case Study in Klang Valley and Pulau Pinang* (2017), a collaborated effort by researchers from Universiti Teknologi MARA (UiTM) and Institute for Infrastructure Engineering and Sustainable Management (IIESM).

In Malaysia, the Solid Waste and Public Cleansing Management Act 2007 (Act 672) had no effects either on the construction industry or on the rate of waste generated from the construction industry (Nagapan et al., 2012) as the Act only provides for solid waste management services to be centralized or federalized from previous decentralized or localized system. Even "Waste Management on Construction Sites" regulation issued by the Environment, Transport and Works Bureau of Hong Kong Government (ETWB, 2004), did not seem to improve the condition of construction waste generated as it only facilitates the sorting of waste. Focus should be put on developing a system to quantify such waste in order to identify areas of potential improvement. In developing a system to quantify such waste, it could also keep track of its source and final destination which will potentially curb the problem of illegal dumping and provide other opportunities for development.

The Legislation

Previously, solid waste management was governed by local authorities. With the Solid Waste and Public Cleansing Management Act 2007 (Act 672), solid waste management is now under federal jurisdiction, governed by the Ministry of Housing and Local Government (MHLG)which acts through the Department of National Solid Waste Malaysia, which is currently termed as Solid Waste Management and Public Cleansing Corporation (SW Corp Malaysia). Initially, Act 672 has no effects, either on the construction industry or on the rate of waste generated from the construction industry as the Act only provides for solid waste management services to be centralized or federalized from the previous decentralized or localized system. Regulation is more focused on avoiding waste from contributing to improper sanitation (Nagapan et al., 2012), thus, consideration weighs more for municipal wastes rather than construction wastes.

With subsequent revisions, Section 5 of the Act imposes responsibility on waste generators to segregate construction wastes on site. Collection of such waste from the site had to be done through a licensed collector, whose responsibility would be to send such waste to its designated location, be it at recycler's, for reuse purposes or to dumpsites. Section 71 of the Act stated fine amounts between RM 10,000.00 up to RM 100,000.00 or imprisonment sentences from 6 months up to 5 years for illegal dumping of construction waste. In The Star Online dated 11 July 2014, Tampin Company was the first to be prosecuted under the said Act.

For the time being, Malaysia does not have sufficient regulation and has no clear responsibilities on how to manage landfill sites (Nagapan et al., 2012). Managing landfill sites would be different from following guidelines on how to operate incinerators and how to conduct proper landfill. Weak enforcement with uncertainty over roles and responsibilities among governing authorities and limited coordination between stakeholders and governing authorities had further contributed to disengagement between policy and practice (Papargyropoulou et al., 2011). Obtained from a study on Construction Waste Management, Malaysian Perspective (Nagapan et al., 2012), there are numerous landfills in Malaysia, 176 which are still operational and 113 which are no longer in operation, as depicted in Table 17.1. Incidents of repeated illegal dumping and illegal dumpsites reoccurring despite media coverage and reports made to the authorities would still be reported in local newspapers (The Star Online, 11 May 2015).

In dealing with construction wastes, there are no clear guidelines for construction companies to adhere to. The only available regulation is to dispose the wastes at landfills. Due to frequent issues of illegal dumping highlighted in local newspapers as well as local researches (Begum, 2005; Pereira et al., 2007; Nagapan et al., 2013; The Star Online, 2010–2015), logical inquiries point at monitoring efforts. Who is monitoring and how is it monitored? Thus, there arose from the issue of uncertainty over roles and responsibilities among governing authorities. A recent study had been conducted to obtain feedbacks from local class G-7 contractors on the implementation of Site Waste Management Plan (SWMP), adopted for the United Kingdom, which aims at minimizing construction wastes (Papargyropoulou et al., 2011). Feedbacks obtained with regards to the barriers of its implementation are the additional costs involved for proper solid waste management practices and the lack of information,

Table 17.1. Landfills in Malaysia

STATE	OPERATIONAL LANDFILLS	NON-OPERATIONAL LANDFILLS
Perlis	1	1
Kedah	11	4
Pulau Pinang	1	2
Perak	20	9
Pahang	19	13
Selangor	7	12
Federal Territory Putrajaya	0	0
Federal Territory KL	0	7
Negeri Sembilan	8	10
Melaka	2	5
Johor	13	21
Kelantan	13	4
Terengganu	9	12
Federal Territory Labuan	1	0
Sabah	20	1
Sarawak	51	12
TOTAL	**176**	**113**
OVERALL LANDFILLS	**289**	

Source: Department of National Solid Waste Management, The Ingenieur, 2009.

guidance, incentives and practical tools to execute SWMP. The main factor of prevention for these contractors from using SWMP in Malaysia is the lack of promotion and encouragement by the Government and The Construction Industry Development Board (CIDB). Thus, it is apparent that the conducts of local contractors are heavily reliant on Government's encouragement.

As with regards to the Malaysian construction industry, The Construction Industry Development Board (CIDB) holds a strong influence on the practice of local contractors. Under Act 520, CIDB has rights to act on enforcements and investigations conferred upon construction sites and construction works. The launch of Industrialised Building System (IBS), Green Building Index (GBI), maintenance of health and safety courses, introduction of skilled workmen certification and many other regulations had been put to effect by CIDB. Therefore, CIDB has a considerably active role with regards to the matter of construction waste management.

Green Building Index (GBI)

Malaysia may be slow in the adaptation of sustainability in construction but it surely is advancing, observed in the launch of Green Building Index (GBI) in 2009 and the gazette of Act 673, Solid Waste and Public Cleansing Management Corporation Act 2007 (Laws of Malaysia). From these two major steps, subsequent revisions and amendments had taken place to regulate the practice.

Progress had been observed over revisions in criteria and rating system for GBI ranging from year 2011 till 2015. GBI provides certification into the category of Platinum, Gold, Silver and Certified. Out of the 29 housing projects being completed and verified as green housing projects, only one project obtained Platinum, S11 House, Selangor by AchiCentre with certification validity from October 2013 till October 2016. In a further list of 135 housing projects which had submitted for design assessment, there are two submissions for Platinum certification, Nadayu 290 Penang—Mulberry Tower and Molek Pine 4which are both high rise residential projects. GBI listing also contains two completed and verified green townships and another 10 more townships in the design assessment list where one township submitted for Platinum certification, TunRazak Exchange (TRX) depicted in Fig. 17.5 below (http:/new.greenbuildingindex.org, July 2017). Achieving Platinum certification in township would definitely require more effort and planning. These increasing numbers are mostly private sector projects, ranging from bungalows to high rise residential projects. The aim of submission to obtain Platinum is a good enough indication on the perceptivity on sustainable construction for housing projects in Malaysia, that there is an obvious demand.

In the development of new housing projects, GBI rating system stressed importance in public transportation access, community connectivity, heat island effect and storm water design under the category of Sustainable Site Planning & Management which holds the highest weightage of 33%, followed by the category of Energy Efficiency with 23%. Indoor Environmental Quality, Material & Resources and Water Efficiency

Figure 17.5. Tun Razak Exchange (TRX). Source: http://trx.my/press-releases/wct-bags-trx-new-infra-contract-buys-a-plot-of-trx-land.

are allotted with similar weightage of 12% each. Waste management is currently being allotted 4% in the rating system namely for the provision of storage and collection of recyclables and to improve waste sorting practice on site. GBI rating system also encouraged adoption of Industrialised Building System (IBS) in residential projects with an allotment of 2% in the score system. While IBS had long been introduced in Malaysia, since the late 60s in the construction of the Tunku Abdul Rahman Public Housing Estate, also known as Pekeliling Flats in Kuala Lumpur, IBS usage is currently reported at the rate of 15 to 20% of overall projects in Malaysia. This percentage is mainly reflected by government building projects from an imposition from year 2008 for all new government building constructions to contain a minimum of 70% IBS composition (http://www.theborneopost.com/2017/04/23/roadblocks-in-adopting-ibs-in-sarawak/, 18th July 2017). Unfortunately, without resolving the geographical and transportation issue, IBS utilization will continue to remain low. Furthermore, a study conducted in Klang Valley concluded that waste generated from residential type of construction produces relatively low amount of waste (Masudi et al., 2011).

In the development of townships, GBI rating system stressed on community planning and design with 26% weightage for consideration on governance, community thrust, greenspaces, amenities and accessibilities. Impact on climate, energy and water carries 20% weightage. Other assessment criteria include environmental and ecology, transportation and connectivity, building and resources, last but not least, business and innovation. These are all physical advancement to a better quality living environment, a produce from sustainable construction.

Conclusion

Even though priority emphasized on preventive actions where in the 3R practice, the order of precedence is to "Reduce, Reuse and Recycle", there is still a need for cure, which is recycling. Despite the possible preventions in reducing and reusing waste, Skyoles (1976) admitted to the existence of an acceptable level of waste or unavoidable waste in the construction industry (Formoso et al., 2002). Having said this, it doesnot imply that such an amount of waste can be ignored. Rather, the acceptable level of construction waste should be apprehended so as to achieve zero waste or zero landfill arising out of construction activities. To achieve zero waste or zero landfill, an establishment of basic initiative to quantify the amount of waste generated would be of absolute importance to provide a basis for future research especially in Malaysia. From a firm quantity of such waste generated, there will be justification for such material to be replaced if its waste quantity is huge. Alternatively, there could also be focused research on recycling and reusing such waste material.

With a standard operating procedure which could quantify waste and track its source and its final destination, it could lead to establishment of supply chain management for construction waste material (CREAM, 2008). With the availability of data on construction waste generated, it enables assessment on the ratio of waste generated against the amount of development. Only with such an assessment that one can conclude on whether or not development had been sustainable. When there is data on quantity or rate of waste generated, one can identify the problem associated with such waste. Knowing its quantity facilitates the feasibility study of construction waste recycling industry as encouraged by CREAM (2012). More importantly, the availability of such data provides specific grounds for studies in waste prevention and to replace natural resources which have high disposal rates (Pereira et al., 2007). If there is a system which is able to track the source, the amount of waste generated and its final destinations or recipients, as opposed to only locate illegal dumpsites, illegal dumping of construction waste should no longer prevail.

Implementing construction waste management opens up possible specializations in its execution. This means more job opportunities as well as career advancement. This is the economic and social sustainability perspective of the construction industry, that with advancement, job opportunities are still being secured. Personnel would be required to manage sorting and transporting of waste on construction sites. The government sector would require personnel in monitoring the practice construction waste management. Such implementation of construction waste management also increases the possibility of generating revenues from conducts of training on the implementation of such new procedures or penalties on non-compliance. Private sector will have boundless innovation in processing and trading such wastes. Hence, the reason for the development of standards, guidelines, policies and best practices to be one of the potential key research areas published in CREAM research focus webpage (2012). Even if current construction waste disposal system in place were adopting published methods, it is still necessary to make relevant certain terms and important information for in-house application, such as detailed steps or greater break down of methods from the published methods in order to obtain a systematic and consistent result (http://www.epa.gov/QUALITY/faq7.html, retrieved 7th April 2013).

The importance of having a systematic practice in place is its chain of effects. Take for example the practice of systematic documentation. In an organization which practices no systematic documentation, production of documents will not be consistent. There are greater chances for the occurrence of internal conflicts from human interaction among members of the organization itself. With a system in place, such as templates in preparing documents, standard flow of processes in generating documents, there is lesser need for human interaction. There will be lesser human interaction because the system provides a guide or reference archive. Not only will there be lesser conflicts but there will be an increased consistency in documents produced, limiting chances or error and lesser time needed in training or explaining to new members in the organization on the documentation practices because the system provides the required information.

References

Act 520, Pembinaan Malaysia Act. 1994. Construction Industry Development Board.

Act 672, Solid Waste and Public Cleansing Management Act. 2007. Ministry of Housing and Local Government.

Adshead, J. 2011. The quest for sustainable buildings: Getting it right at the planning stage. Green Buildings and the Law 76–93, Retrieved from www.routledge.com/books/details/9780203866801/ [accessed 21st April 2013].

Asnani, P. U. 2000. Modernization of Solid Waste Management Practices in India Through Hon'ble Supreme Court's Intervention, Proceedings of the Second International Conference on Solid Waste Management, 22nd–25th March, 2000.

Azman, M. N. A., M. S. S. Ahamad and N. D. Hilmi. 2012. The perspective view of Malaysian industrialized building system (IBS) under IBS precast manufacturing. *In*: The 4th International Engineering Conference—Towards Engineering of 21st Century.

Bai, R. and M. Sutanto. 2002. The practice and challenges of solid waste management in Singapore. Waste Management 22(5): 557–567.

Begum, R. A. 2005. Economic analysis on the potential of construction waste minimisation and recycling in Malaysia. Unpublished PhD Thesis. Universiti Kebangsaan Malaysia (UKM). Bangi, Selangor.

Begum, R. A., C. Siwar, J. J. Pereira and A. H. Jaafar. 2007. Implementation of waste management and minimisation in the construction industry of Malaysia. Resources, Conservation and Recycling 51(1): 190–202.

Begum, R. A., S. K. Satari and J. J. Pereira. 2010. Waste generation and recycling: Comparison of conventional and industrialized building systems. American Journal of Environmental Sciences 6(4): 383.

Bossink, B. A. G. and H. J. H. Brouwers. 1996. Construction waste: Quantification and source evaluation. Journal of Construction Engineering and Management 122(1): 55–60.

Cha, H. S., J. Kim and J. Y. Han. 2009. Identifying and assessing influence factors on improving waste management performance for building construction projects. Journal of Construction Engineering and Management 135(7): 647–656.

Chan, H. C. Y. and W. F. K. Fong. 2002. Management of construction and demolition materials and development of recycling facility in Hong Kong. *In*: Proceeding of International Conference on Innovation and Sustainable Development of Civil Engineering in the 21st Century.

Chang, N. B., K. S. Lin, Y. P. Sun and H. P. Wang. 2001. An engineering assessment of the burning of the combustible fraction of construction and demolition wastes in a redundant brick kiln. Environmental Technology 22(12): 1405–1418.

Cole, R. J. 2000. Building environmental assessment methods: assessing construction practices. Construction Management & Economics 18(8): 949–957.

CREAM. 2008 and 2012. About CREAM Priority Area. http://www.cream.my/main/index.php/research-development-r-d/environmental-sustainability [accessed 12th August 2008, accessed 23rd April 2012].

Dainty, A. R. and R. J. Brooke. 2004. Towards improved construction waste minimisation: a need for improved supply chain integration? Structural Survey 22(1): 20–29.
Department of Statistics Malaysia, https://www.dosm.gov.my [accessed 10th April 2016].
Enshassi, A. 2000. Environmental concerns for construction growth in Gaza Strip. Building and Environment 35(3): 273–279.
Esin, T. 2007. A study regarding the environmental impact analysis of the building materials production process (in Turkey). Building and Environment 42(11): 3860–3871.
Ferguson, J. 1995. Managing and minimizing construction waste: A practical guide. Thomas Telford.
Formoso, C. T., L. Soibelman, C. De Cesare and E. L. Isatto. 2002. Material waste in building industry: main causes and prevention. Journal of Construction Engineering and Management 128(4): 316–325.
Gavilan, R. M. and L. E. Bernold. 1994. Source evaluation of solid waste in building construction. Journal of Construction Engineering and Management 120(3): 536–552.
Hao, J. L., M. J. Hills and V. W. Tam. 2008. The effectiveness of Hong Kong's construction waste disposal charging scheme. Waste Management & Research, 26(6): 553–558. Retrieved from http://wmr.sagepub.com/cgi/content/abstract/26/6/553 [accessed 21st April 2013].
Hickman, J. H. L. 2000. People, the key to successful solid waste management. Proceedings of the Second International Conference on Solid Waste Management, 22nd–25th March 2000.
Inyang, H. I. 2003. Framework for recycling of wastes in construction. Journal of Environmental Engineering. 129(10): 887–898.
Ishak, N. H., A. R. M. Ariffin, R. Sulaiman and M. N. M. Zailani. 2016. Rethinking Space Design Standards Toward Quality Affordable Housing In Malaysia. In MATEC Web of Conferences (Vol. 66, p. 00112). EDP Sciences.
Jones, P. 2002. Integrated Solid Waste Management: A Life Cycle Inventory: Forbes McDougall, Peter White, Marina Franke, Peter Hindle, Blackwell Science, Oxford, UK, 2001, ISBN 0632058897, 530 pages,(plus CD Rom)£ 99, www. Blackwell-science. com. Corporate Environmental Strategy, 9(3): 320–321.
Key Data, Property Market Report. 2016. Valuation and Property Services Department, Ministry of Finance Malaysia.
Kulatunga, U., D. Amaratunga, R. Haigh and R. Rameezdeen. 2006. Attitudes and perceptions of construction workforce on construction waste in Sri Lanka. Management of Environmental Quality: An International Journal 17(1): 57–72.
Land Use Report. 2012. National Land Use Information Centre, Federal Department of Town and Country Planning.
Lawson, N., I. Douglas, S. Garvin, C. McGrath, D. Manning and J. Vetterlein. 2001. Recycling construction and demolition wastes—a UK perspective. Environmental Management and Health 12(2): 146–157.
Liu, B. Y., J. H. J. Chang and J. F. Shih. 2000. A Case Study of the Mandatory Household Waste Sorting Program and the Performance of Resource Recycling, Proceedings of the Second International Conference on Solid Waste Management, 22nd–25th March, 2000.
Lu, M., C. S. Poon and L. C. Wong. 2006. Application framework for mapping and simulation of waste handling processes in construction. Journal of Cnstruction Engineering and Management 132(11): 1212–1221.
Mallak, S. K., M. B. Ishak and A. F. Mohamed. 2014. Waste Minimization Benefit and Obstacles for Solid Industrial Waste in Malaysia. IOSR Journal of Environmental Science, Toxicoogy and Food Technology.
Masudi, A. F., C. R. C. Hassan, N. Z. Mahmood, S. N. Mokhtar and N. M. Sulaiman. 2011. Construction waste quantification and benchmarking: A study in Klang Valley, Malaysia. Journal of Chemistry and Chemical Engineering 5(10): 909–916.
McDougall, F. R. and J. P. Hruska. 2000. Report: The use of Life Cycle Inventory tools to support an integrated approach to solid waste management. Waste Management and Research 18(6): 590–594.
Nagapan, S., I. Abdul Rahman and A. Asmi. 2012. Construction waste management: Malaysian perspective.
Papargyropoulou, E., C. Preece, R. Padfield and A. A. Abdullah. 2011. Sustainable construction waste management in Malaysia: A contractor's perspective. In Management and Innovation for a Sustainable Built Environment MISBE 2011, Amsterdam, The Netherlands, June 20–23, 2011. CIB, Working Commissions W55, W65, W89, W112; ENHR and AESP. Retrieved from misbe2011.fyper.com/proceedings/documents/224.pdf [accessed 21st April 2013].
Pereira, J.J., M.D. Jantan and G. Sundaraj. 2007. Managing Construction Waste in Malaysia. LESTARI, UKM.

R&D Themes and Titles of the Construction Research Institute of Malaysia (CREAM) 2008/2009 Research Fund List.

R&D Themes and Titles of the Construction Research Institute of Malaysia (CREAM) 2012 Research Areas.

Ramli, N. H. M. 2010. A Computerized Construction Waste Exchange System for Malaysia (Doctoral dissertation, Jabatan Kejuruteraan Perisian, Fakulti Sains Komputer dan Teknologi Maklumat, Universiti Malaya).

Redzuan, A. R. M., A. C. Hassan and A. H. Jamaludin. 2017. Quantification of Construction Waste Generated in Residential Housing Projects via Heap Survey Sampling with the Method of Visual Estimation: A Case Study in Klang Valley and Pulau Pinang. Raja Nor Husna Raja Mohd Noor, Siti Akmalina Rahmat,"Intan Rohani Endut. Journal of Engineering and Applied Sciences 12(4): 792–796.

Shakantu, W., M. Muya, J. Tookey and P. Bowen. 2008. Flow modelling of construction site materials and waste logistics: A case study from Cape Town, South Africa. Engineering, Construction and Architectural Management 15(5): 423–439.

Shen, L. Y., V. W. Y. Tam, C. M. Tam and S. Ho. 2000. Material wastage in construction activities—a Hong Kong survey. pp. 125–31. *In*: Proceedings of the first CIB-W107 international conference—creating a sustainable construction industry in developing countries.

Skoyles, E. R. 1976. Materials Wastage—A Misuse of Resources.

Taiichi, O. 1988. Workplace Management. Trans. Andrew P. Dillon. Cambridge: Productivity Press.

Tam, V. W., L. Y. Shen, I. W. Fung and J. Y. Wang. 2007. Controlling construction waste by implementing governmental ordinances in Hong Kong. Construction Innovation 7(2): 149–166.

Tam, V. W. 2011. Rate of reusable and recyclable waste in construction. Open Waste Management Journal 4(1): 28–32.

Tan, R. B. and H. H. Khoo. 2006. Impact assessment of waste management options in Singapore. Journal of the Air & Waste Management Association 56(3): 244–254.

Wan-A-Kadir, W. R. 1997. The Development of a Framework for Sustainable Waste Management Policy and Strategy for Malaysia (Doctoral dissertation, University of Salford).

Wan, S. K., M. M. Kumaraswamy and D. T. Liu. 2009. Contributors to construction debris from electrical and mechanical work in Hong Kong infrastructure projects. Journal of Construction Engineering and Management 135(7): 637–646.

Zainun, N. Y., I. A. Rahman and R. A. Rothman. 2016. Mapping of Construction Waste Illegal Dumping Using Geographical Information System (GIS). In IOP Conference Series: Materials Science and Engineering (Vol. 160, No. 1, p. 012049). IOP Publishing.

Index

R

S

T

U

W

About the Editors

AbdulLateef Olanrewaju, Ph.D. is with Universiti Tunku Abdul Rahman, Malaysia. His academic activities center on lecturing and research associated with building maintenance, building economics, housing, sustainable buildings and value management. He has authored and co-authored several scientific papers. Olanrewaju is a referee for leading journals. He has served as a panelist and keynote speaker on sustainable housing, sustainable building maintenance and the construction sector.

Dr. Zalina Shari is an Associate Professor at Universiti Putra Malaysia (UPM). She received her PhD in Architecture (specialising in building sustainability assessment) from the University of Adelaide. She has published more than 60 book chapters and articles related to green building and sustainability in the built environment. She serves as a member of editorial boards for two international journals and in 2017, she received the "Vice Chancellor Fellowship Award for Excellent Educator Category" from UPM.

Dr. Zhonghua Gou is a Senior Lecturer in Architecture at Griffith University, Australia. His research interests focus on sustainable urban and architectural design in tropical and subtropical climates. He has published more than 50 peer-reviewed journal articles. He is a guest editor for the Journal of Buildings and a regular reviewer for many journals. Zhonghua is a LEED Accredited Professional and Certified Carbon Auditor, and he practices as a green building consultant in Hong Kong and mainland China.

9781138102606

9781138102606